NONLINEAR PARTIAL DIFFERENTIAL
EQUATIONS IN APPLIED SCIENCE
PROCEEDINGS OF THE U.S. – JAPAN
SEMINAR, TOKYO, 1982

NORTH-HOLLAND
MATHEMATICS STUDIES 81

Lecture Notes in Numerical and Applied Analysis Vol. 5
General Editors:
H. Fujita (University of Tokyo) and M. Yamaguti (Kyoto University)

Nonlinear Partial Differential Equations in Applied Science; Proceedings of The U.S.–Japan Seminar, Tokyo, 1982

Edited by
HIROSHI FUJITA (University of Tokyo)
PETER D. LAX (New York University)
GILBERT STRANG (Massachusetts Institute of Technology)

1983
NORTH-HOLLAND PUBLISHING COMPANY
AMSTERDAM · NEW YORK · OXFORD

KINOKUNIYA COMPANY LTD.
TOKYO JAPAN

NORTH-HOLLAND PUBLISHING COMPANY - AMSTERDAM · NEW YORK · OXFORD
KINOKUNIYA COMPANY - TOKYO

ISBN: 0 444 86681 7

Publishers
NORTH-HOLLAND PUBLISHING COMPANY
AMSTERDAM · OXFORD · NEW YORK

* * *

KINOKUNIYA COMPANY LTD.
TOKYO JAPAN

Sole distributors for the U.S.A. and Canada
ELSEVIER SCIENCE PUBLISHING COMPANY. INC.
52 VANDERBII.T AVENUE
NEW YORK. N.Y. 10017

Distributed in Japan by KINOKUNIYA COMPANY LTD.

Lecture Notes in Numerical and Applied Analysis Vol. 5

PRINTED IN JAPAN

PREFACE

Nonlinear equations come to us in tremendous variety, each with its own questions and its own difficulties. At one extreme are the completely integrable equations, with constants of the motion and a rich algebraic structure. At the other extreme is chaos, with turbulent solutions and statistical averages. Between these two possibilities, algebraic and ergodic, lies the full range of nonlinear phenomena. There are smooth solutions which develop shocks, or bifurcate, or maintain slow and nearly periodic variations that imitate the linear theory. Each of these questions requires a separate treatment, and the subject would be simpler if we know for every equation which behavior to expect.

Nevertheless these equations, the nonlinear partial differential equations which arise in applications, share one crucial property. They are all vulnerable when the right pattern in found. It is a slow process, to uncover and reveal their structure, but it is moving forward.

The papers in this volume reflect a part of that progress. They were presented at the U.S.-Japan Seminar in Tokyo in July 1982.

One goal of the seminar was to establish personal contact among those mathematicians who are actively working for these difficult but fascinating equations in the U.S. and in Japan. The other goal was a wider one, that is, to invoke most advanced scientific talks and discussions on major topics in this developing field of applied analysis.

Thanks to the cooperation of all participants from the U.S., Japan, and some third countries including China, the seminar was successful in both sense mentioned above and we believe that these proceedings of the seminar which contain all papers delivered there will contribute much to the progress of the study of nonlinear problems.

Finally, we, who served also as the coordinators of the seminar, wish to express our gratitude to the governmental agencies, i.e., National Science Foundation and Japan Society for the Promotion of Science, for their support and to industrial companies in Japan for practical assistances which they gave as institutional participants. Last but not least, our gratitudes go to all of our committee members and staff members of the secretariat of the seminar for their enthusiasm and devotion.

September 15, 1983

H. FUJITA
P. D. LAX
G. STRANG

CONTENTS

PROGRAM

MONDAY, JULY 5

8:45 Opening of Seminar

Session 5–1

9:00–10:00 Prof. T. Kato (University of California, Berkeley)
"Quasi-Linear Equations of Evolution in Nonreflexive Banach Spaces"

10:15–11:00 Prof. K. Masuda (Tôhoku University)
"Some Remarks on Blow-up of Solutions of Nonlinear Diffusion Equations"

11:05–11:50 Dr. T. Suzuki (University of Tokyo)
"Deformation Formulas and their Applications to Spectral and Evolutional Inverse Problems"

Session 5–2

13:45–14:30 Mr. H. Okamoto (University of Tokyo)
"Stationary Free Boundary Problems for Circular Flows with or without Surface Tension"

14:35–15:20 Prof. Lin Qun (Institute of Systems Science, Academia Sinica) and Prof. Jiang Li-shang (Peking University)
"The Existence and the Finite Element Approximation for the System $\Delta u=\sum_{j=1}^{N} u_j \frac{\partial u}{\partial x_j}+f$"

15:20–16:00 Coffee Break

16:00–17:00 Prof. H. F. Weinberger (University of Minnesota)
"A Simple System with a Continuum of Stable Inhomogeneous Steady States"

TUESDAY, JULY 6

Session 6–1

9:00–10:00 Prof. R. J. DiPerna (Duke University)
"Shock Waves and Entropy"

10:15–11:00 Prof. S. Ukai (Osaka City University) and Prof. K. Asano (Kyoto University)
"Stationary Solutions of the Boltzmann Equation"

11:05–11:50 Prof. T. Nishida (Kyoto University) and Dr. A. Matsumura (Kyoto University)
"Initial Boundary Value Problems for the Equations of Compressible Viscous and Heat-Conductive Fluid"

Session 6–2

13:45–14:30 Prof. S. Miyatake (Kyoto University)
"Integral Representation of Solutions for Equations of Mixed Type in a Half Space"

14:35–15:20 Prof. H. Fujii (Kyoto Sangyo University) and Prof. Y. Nishiura (Kyoto Sangyo University)
"Global Aspects in Bifurcation Problems for Nonlinear Diffusion Systems" (tentative)

15:20–16:00 Coffee Break

16:00–17:00 Prof. A. C. Newell (University of Arizona)
"Two-Dimensional Convection Patterns in Large Aspect Ratio Systems"

WEDNESDAY, JULY 7

Session 7–1

9:00–10:00 Prof. M. Sato (Research Institute for Mathematical Sciences, Kyoto University)
"Soliton Equations as Dynamical Systems on Infinite Dimensional Grassmann Manifold"

10:15–11:00 Prof. A. C. Newell (University of Arizona)
"The Connection between Wahlquist-Estabrook, Hirota, τ Function, and Inverse Scattering Methods for the AKNS Hierarchy"

11:05–11:50 Prof. Zhou Yu-lin (Peking University)
"Some Problems for Nonlinear Evolutional Systems of Partial Differential Equations"

THURSDAY, JULY 8

Session 8–1

9:00–10:00 Prof. M. Yamaguti (Kyoto University)
"'Chaos' Caused by Discretization"

10:15–11:00 Prof. H. Matano (Hiroshima University)
"Asymptotic Behavior of the Free Boundaries Arising in One Phase Stefan Problems in Multi-Dimensional Spaces"

11:05–11:50 Dr. E. Hanzawa (Hokkaido University)
"Nash's Implicit Function Theorem and the Stefan Problem"

Session 8–2

13:45–14:30 Prof. H. Kawarada (University of Tokyo)
"New Penalty Method and its Application to Free Boundary Problems"

14:35–15:20 Prof. T. Ushijima (University of Electro-Communications)
"On the Linear Stability Analysis of Magnetohydrodynamic System"

15:20–16:00 Coffee Break

16:00–17:00 Prof. G. Papanicolaou (Courant Institute of Mathematical Sciences, New York University)

"Modulation Theory for the Cubic Schrödinger Equation in Random Media"

FRIDAY, JULY 9

Session 9–1

9:00–10:00 Prof. R. V. Kohn (Courant Institute of Mathematical Science, New York University)
"Partial Regularity for the Navier-Stokes Equations"

10:15–11:00 Mr. Y. Giga (Nagoya University)
"The Navier-Stokes Initial Value Problem in L^p and Related Problems"

11:05–11:50 Prof. J. G. Heywood (University of British Columbia)
"Stability, Regularity and Numerical Analysis of the Nonstationary Navier-Stokes Problem"

Session 9–2

13:45–14:30 Prof. Ying Lung-an (Peking University) and Prof. Teng Zhen-huan (Peking University)
"A Hyperbolic Model of Combustion"

14:35–15:20 Prof. T. Miyoshi (Kumamoto University)
"Yielding and Unloading in Semi-Discrete Problem of Plasticity"

15:20–16:00 Coffee Break

16:00–17:00 Prof. G. Strang (Massachusetts Institute of Technology)
"Optimization Problems for Partial Differential Equations"

17:05 Closing of Seminar

Lecture Notes in Num. Appl. Anal., **5**, 1–15 (1982)
Nonlinear PDE in Applied Science. U.S.-Japan Seminar, Tokyo, 1982

CONSERVATION LAWS AND THE WEAK TOPOLOGY

Ronald J. DiPerna

Duke University
Durham, North Carolina 27706

We shall discuss some results concerning the convergence of approximate solutions to hyperbolic systems of conservation laws. The general setting is provided by a system of n conservation laws in one space dimension,

$$u_t + f(u)_x = 0 \tag{1}$$

where $u = u(x,t) \in R^n$ and f is a smooth nonlinear map from R^n to R^n. We assume that f is strictly hyperbolic in the sense that its Jacobian has n real and distinct eigenvalues

$$\lambda_1(u) < \lambda_2(u) < \ldots < \lambda_n(u).$$

With regard to approximation, one is interested in sequences of approximate solutions generated by parabolic systems

$$u_t + f(u)_x = \varepsilon D u_{xx}, \quad u = u_\varepsilon(x,t)$$

and by finite difference schemes

$$\partial_t u + \partial_x f(u) = 0, \quad u = u_{\Delta x}(x,t),$$

which are conservative in the sense of Lax and Wendroff [8]. A standard strategy for convergence seeks to establish uniform estimates on both the amplitude and derivatives of the approximate solutions in appropriate metrics and then appeal to a compactness argument to produce a subsequence that converges in the strong topology. One may regard convergence of the entire sequence as a question of uniqueness of the limit. We recall that in the setting of hyperbolic conservation laws the maximum norm and the total variation norm yield a natural pair of metrics in which to investigate the stability of the solution. The L^∞ norm measures the solution amplitude and the total variation norm measures the solution gradient. Their relevance for conservation laws is established by the following theorem of Glimm [5] dealing with the stability and convergence of the approximate solutions generated by his random choice method applied to the Cauchy problem.

<u>Theorem 1</u>. If the total variation of the initial data $u_0(x)$ is sufficiently small then a sequence of random choice approximations $u_{\Delta x}$ converges pointwise almost everywhere to a globally defined distributional solution u maintaining uniform control on the amplitude and spatial variation:

$$|u_{\Delta x}(\cdot,t)|_\infty \leq \text{const. } |u_0|_\infty$$

$$\text{TV } u_{\Delta x}(\cdot,t) \leq \text{const. TV } u_0 \ .$$

The constants are independent of the mesh length and depend only on the flux function f.

The proof is based on a general study elementary wave interactions in the exact solution and in the random choice approximations $u_{\Delta x}$. It remains an open problem to prove or disprove the corresponding estimates for conservative finite difference schemes and parabolic systems. In the latter direction we refer the reader to [3] which contains an analysis of discrete wave interactions in conservative schemes together with a stability and convergence theorem for a class of methods involving the hybridization of the random choice method with first order accurate conservative methods.

Here we shall discuss new compactness theorems for sequences of approximate solutions generated by diffusive systems and conservative difference schemes. The proof involves the theory of compensated compactness which originates in the work of Tartar [11] and Murat [9,10] and the main step provides a proof of a conjecture of Tartar [11]. The analysis appeals to the weak topology and averaged quantities rather than the strong topology and the fine scale features. Regarding the weak topology and the elliptic conservation laws of elasticity we refer the reader to the work of Ball [1]. The principle statement is that for a class of approximation methods, which respect the entropy condition, L^∞ stability alone implies convergence. Gradient estimates are not required to pass to the limit in the nonlinear functions.

We shall first recall some background involving Tartar's work on weak convergence and compensated compactness. Consider a sequence of functions

$$u_n(y): R^m \to R^n$$

which is uniformly bounded in L^∞. It is well-known that one may extract a subsequence which converges in the weak-star topology of L^∞:

$$\lim \int_B u_n(y)dy = \int_B u(y)dy$$

for all bounded $B \subset R^n$. We recall that in general the sequence u_n need not contain a strongly convergent subsequence, i.e. a subsequence converging pointwise a.e. to u. In particular, if g is a real-valued map on R^m

$$\lim g(u_n(y)) \neq g(u).$$

However, after passing to subsequence, composite weak limits may be represented as expected values of associated probability measures in the following sense. There exists a subsequence of u_n (still denoted here by u_n) and a family of probability measures over the range space R^n,

$$\{\nu_y : y \in R^m\}$$

such that for all continuous $g: R^m \to R$,

$$\lim_{n\to\infty} g(u_n(y)) = \int_{R^n} g(\lambda) d\nu_y(\lambda) .$$

The limit on the left hand side is taken in the weak-star topology of L^∞ and equality holds for almost all y in R^m. Here λ denotes a generic point in the range space R^n. This result stems from the work of L. C. Young and was first used in the setting of conservation laws by Tartar [11]. It is not difficult to show that strong convergence corresponds to the case where the representing measure ν_y reduces to a point mass concentrated at $u(y)$:

$$\nu_y = \delta_{u(y)}$$

More generally, the deviation between weak and strong convergence is measured by the spreading of the support of ν_y. If g is Lipschitz then

$$|g(\lim u_n) - \lim g(u_n)|_\infty \leq \text{const.} \max_y \text{diam spt } \nu_y .$$

In the framework of conservation laws, the goal is to show that the representing measures associated with a family of exact or approximate solutions reduces to a point mass or is contained in a set whose geometry allows one to deduce the continuity of the special nonlinear maps appearing in the equations. In the case of a scalar conservation law Tartar [11] has shown that ν_y reduces to a point mass if f is convex and that, in general, ν_y is supported on an interval where f is affine. Here we shall discuss the reduction of ν for strictly hyperbolic systems of two equations with non-degenerate eigenvalues. The analysis is based on a study of the Lax progressing entropy waves in state space [7], specifically on connections between their structure and the structure of wave patterns in the physical space, cf. [2] for details and additional references. We also refer the reader to Lax [6] which contains a discussion of the scalar conservation law and the viscosity method in the setting of the weak topology.

Before discussing the general case we shall cite an example. Consider the equations of elasticity in Lagrangian form with artificial viscosity

$$u_t - \sigma(v)_x = \varepsilon\, u_{xx}$$

$$v_t - \quad u_x = \varepsilon\, v_{xx} ,$$

and assume that $\sigma' > 0$ while $\operatorname{sgn} v\, \sigma'' > 0$. Given initial data in L^∞, there exists for each fixed ε a globally defined solution the amplitude of which remains uniformly bounded as the viscosity parameter ε vanishes,

$$|u(\cdot,t)| + |v(\cdot,t)|_\infty \le \text{const.}$$

Here the constant depends only on σ and in the L^∞ norm of the data. The bound follows from the presence of invariant regions in the state space [14]. We claim that by appealing only to the L^∞ stability and the entropy condition, one may extract a subsequence

$(u_{\varepsilon_k}, u_{\varepsilon_k})$ which converges pointwise a.e. to a globally defined distributional solution of the associated hyperbolic system

$$u_t - \sigma(v)_x = 0 \tag{2}$$
$$v_t - u_x = 0 \ .$$

A similar result can be established for a class of first order finite difference schemes which are based on averaging the Riemann problem, e.g. the Lax-Friedrichs scheme and Godunov's scheme.

The source of the compactness in the strong topology lies in the nonlinear structure of the wave speeds and in the dissipation of generalized entropy along propagating shocks. We shall first recall the notion and some basic properties of generalized entropy as for mulated by Lax [7]. Consider a system of n conservation laws (1). A pair of real-valued mappings on the state space R^n

$$\eta: R^n \to R; \quad q: R^n \to R$$

is called an entropy pair if all smooth solutions of (1) satisfy an addition conservation law of the form

$$\eta(u)_t + q(u)_x = 0 \ . \tag{3}$$

For the purposes at hand we shall restrict our attention to the class of systems having an entropy pair with η strictly convex. As observed by Lax and Friedrichs [15] this class includes the basic systems of continuum mechanics. Furthermore, Lax [7] showed that all strictly hyperbolic systems of two equations has at least a locally defined strictly convex entropy and that a broad class has a globally defined strictly convex entropy. The basic compatibility condition which links the entropy η to its flux g may be derived as follows. Suppose $u(x,t)$ is a C^1 solution and consider the quasilinear forms of the systems of conservation laws (1) together with the extension (3):

$$(4)\qquad \begin{aligned} u_t + \nabla f(u) u_x &= 0 \\ \nabla\eta(u) u_t + \nabla q(u) u_x &= 0 . \end{aligned}$$

By replacing the time derivative of u by the spatial derivative we find that (3) is equivalent to

$$\{-\nabla\eta(u)\ \nabla f(u) + \nabla q(u)\} u_x = 0 .$$

Hence the condition

$$(5)\qquad \nabla\eta(u)\ \nabla f(u) = \nabla q(u), \quad u \in R^n$$

is a necessary and sufficient for the existence of an entropy pair. We observe that (5) represents a system of n linear, variable coefficient partial differential equations in two unknowns η and q. If $n > 2$ it is formally over determined but fortunately has a (convex) solution in the setting of mechanics. Concerning the structure of (5) we recall the observation of Loewner that the compatibility condition (5) retains the same classification as the original system (1). In our setting the demonstration that (5) is hyperbolic is straightforward: consider the right eigenvectors of the Jacobian of f

$$\nabla f(u)\ r_j(u) = \lambda_j(u)\ r_j(u) .$$

Taking the inner product of (5) with r_j immediately yields the characteristic form of (4):

$$(\lambda_j \nabla\eta - \nabla q) \cdot r_j = 0, \quad j = 1,2$$

In the following discussion we shall be mainly interested in the determinate case $n = 2$ which can be illustrated with a variety of examples. In particular, it is useful to keep in mind that the smooth motion of an elastic medium which conserves mass and momentum also conserves mechanical energy. For system (2) one may take

$$\eta = \frac{1}{2} u^2 + \Sigma(v), \quad \Sigma'(v) = \sigma(v) .$$

The convex function η serves as a generalized entropy for (2) with generalized entropy flux

$$q = u\,\Sigma(v)$$

The identity (5) states the time rate of change of mechanical energy is balanced by the rate at which the stress tensor performs work.

Within the class of conservation laws with a convex extension it is standard to impose the Lax entropy inequality

$$\eta(u)_t + q(u)_x \leq 0 \tag{6}$$

on weak solutions $u(x,t)$ for the purpose of distinguishing the physically relevant weak solutions from the set of all possible weak solutions. Solutions satisfying (6) are called <u>admissible</u>. We note that the distributional inequality is meaningful if u is merely a locally bounded function. For our current purposes we shall restrict our attention to weak solutions which lie in the space $L^\infty \cap BV$. Here BV denotes the class of functions of several variables which have bounded variation in the sense of Cesari, i.e. first order partial derivatives representable as locally bounded Borel measures [4,12]. Experience with conservation laws has shown that $L^\infty \cap BV$ is a natural function space for the solution operator. In this connection we note that solutions constructed by the random choice method lie in the space $L^\infty \cap BV$ by virtue of the stability estimates of theorem 1. Within $L^\infty \cap BV$ one can demonstrate that the measure

$$\theta_u \overset{\text{def.}}{=} \eta(u)_t + q(u)_x$$

is concentrated on the shock set of $\Gamma(u)$ the solution u, i.e. the set of points of discontinuity and consequently that the entropy inequality (5) holds if and only if all shock waves in u dissipate generalized entropy:

$$\theta_u(E) \leq 0$$

for all Borel $E \subset \Gamma(u)$. This inequality reduces to the second law of thermodynamics in the setting of fluid flow. Finally, we shall restrict attention to systems with non-degenerate eigenvalues, i.e. systems for which the wave speeds are monotone functions of the wave amplitudes. Technically we assume that λ_j is monotone in the corresponding eigendirection:

$$r_j \cdot \nabla\lambda_j \neq 0 \ . \tag{7}$$

We note that the genuine nonlinearity condition (7) introduced by Lax [16] is satisfied by several systems of interest: the isentropic equations of gas dynamics for a polytropic gas, the equations of shallow water waves, the equations of elasticity if $\sigma'' \neq 0$.

<u>Theorem 2</u>. Consider a strictly hyperbolic genuinely nonlinear system of two conservation laws with a strictly convex entropy. Suppose u_n is a sequence of admissible solutions in $L^\infty \cap BV$. If

$$|u_n|_\infty \leq M$$

where the constant M is independent of n, there exists a subsequence that converges pointwise a.e. to an admissible solution.

Thus the exact solution operator restricted to admissible solutions forms a compact mapping from L^∞ to L^1_{loc}. The source of the compactness lies in the loss of information associated with admissible shock waves and in the nonlinear structure of the eigenvalues. We emphasize that the compactness is established without derivative estimates.

Next, we shall discuss the compactness of solution sequences generated by diffusion processes

$$u_t + f(u)_x = \varepsilon\, Du_{xx}, \tag{8}$$

where for simplicity D is a constant $n \times n$ matrix. In order to ensure correct entropy production in the limit as ε vanishes, it is sufficient (and nearly necessary) to require that the diffusion matrix D be non-negative with respect to the second derivative of η, i.e.

$$\nabla^2 \eta\, D \geq 0 .$$

With regard to the general question of admissibility shock structure and proper diffusion matrices we refer the reader to R. Pego [17,18].

Theorem 3. Suppose f is a strictly hyperbolic genuinely nonlinear map on R^2 with a strictly convex entropy η and suppose that the diffusion matrix D is positive definite with respect to $\nabla^2\eta$. If u_ε is a sequence of smooth solutions to (8) satisfying

$$|u_\varepsilon|_\infty \leq M ,$$

there exists a subsequence which converges pointwise a.e. to an admissible solution u of the associated hyperbolic system (1).

Hence the solution operators S_ε of the parabolic system (8) provide a family of mappings which is compact from L^∞ to L^1_{loc} uniformly with to ε. The compactness present at the hyperbolic level is preserved uniformly in ε provided that the diffusion matrix enduces favorable entropy production in the limit.

In the setting of continuum mechanics we recall that the standard diffusion matrices are merely positive semi-definite because mass diffusion is neglected. However with additional work one can establish the corresponding result.

Theorem 4. Suppose that $(\rho_\varepsilon, u_\varepsilon)$ is a sequence of smooth solutions of compressible Navier-Stokes

$$\rho_t + (\rho u)_x = 0$$

$$(\rho u)_t + (\rho u^2 + p(\rho))_x = \varepsilon u_{xx}$$

for a polytropic gas $p = A\rho^\gamma$, $\gamma > 1$. If the flow is uniformly bounded and avoids the vacuum state, i.e.

$$0 < m \leq \rho_\varepsilon \leq M \quad \text{and} \quad |u_\varepsilon|_\infty \leq M ,$$

then there exists a subsequence which converges pointwise a.e. to an admissible solution of the compressible Euler equations. We note that the compressible Euler equations losses its strict hyperbolicity at the vacuum state. It is an interesting open problem to establish the corresponding result without the hypothesized uniform lower bound on the density ρ. At a more fundamental level, it remains an open problem to prove uniform L^∞ estimates in general circumstances. For example in the case of hyperbolic systems (1), it remains an open problem to prove that

$$|u(\cdot,t)|_\infty \leq \text{const.}\ |u_0|_\infty \tag{9}$$

for admissible solutions in $L^\infty \cap BV$ with small data. The estimate (9) is motivated by physical considerations but has only been verified for solutions constructed by the random choice method.

The proof of the theorems described above utilizes the theory of compensated compactness. In this connection we refer the reader to the work of Tartar [11] and Murat [9,10] and to the forthcoming Proceedings of the NATO/LMS Advanced Study Institute on Systems of Nonlinear Partial Differential Equations held at Oxford 1982 and organized by J. Ball et al. Here we shall simply mention one of the problems which the theory addresses: characterize the nonlinear functions $g(u): R^n \to R$ which are continuous in the weak topology when restricted to sequences of functions $u_n(y): R^m \to R^n$ which satisfy linear constant coefficient partial differential constraints. As an example we mention a result from electrostatics which historically motivated the general theory. Consider vector fields

$$z_n: R^3 \to R^3 \quad \text{and} \quad w_n: R^3 \to R^3$$

converging weakly in L^2

$$z_n \to z; \quad w_n \to w .$$

Suppose that the expansion in z_n is controlled as well as the rotation in w_n to the extent that both the sequences of distributions

$$\text{div } z_n \quad \text{and} \quad \text{curl } w_n$$

lie in a compact subset of the negative Sobolev space H^{-1}_{loc}. Here distinguished linear combinations of partial derivatives are compact after the loss of one derivative. Under these circumstances there is precisely one smooth real-valued function $\phi(z,w)$ which is continuous in the weak topology, i.e. satisfies

$$\phi(z,w) = \lim \phi(z_n,w_n) ,$$

and its given by the inner product

$$\phi(z,w) = \langle z,w\rangle .$$

Although, in general, the individual terms $z^j w^j$ of the inner product are not weakly continuous, there exists compensation among the terms of the sum

$$\langle z,w\rangle = \sum_{j=1}^{3} z^j w^j$$

which allows for weak continuity. From the point of view of electrostatics, there is precisely one quantity, the electrostatic energy density which can be measured, provided one agrees that the process of measurement is modeled by averaged quantities.

For the purpose of applications to conservation laws, let us recall the duality between the divergence and the curl in the plane,

$$\text{div } z = \text{curl } z^*$$

where z^* denotes the orthogonal complement of z and consider the basic entropy inequality formulated with respect to two distinct entropy pairs (η_j,q_j) $j = 1,2$. If, for example, u_n is a sequence of admissible weak solutions in $L^\infty \cap BV$ it can be shown, by appealing to Sobolev embedding, that the sequence of distributions

$$\eta(u_n)_t + q(u_n)_x$$

lies in a compact subset of H^{-1}_{loc} if η is convex and consequently that

$$\eta_j(u_n)_t + q_j(u_n)_x$$

lies in a compact subset of H^{-1}_{loc} for arbitrary (η_j,q_j). Thus the divergence of the entropy field (η_1,q_1) and the curl of the entropy field $(-q_2,\eta_2)$ both lie in a compact subset of H^{-1}_{loc}.

The continuity of the inner product yields a commutativity relation for the representing measure ν. For all entropy pairs (η_j, q_j), $j = 1,2$ we have

$$(9) \qquad \langle \nu, \eta_1 q_2 - \eta_2 q_1 \rangle = \langle \nu, \eta_1 \rangle \langle \nu, q_2 \rangle - \langle \nu, \eta_2 \rangle \langle \nu, q_1 \rangle$$

where ν denotes representing measure at an arbitrary point, i.e. $\nu = \nu_{(x,t)}$. Tartar showed that (9) implies ν is a Dirac measure for a genuinely nonlinear scalar conservation law. In [2] it is shown using the Lax progressing entropy waves that (9) implies that ν is a point mass for general genuinely nonlinear systems of two equations and for the special case of elasticity which has a linear degeneracy alone an isolated curve.

References

[1] Ball, J. M., Convexity conditions and existence theorems in nonlinear elasticity, Arch. Rational Mech. Anal. 63 (1977) 337-403.

[2] DiPerna, R. J., Convergence of approximate solutions to conservation laws, Arch. Rational Mech. Anal., to appear, (1983).

[3] DiPerna, R. J., Finite difference schemes for conservation laws, Comm. Pure Applied Math. 25 (1982) 379-450.

[4] Federer, H., Geometric Measure Theory (Springer, New York, 1969).

[5] Glimm, J., Solutions in the large for nonlinear hyperbolic systems of equations, Comm. Pure Appl. Math. 18 (1965) 697-715.

[6] Lax, P. D., Weak solutions of nonlinear hyperbolic equations and their numerical computation, Comm. Pure Appl. Math. 7 (1954) 159-193.

[7] Lax, P. D., Shock waves and entropy, in Contributions to nonlinear functional analysis, e.d. E. A. Zarantonello, Academic Press, (1971) 603-634.

[8] Lax, P. D. and B. Wendroff, Systems of conservation laws, Comm. Pure Appl. Math. 13 (1960) 217-237.

[9] Murat, F., Compacité par compensation, Ann. Scuola Norm. Sup. Pisa 5 (1978) 489-507.

[10] Murat, F., Compacité par compensation: Condition necessaire et suffisante de continuite faible sous une hypotheses de rang constant, Ann. Scula Norm. Sup. 8 (1981) 69-102.

[11] Tartar, L., Compensated compactness and applications to partial differential equations, in Research Notes in Mathematics, Nonlinear Analysis and Mechanics: Heriot-Watt Symposium, Vol. 4, Ed. R. J. Knops, Pitman Press, 1979.

[12] Vol'pert, A. I., The spaces BV and quasilinear equations, Math. USSR, Sb. 2 (1967) 257-267.

[13] Dacorogna, B., Weak continuity and weak lower semicontinuity of nonlinear functionals, Lefschetz Center for Dynamical Systems Lecture Notes #81-77, Brown University, (1981).

[14] Chueh, K. N., Conley, C. C. and J. A. Smoller, Positivity invariant regions for systems of nonlinear diffusion equations, Indiana Math. J. 26 (1977) 373-390.

[15] Friedrichs, K. O. and P. D. Lax, Systems of conservation laws with a convex extension, Proc. Mat. Acad. Sci. USA 68 (1971) 1686-1688.

[16] Lax, P. D., Hyperbolic systems of conservation laws II, Comm. Pure Appl. Math. 10 (1957) 537-566.

[17] Pego, R., Viscosity matrices for a system of conservation laws, Center for Pure and Applied Mathematics, University of California, Berkeley, preprint.

[18] Pego, R., Linearized stability of shock profiles, CPAM, University of California, Berkeley, preprint.

Lecture Notes in Num. Appl. Anal., **5**, 17–35 (1982)
Nonlinear PDE in Applied Science. U.S.-Japan Seminar, Tokyo, 1982

Global Bifurcation Diagram in Nonlinear Diffusion Systems

Hiroshi FUJII and Yasumasa NISHIURA
Department of Computer Sciences
Faculty of Science
Kyoto Sangyo University
Kyoto 603, JAPAN.

§1. Introduction

Global phenomena of pattern formation in systems of reaction-diffusion equations is the main theme of the present paper. The system is written as

$$(P)\qquad \begin{aligned} u_t &= d_1 u_{xx} + f(u,v) \\ v_t &= d_2 v_{xx} + g(u,v) \end{aligned} \qquad \text{in } (t,x) \in (0,+\infty) \times I,$$

$$u_x = v_x = 0 \qquad \text{on } (t,x) \in (0,+\infty) \times \partial I,$$

where $I = (0,\pi)$, and ∂I its boundary. The system (P) is assumed to possess Turing's *diffusion induced instability*, which appears typically in mathematical biology [7]. In other words, we are interested in the structure of global bifurcation diagram - "global" with respect to the two diffusion parameters $(d_1,d_2) \in \mathbb{R}_+^2$ - of the following stationary system :

$$(SP)\qquad \begin{aligned} d_1 u_{xx} + f(u,v) &= 0, \\ \frac{1}{\alpha} v_{xx} + g(u,v) &= 0, \end{aligned} \qquad \text{in } I,$$

with the boundary conditions $(P)_3$ on ∂I; here, we put $d_2 = 1/\alpha$.

The system (P) has been studied by a number of authors from various kind of viewpoints. In particular, the bifurcation theoretic work of Mimura, Nishiura and Yamaguti [3] has motivated the studies which succeed, such as Mimura, Tabata and Hosono [4] who studied the singular limit $d_1 \downarrow 0$ of (SP) using the singular perturbation technique; the second author [5] has obtained a complete bifurcation diagram with respect to d_1 of (SP) in the limit case $d_2 \uparrow +\infty$(i.e., $\alpha \downarrow 0$). His limit system is called the *shadow system*. The first author has developed a new numerical algorithm to detect and trace all bifurcating branches using a group theoretic method [2]. Fujii, Mimura and Nishiura [1] studied local structures of (SP) near double bifurcation points (adopting a group theoretic argument), and drew a global picture of bifurcation diagram, integrating the above analytical and numerical results.

The purpose of this paper is, in part, to give a survey of those works, and in part, to describe new results which have been obtained after the publication of [1]. Our method is based on the study of:

(1) local structure at double bifurcation points introducing the Lie group D_∞[1],

(2) the complete bifurcation analysis of the shadow system [5],

(3) the singular-shadow limit of $d_2 \uparrow +\infty$, $d_1 \downarrow 0$ - which we call the *singular-shadow edge*,

(4) the structure of "singular solutions" at the singular limit $d_1 \downarrow 0$, and

(5) an integration of these analytical results to have a global picture of bifurcating branches.

A key in our paper is the discovery of singular branches which possess both boundary and interior transition layers, and of singular limit points as its consequence. We shall see in the present paper that *the structure of solutions at the singular-shadow edge seems to play the role of "organizing centre" of the whole global structure.*

The solution space for the system (SP) will be $\mathbb{R}^2_+ \times X$ ($\ni ((d_1,\alpha),U)$, where X is the Hilbert space $\mathbb{H}^2_N = (H^2_N)^2 = (H^2(I) \cap$ (the boundary conditions $(P)_3$)).

We state the assumptions on the nonlinearities f and g.

(A.1)(i) There exists a unique constant solution $U = \bar{U} = (\bar{u},\bar{v}) > 0$ of (SP). See, Fig. 0.1.

(ii) $\bar{U}$ is a stable solution of the kinetic system of (P). I.e., the Jacobian matrix at $\bar{U}$, $B = \{\partial(f,g)/\partial(u,v)\}|\bar{U}$, satisfies $\mathrm{tr}(B) < 0$ and $\det(B) > 0$.

(iii) (P) is an activator-inhibitor system, i.e., the elements of B have the sign

$$B = \begin{bmatrix} b_{11} & b_{12} \\ b_{21} & b_{22} \end{bmatrix} = \begin{bmatrix} + & - \\ + & - \end{bmatrix}$$

(A.2) The zero level curve of $f(u,v)$ is S-shaped and $f < 0$ in the upper region of the sigmoidal curve. Fig.0.1; $f = 0$ has three real roots $u_{-1}(v) \leqq u_o(v) \leqq u_{+1}(v)$ for $v \in \Lambda$. When it is solved with respect to u, it has three branches $h_{-1}(v) \leqq h_o(v) \leqq h_{+1}(v)$.

(A.3) Define $J(v)$ by

$$J(v) = \int_{u_{-1}(v)}^{u_{+1}(v)} f(s,v)ds. \tag{0.1}$$

Then, $J(v) = 0$ holds if and only if $v = \eta^* \in \Lambda$, and $d(J(v))/dv < 0$ at $v = \eta^*$.

(A.4) Let $G_{\pm}(v) = g(h_{\pm 1}(v),v) \in C^1(\Lambda)$. Then, $dG_{\pm}(v)/dv < 0$ for all $v \in \Lambda$.

There are a number of examples within the setting (A.1)-(A.4). See, [1].

The *May-Mimura model* for diffusive prey and predator system is an example, in which

$$f(u,v) = \{f_o(u) - v\}u, \quad \text{and} \quad g(u,v) = -\{g_o(v) - u\}v, \tag{0.2}$$

where $\quad f_o(u) = (35 + 16u - u^2)/9, \quad$ and $\quad g_o(v) = 1 + (2/5)v.$

This model has been used for numerical tracing of bifurcating branches in [1].

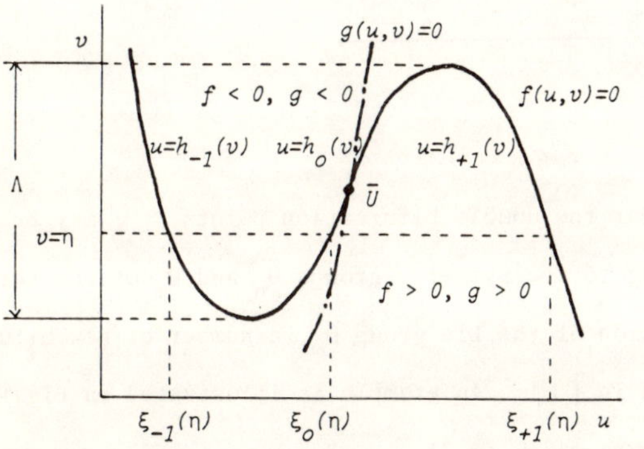

Fig. 0.1.

§1. View near double bifurcation points

The constant state $U = \bar{U}$ has an infinite number of hyperbolic curves Γ_n ($n = 1,2,\ldots$) on the (d_1,d_2)-plane, defined by

(1.1) $$\Gamma_n = \{(d_1,d_2) \in \mathbb{R}_+^2 \mid d_2 = \frac{1}{n^2}(\frac{b_{12}b_{21}}{n^2 d_1 - b_{11}} + b_{22})\}, \quad (n=1,2,\ldots).$$

Γ_n $(n=1,2,\ldots)$ is the set of $(d_1,d_2) \in \mathbb{R}_+^2$ where the constant state $\bar{U}$ has zero eigenvalues, and hence correspond to primary bifurcation points of $\bar{U}$. See, [3] and [1]. Each Γ_n has intersections with the other Γ_m's, $m \neq n$, $m = 1,2,\ldots$ Namely, $\Gamma_{n,m} = \Gamma_n \cap \Gamma_m$ $(n, m = 1,2,\ldots, n \neq m)$ are double bifurcation points of $\bar{U}$, where the linearized operator of (SP) has two dimensional kernel. Let $\Gamma_n' = \Gamma_n / \bigcup_{m \neq n} \Gamma_{n,m}$. From each points of Γ_n', there appears a bifurcating sheet of solutions of (SP), which consists of functions with D_n-symmetry ([1]). See, Fig.1.1. Here, by a function $U \in X$ is D_n-symmetric we mean that its even extension $\tilde{U}$ to $[-\pi,\pi]$, considered as a periodic function defined on the circle $[-\pi, \pi]$, is invariant under the group D_n; D_n is the dihedral group which sends a regular n-polygon onto itself.

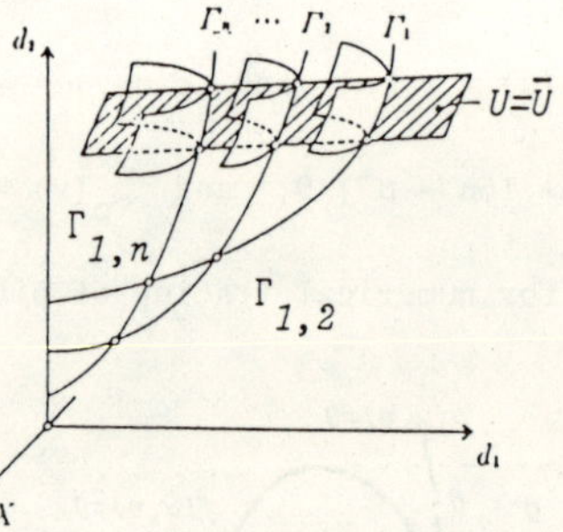

Fig. 1.1.

The local structure near the double bifurcation points $\Gamma_{n,m}$ may be classified into several types according to the symmetry groups D_n and D_m of the kernel at $\Gamma_{n,m}$. Using the group representation of the Lie group D_∞, a number of new bifurcation diagrams have been obtained in [1]. An example is illustrated in Fig.1.2, which is the case for the May-Mimura model (0.2).

The double singularity at $\Gamma_{1,2}$ may deserve a further investigation. Since the bifurcation equations at $\Gamma_{1,2}$ take the form (due to a group theoretic argument):

$(1.2)_1$ $$\rho_1\{-(a_1\sigma_1 - a_2\sigma_2) + p_{11}\rho_2 + p_{30}\rho_1^2 + p_{12}\rho_2^2 + (\text{higher order terms})\} = 0,$$

$$(1.2)_2 \qquad -(b_1\sigma_1 - b_2\sigma_2)\rho_2 + q_{20}\rho_1^2 + q_{21}\rho_1^2\rho_2 + q_{03}\rho_2^3 + (\text{higher order terms}) = 0,$$

where $\underline{\sigma} = \underline{d} - \Gamma_{1,2}$, the local bifurcation parameters, and a_i, b_i, $p_{i,j}$ and $q_{i,j}$'s are all constants; $\underline{\rho} = (\rho_1, \rho_2) \in \mathbb{R}_+^2 \simeq$ the kernel space at $\Gamma_{1,2}$, is the amplitude vector of the bifurcation equations. The bifurcation diagram near $\Gamma_{1,2}$ is shown in Fig.1.3 for the case $p_{11}q_{20} > 0$ ([1]). Note, however, that the destination of the primary D_1-branch is *not* indicated in this diagram, as well as those of the secondary branches of the D_2-branch.

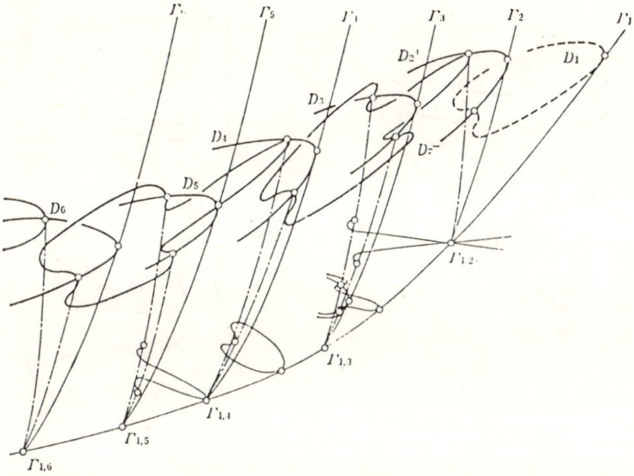

Fig. 1.2. Schematic Bifurcation Diagram near Γ_1.

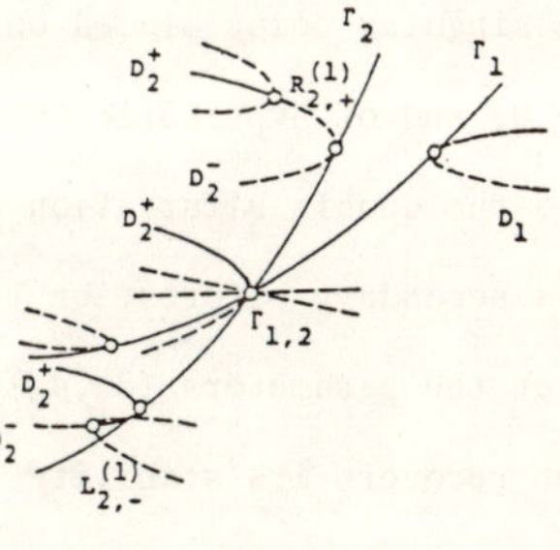

Fig. 1.3. Local Bifurcation Diagram near $\Gamma_{1,2}$ $(p_{11}q_{20} > 0)$.

If one unfolds (1.2) near the degenerate parameter values $p_{11}q_{20} = 0$, an interesting local bifurcation structure near $\Gamma_{1,2}$ reveals itself. In fact, as may be the case for the May-Mimura model (0.2), let us suppose that $q_{20} = q^o_{20} \neq 0$, and let p_{11} be the unfolding parameter as $|p_{11}| < \delta_o$ (δ_o: sufficiently small). The bifurcation diagram thus obtained are shown in Figs. $(1.4)_a$ and $(1.4)_b$, which correspond respectively to the case $q^o_{20}p_{11} > 0$ and < 0, for $|p_{11}| < \delta_o$.

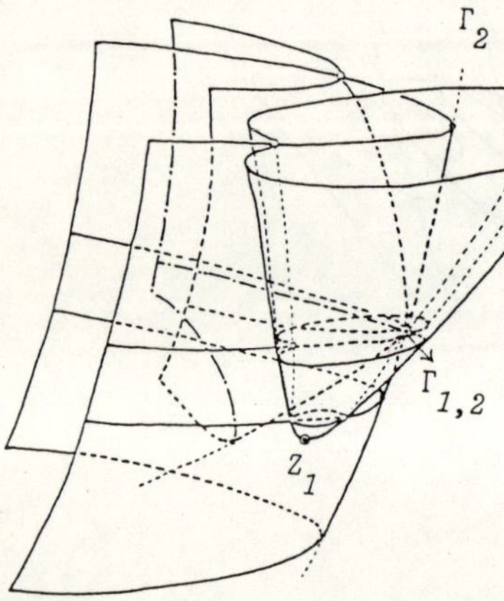

Fig. 1.4$_a$

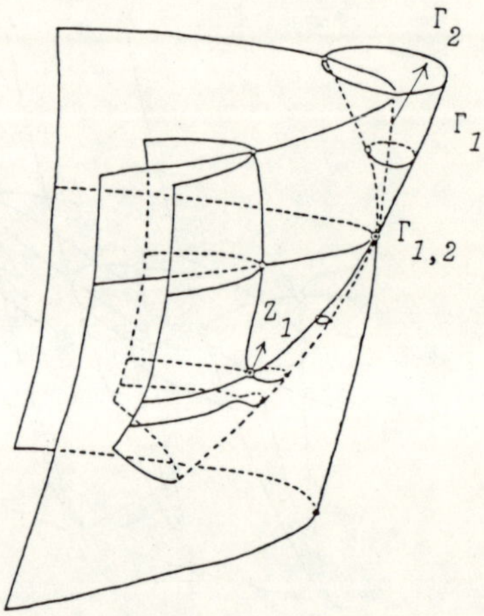

Fig. 1.4$_b$

The "D_1 pot" near $\Gamma_{1,2}$; Z_1 is a simple degenerate singular point of elliptic type ((a), $p_{11}q_{20} > 0$); of hyperbolic type ((b), $p_{11}q_{20} < 0$).

Z_1 is a simple degenerate singular point placed on the D_2-sheet, respectively of elliptic type when $q^o_{20}p_{11} > 0$, and of hyperbolic type when $q^o_{20}p_{11} < 0$. As p_{11} tends to zero, Z_1 approaches to the double bifurcation point $\Gamma_{1,2}$. Thus, one sees that the primary sheet D_2 has a secondary bifurcation line, which passes Z_1 and $\Gamma_{1,2}$. See, Fig.1.5. If one let the parameters (d_1,d_2) cross this line from the right to the left, the D_2-sheet recovers its stability - hence, this secondary line is called the *recovery line of the* D_2*-sheet*.

A remarkable fact in these diagrams is that in both cases there appears a " pot-like" structure due to the existence of $\Gamma_{1,2}$. We note that this pot-like structure has been predicted numerically in [1]. A remark is that due to the *fold-up principle* ([1]), every D_n-primary sheet takes the pot-like form, the origin of which is a degenerate simple singularity Z_n located on the D_{2n}-sheet.

The basic question is the global behavior of this "pot", and also of the other secondary branch born at $\Gamma_{1,2}$. It is also worthy of noting that from $\Gamma_{2,3}$, there appears a secondary bifurcation line of D_2 - which actually corresponds to the points where D_2 loses its stability again. Hence, this line may be called the *losing line of* D_2. Between $\Gamma_{1,2}$ and $\Gamma_{2,3}$, one sees an outcrop of the stable region of the primary D_2-sheet. See, Fig.1.5.

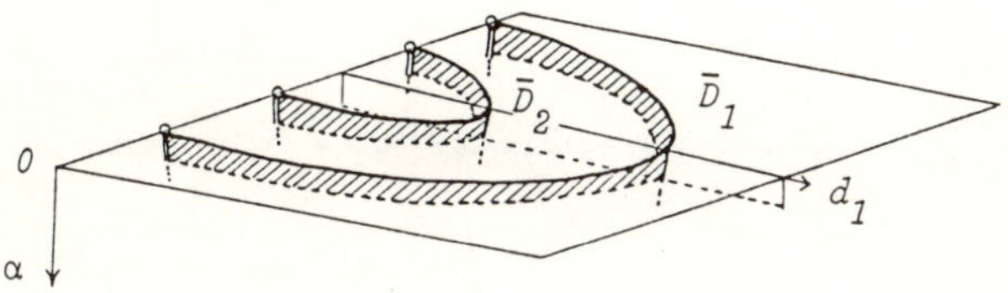

Fig. 2.1. The Shadow Branches and their extension to α > 0.

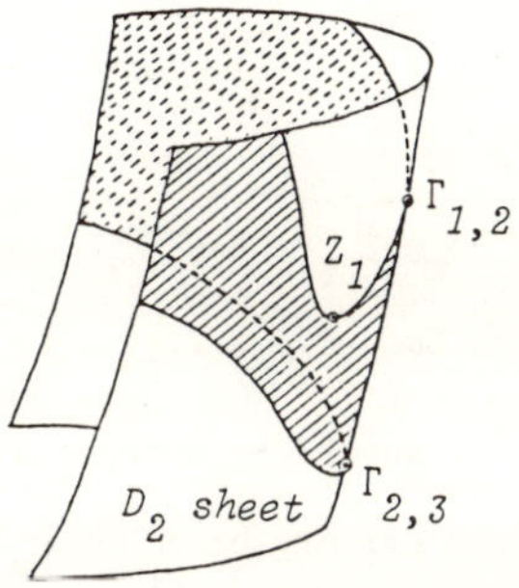

Fig. 1.5. The recovery and losing lines of the D_2 sheet; the shaded part shows the stable region.

§2. View on and near the shadow ceiling - global existence of bifurcating branches

By the *shadow ceiling* we mean the limit space $\mathbb{R}_+ \times X$ of $\alpha \downarrow 0$. If solutions of (SP) are uniformly L_∞-bounded with respect to d_1 and α, one may have the limit system:

$$\text{(SS)} \quad \begin{cases} d_1 u_{xx} + f(u,\eta) = 0, & \text{in } I, \\ \int_I g(u,\eta)dx = 0, \end{cases}$$

with the boundary condition $u_x = 0$ on ∂I, where $v = \eta$ is a *constant* function. The second equation comes from the integration of $(SP)_2$ over the interval I. The system (SS) is called the *shadow system* for (SP).

The global behavior of solutions of (SS) with respect to $d_1 \downarrow 0$ is the first object of the study here, which is expected to approximate the global behavior of solutions of (SP) for sufficiently small $\alpha > 0$. In fact, a complete bifurcation diagram for (SS) has been obtained by the second author [6].

<u>*Theorem 2.1.*</u> *(i) The one-mode bifurcating branch $\bar{D}_1$ emanating from $(\bar{d}_1^c, \bar{U})$ continues to exist as $d_1 \downarrow 0$. By the fold-up principle, the same conclusion holds to the n-mode branches $\bar{D}_n$ emanating from $(\bar{d}_n^c, \bar{U})$. Fig.2.1. (ii) These shadow branches $\bar{D}_n$ $(n = 2,3,\ldots)$ do not recover their stability on the way to the limit $d_1 \downarrow 0$, and consequently, they have* no secondary branches *in a generic sense. Hence, only the $\bar{D}_1$-branch is the stable one among the branches on the shadow ceiling.*

<u>*Theorem 2.2.*</u> *The global existence result, i.e., the statement (i) of Theorem 2.1 holds as well for the bifurcating branches D_n's $(n = 1,2,..)$ of (SP) for small $\alpha > 0$. Namely, every bifurcating branch hits the singular wall $d_1 = 0$ for small $\alpha \geqq 0$. See, Fig.2.1.*

It should be noted that the statement (ii) of Theorem 2.1 does no more hold for the primary branches D_n's of (SP) even for sufficiently small $\alpha > 0$, except for $n = 1$.

<u>§3. The singular shadow edge - Edge continua</u>

The study of the limit $d_1 \downarrow 0$ of the shadow system (SS) plays a key role in subsequent discussions. The reduced shadow system as $d_1 \downarrow 0$ is defined by

$$\text{(RSS)} \qquad \begin{cases} f(u,\eta) = 0, \\ \int_I g(u,\eta)\, dx = 0. \end{cases}$$

A solution $(u,\eta) \in X_o$ (where $v = \eta$ is a *constant* function) of (RSS) is called *a reduced shadow solution*, where $X_o = L^2(I) \times H_N^2(I)$. The system (RSS) has a vast of solutions as compared with (SS) or (SP). In fact, one takes $\eta \in \Lambda$ arbitrarily. Suppose $\xi_\ell(\eta)$, $\ell = 0, \pm 1$, are the three solutions of $(RSS)_1$. See, (A.2). Let $u(\eta;x)$ be any step function in which $u(\eta;x)$ takes either of $\xi_\ell(\eta)$, $\ell = 0, \pm 1$, for almost all $x \in I$. Let $I_\ell(\eta) = \{ x \in I \mid u(\eta;x) = \xi_\ell(\eta)\}$, $\ell = 0, \pm 1$. Then, $(RSS)_2$

reduces to

$$(3.1)\qquad \sum_{\ell=0}^{\pm 1} g(\xi_\ell(\eta),\eta)\,|I_\ell(\eta)| = 0,$$

where $|I_\ell(\eta)|$ = measure of $I_\ell(\eta)$, $\ell = 0, \pm 1$.

Thus, for any $\eta \in \Lambda$, if one chooses $I_\ell(\eta)$, $\ell = 0, \pm 1$, so as to satisfy (3.1), the corresponding step function is a reduced shadow solution of (RSS). See, Fig. 3.1. Note that there are many such step functions, since only the ratio of $|I_\ell(\eta)|$'s has the meaning in (3.1).

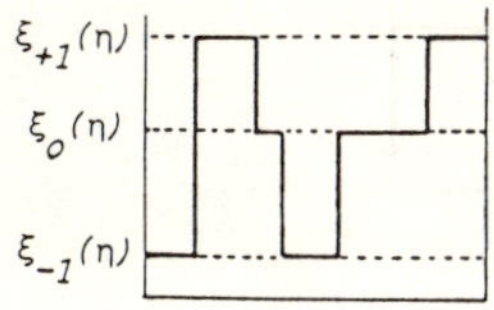

Fig. 3.1.

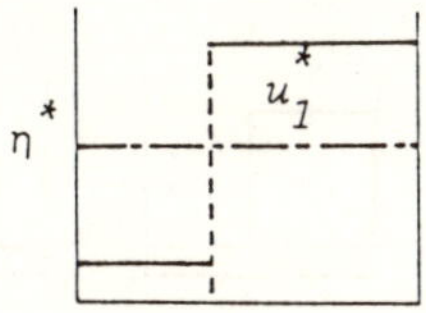

Fig. 3.2.
The Singular Shadow Limit Solution $(n = 1)$.

Among the reduced shadow solutions, we pick up those which satisfy the relation:

$$(3.2)\qquad g(\xi_{+1}(\eta^*),\ \eta^*)\,|I^*_{+1}| + g(\xi_{-1}(\eta^*),\ \eta^*)\,|I^*_{-1}| = 0,$$

and write them as $(\tilde{u}^*(x),\eta^*)$. Let $\tilde{u}^*_n(x)$ be a function of $\{\tilde{u}^*(x)\}$, the even extension of which (considered as a periodic function on the circle $[-\pi,\pi]$) has exactly n intervals of I^*_{+1} (and of I^*_{-1}). (For the definition of η^*, see (0.1).) Namely, $\tilde{u}^*_n(x)$ is a function of $\{\tilde{u}^*(x)\}$ which has n boundary discontinuities. Finally, let $u^*_n(x)$ be a function of $\{\tilde{u}^*_n(x)\}$, the even extension of which is invariant under the group action D_n. We have the following

<u>Theorem 3.1.</u> The shadow branch $\bar{D}_n$ converges to $(u^*_n(x),\eta^*)$ as $d_1 \downarrow 0$ in the punctuated sense. See, Fig.3.2 for $n = 1$. Namely, it converges to $u^*_n(x)$ uniformly on the interval $\bar{I} - \sum_{i=1}^{n}(x^*_i-\kappa, x^*_i+\kappa)$ for any $\kappa > 0$, where $\{x^*_i\}_{i=1}^{n}$ are the points of discontinuity of $u^*_n(x)$ and the location of each discontinuity is determined by (3.2).

One may thus call $(u^*_n(x),\eta^*)$ the $\bar{D}_n$-limit solution.

The set of functions $\{\tilde{u}^*_n(x)\}$ can be obtained from the limit solution $u^*_n(x)$ by *a translation, an extension,* or *a contraction* of intervals of the blocks of $u^*_n(x)$, so long as such an operation keeps the ratio $|I^*_{+1}| / |I^*_{-1}|$. (Note that a *division* of a block of $u^*_n(x)$ yields a function of $\{\tilde{u}^*_{n+2k}(x)\}$ (for some $k \geqq 1$).) Thus, the set $\{\tilde{u}^*_n(x)\}$ may consist of a set of one-parameter families of functions, including the limit state $u^*_n(x)$ - which we call the *edge continua*. An example is illustrated in Fig.3.3, where the continua of the $u^*_2(x)$ and $u^*_4(x)$ are shown.

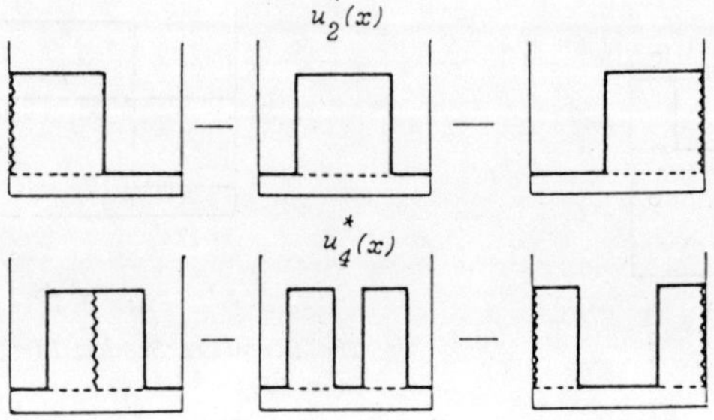

Fig.3.3. Edge continua of $u^*_2(x)$ and $u^*_4(x)$, formed by translations of blocks. Note that the terminal states (the right and left pictures) are different from $u^*_1(x)$ (upper), and $u^*_2(x)$ (lower), since they contain "slits" at $x = 0, \frac{\pi}{2}$ or π.

§4. View on the singular wall

By the singular wall, we mean the limit space $\mathbb{R}_+ \times X_0$ ($\ni (\alpha, (u,v))$) of $d_1 \downarrow 0$. The goal of this section is the study of the structure of solutions on the singular wall of:

$$\text{(RP)} \qquad \begin{cases} f(u,v) = 0, & \\ \dfrac{1}{\alpha} v_{xx} + g(u,v) = 0, & \text{in } I, \\ v_x = 0, & \text{on } \partial I. \end{cases}$$

The system (RP) is called the reduced problem of (SP), and its solutions *reduced solutions*. However, we are only interested in such reduced solutions that from which we can extract smooth solutions of (SP) for (small) positive $d_1 \downarrow 0$. Such limit solutions will be called *singular solutions* (,and a *singular branch* if it consists of a one-parameter family of singular solutions).

Let S_o denote the set of singular solutions. We associate to S_o the following *asymptotic norm*. For $U = (u,v) \in S_o$, the asymptotic norm $\|U\|_s$ is defined by

$$\|U\|_s = \lim_{\varepsilon \downarrow 0} \| (u(\varepsilon;x), v(\varepsilon;,x)) \|_{2,\varepsilon},$$

where $(u(\varepsilon;,x), v(\varepsilon;,x))$ is a family of solutions of (SP) which converges to U as $\varepsilon = \sqrt{d_1}$ in the punctuated sense, and

$$\|(u,v)\|_{2,\varepsilon} = \sum_{k=0}^{2} \max_{x \in \bar{I}} |(\varepsilon \frac{d}{dx})^k u| + \max_{x \in \bar{I}} |(\frac{d}{dx}) v|.$$

Mimura, Tabata and Hosono [4] have found a family of singular branches with interior transition layers for sufficiently small $\alpha > 0$. On the singular wall, these singular branches correspond to the double solid lines in Fig.4.1. At $\alpha = 0$, they start from the *singular-shadow solutions* (u_n^*, η^*) (n=1,2,...), and hence connect to the shadow branches $\bar{D}_n$ (n=1,2,...) at the edge.

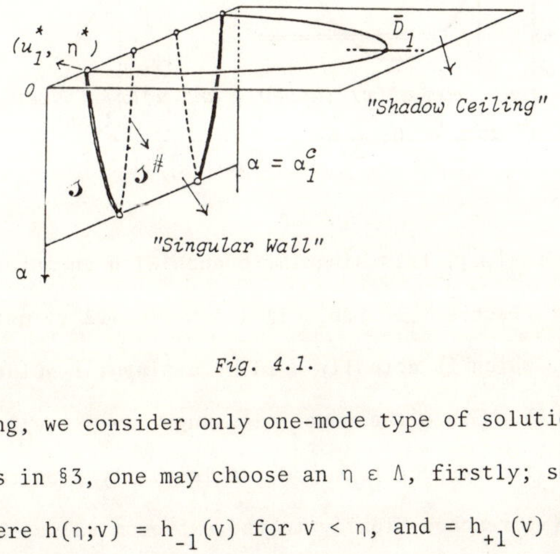

Fig. 4.1.

In the following, we consider only one-mode type of solutions for simplicity of presentation. As in §3, one may choose an $\eta \in \Lambda$, firstly; solve $f(u,v) = 0$ to have $u = h(\eta;v)$, where $h(\eta;v) = h_{-1}(v)$ for $v < \eta$, and $= h_{+1}(v)$ for $v > \eta$(Fig.4.2). Substitution of $u = h(\eta;v)$ into $(RP)_2$ leads to a scalor equation for v:

$$(4.1) \qquad \frac{1}{\alpha} v_{xx} + G(\eta;v) = 0, \qquad x \in I,$$

where $G(\eta;v) = g(h(\eta;v),v)$. Note that $G(\eta;v)$ has a discontinuity in v at $v = \eta$. Eq.(4.1) with $v_x = 0$ on ∂I, has an α-family of strictly increasing solutions $V_1^{\eta,\alpha}(x) \in C^1(\bar{I})$, $0 < \alpha < \alpha_1$ for each $\eta \in \Lambda$. Let $U_1^{\eta,\alpha}(x) = h(\eta;V_1^{\eta,\alpha}(x))$. According to the construction, the function $U_1^{\eta,\alpha}(x)$ has a discontinuity at $x = x^* \varepsilon I$. Then, the couple $(U_1^{\eta,\alpha}, V_1^{\eta,\alpha})$, $0 < \alpha < \alpha_1$, is an α-family of reduced solutions for each $\eta \varepsilon \Lambda$.

A question is that when a small diffusion $d_1 > 0$ is introduced to $(RP)_1$, under what conditions the discontinuity of $U_1^{\eta,\alpha}$ can be smoothed out by an interior transition layer. The following result has been obtained in [4]:

If the Fife condition

$$\int_{\xi_{-1}(\eta)}^{\xi_{+1}(\eta)} f(s,\eta)\, ds = 0, \tag{4.2}$$

is satisfied, then the couple $(U_1^{\eta,\alpha}, V_1^{\eta,\alpha})$ *is a singular solution.* Thanks to the assuumption (A.3), there exists a unique separation point $\eta = \eta^*$, which satisfies (4.2). Hence, follows an α-family of singular solutions with an interior transition layer $\mathcal{J} = \{(U_1^{\eta,\alpha}, V_1^{\eta,\alpha}) \varepsilon X_0,\ 0 < \alpha < \alpha_1^c\}$.

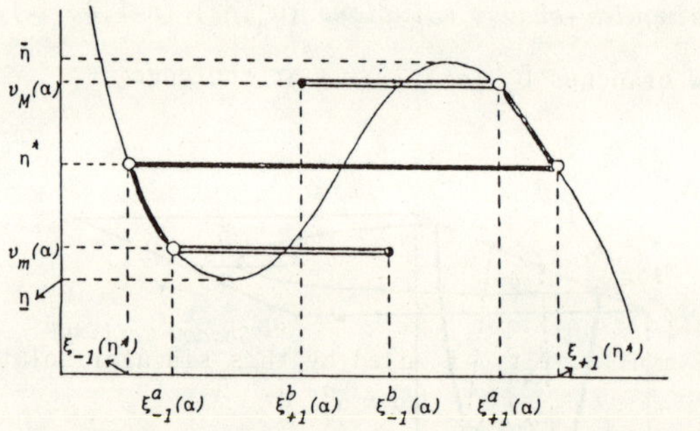

Fig. 4.2.

As is remarked in [1], this singular branch with an interior transition layer ceases to exist as α reaches α_1^c. See, Fig.4.1. Fig.4.2 illustrates that part of the nonlinearity f which is actually used by a singular solution $V_1^{\eta^*,\alpha}(x)$ of (4.1) (the solid line). Since the numerical range $(v_m(\alpha), v_M(\alpha))$ of $V_1^{\eta^*,\alpha}(x)$ is monotone increasing with respect to $\alpha \geq 0$, it is easily seen that, as α increases, one of $v_m(\alpha)$ and $v_M(\alpha)$ reaches finally to the extremum values $\underline{\eta}$ or $\bar{\eta}$ of f. This is the reason why the singular branch ceases its existence at $\alpha = \alpha_1^c$.

It is wondered whether this is all the existing singular branches, and what happens at the "critical point" $\alpha = \alpha_1^c$ on the singular wall. This is a question which has been left open for a long time. The answer is that *there appears another singular branch from* α_1^c *upwards to* $\alpha \downarrow 0$. See, broken lines in Fig.4.1.

This *new* branch is characterized by that it has both *boundary and interior* transition layers. Hence, we shall call it the *singular branch with boundary and interior transition layers*. The critical point α_1^c will be called the *singular limit point*. The detailed construction and proofs will be published elsewhere. However, it should be remarked that such a singular solution can *not* be constructed within the *Fife setting* as in [4]. Moreover, the "boundary layer" thus constructed is completely different in nature from boundary layers observed in Dirichlet boundary-value problems.

The construction on the singular wall of a singular branch with boundary and interior transition layers can be performed simply by adding boundary layers (actually, "boundary slits") to $(U_1^{\eta^*,\alpha}, V_1^{\eta^*,\alpha})$. There are three such branches, corresponding to where the slit exists: at the left ($^{\#}\mathfrak{s}$), or right end ($\mathfrak{s}^{\#}$), or at the both ends ($^{\#}\mathfrak{s}^{\#}$). See, Fig.4.4. The essential point in our construction is that *the depths of the slit* $(\xi_{-1}^b - \xi_{-1}^a)$ and $(\xi_{+1}^b - \xi_{+1}^a)$ *are determined by the generalized Fife condition* (See, Fig.4.2):

$$(4.3) \qquad \int_{\xi_{-1}^a}^{\xi_{-1}^b} f(s,v_m)\,ds = 0\,, \quad \text{and} \int_{\xi_{+1}^a}^{\xi_{+1}^b} f(s,v_M)\,ds = 0.$$

Note that the part of nonlinearity f used by this singular solution is the solid line plus (one or both of)the double solid lines in Fig.4.2.

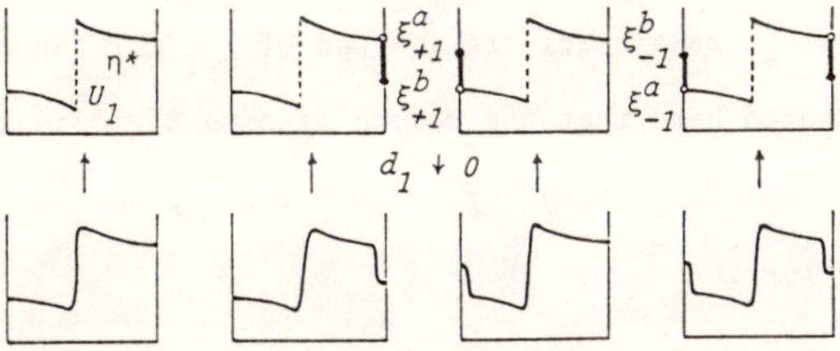

Fig.4.3. The Singular Solutions with Boundary- and Interior-Transition Layers.

It is noted that the four branches constructed in this way are different each other when they are measured by the asymptotic norm $||\cdot||_s$. (They have the same X_o-norm.)

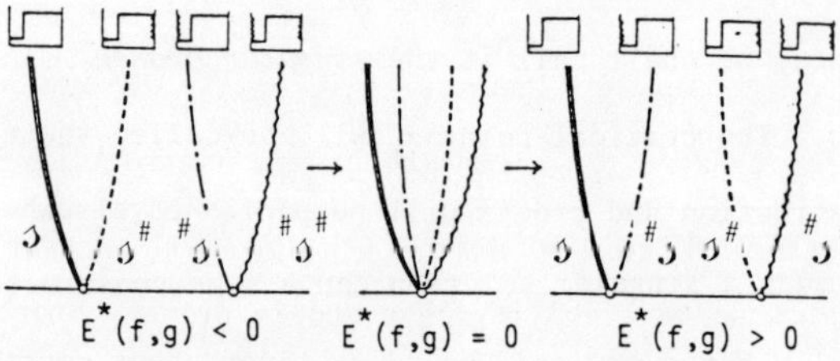

Fig. 4.4. The Dependency on Nonlinearities of Singular Branches.

To study the interrelation of these branches, one may need the quantities:

$$E^*(f,g) = \int_{\underline{\eta}}^{\bar{\eta}} G(\eta^*;v)dv.$$

Suppose $E^*(f,g) < 0$. Then, as α tends to α_1^c, the depth of the slit at ξ_{+1} tends to zero, while the depth at ξ_{-1} remains bounded away from zero. (And, *vice versa* if $E^*(f,g) > 0$.) Thus, the four singular branches form two "wedges" on the singular wall as in Fig.4.1, since the asymptotic norms of the branches $\mathcal{S}$ and $\mathcal{S}^{\#}$ take the same value at $\alpha = \alpha_1^c$. The same is true for the branches ${}^{\#}\mathcal{S}^{\#}$ and ${}^{\#}\mathcal{S}$.

What happens when one deforms the nonlinearity (f,g) so that $E^*(f,g)$ changes smoothly ? See, Fig.4.4. The four singular branches move smoothly, and when $E^*(f,g) = 0$, the tops of the two wedges meet together. Then, they split again for $E^*(f,g) > 0$. Note that an exchange of branches occurs in this process, since the depth of the slit at ξ_{+1} remains finite instead of ξ_{-1} for the case $E^*(f,g) > 0$.

It should be noted here that *the wedges of singular branches on the wall are traces of "hitting" and "splitting" of limit points of some branches of the stationary problem (SP).*

§5. Discussions - the Global View

The purpose of discussions here is to integrate the analytical results in §1 - §4 into a unified view to the global bifurcation structure for our nonlinear diffusion systems. One of the main interests is to see the mechanism of successive recovery and losing of stability observed in primary branches of (SP). (See, [1].)

The local structure of double singularities placed on the trivial sheet $((d_1, d_2), \bar{U}) \in \mathbb{R}_+^2 \times X$ (§1), the global structure on the shadow ceiling (§2), and the somewhat complex structure of singular branches on the singular wall (§4) - they are all expected to reflect the real existing bifurcation structure of the nonlinear diffusion system.

The first key seems to be the structure of continua at the singular-shadow edge. In fact, an integration of all the above results, together with the numerical evidences reported in [1], may lead us to a working hypothesis on the edge continua.

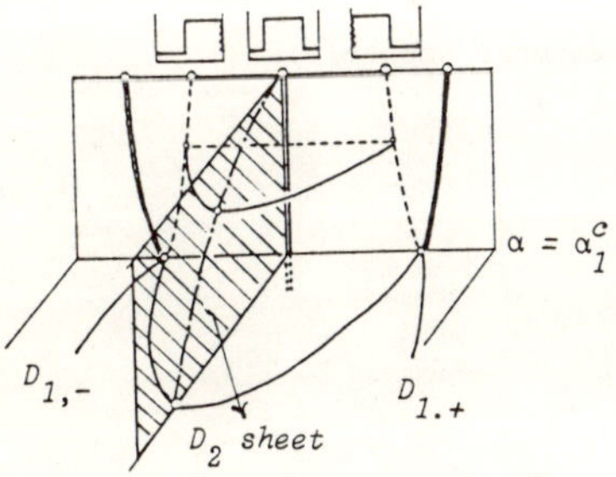

Fig. 5.1$_a$

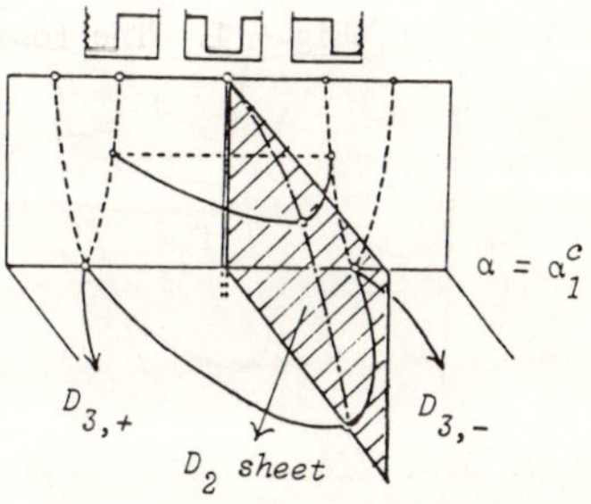

Fig. 5.1$_b$

Extension of the Edge Continuum; The line —·——·- shows the recovery line of $D_{2,-}$ and $D_{2,+}$ which tend to the Shadow Singular Limits. The two Singular Limit Points of the D_1 sheet appear here.

Two singular branches with boundary slits are said to be *terminal branches* of $D_{n,+}$ (or $D_{n,-}$) if at $\alpha = 0$, they are connected by a (one-parametrized) edge continuum which includes the limit state $D_{n,+}$(or $D_{n,-}$). Then, *for a pair of two terminal singular branches of $D_{n,\pm}$, there may exist a sheet of solutions of (SP)* *which connect the two terminal branches, and to which intersects the primary sheet* $D_{n,\pm}$ *transversally*. See, Fig.5.1 (a) for the case of $D_{2,-}$, and Fig.5.1 (b) for $D_{2,+}$.

Assuming this, the global picture of the D_1-sheet looks like Fig.5.2. The pot-like structure in Fig.1.4 which begins at Z_1 on the D_2 sheet extends, and as α becomes snmaller, the "loop" expands until it hits the singular wall at $\alpha = \alpha_1^c$, where it yields two singular limit points. As $\alpha < \alpha_1^c$, the loop splits into two arcs, one is, of course, a cross-section of the primary D_1-sheet, and the other is the branch connecting the two terminal singular states of D_2, which have a boundary- and interior transition layers. As α tends to zero, the latter arc shrinks to the edge continuum of D_2, while the former remains as the primary $\bar{D}_{1,\pm}$ shadow branches.

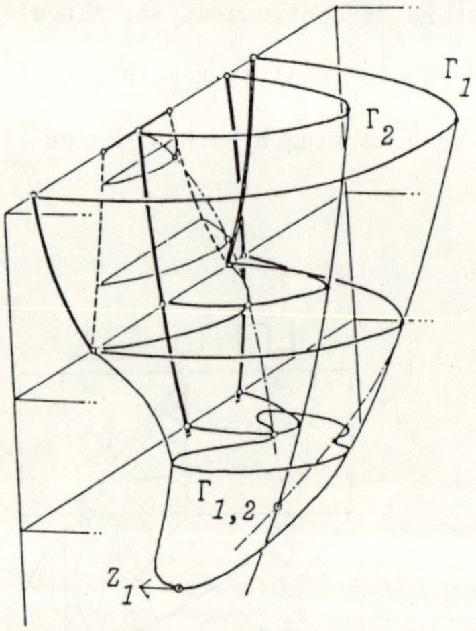

Fig. 5.2. The Global Picture of the D_1 *pot.*

An important consequence of this picture is that the outer surface of this D_1-pot is the *stable region* of (P), while the inner surface corresponding to the boundary- and interior-layered solutions is the *unstable region*. A remark should be made here. There remains a possibility that this *stable* region may have some isolated *"unstable islands"* encircled by a Hopf secondary bifurcation line. However, it is shown in [5] by an *a priori* estimate that such Hopf points, if exists, cannot exist for sufficiently small $\alpha > 0$.

We note that such a global picture is supported by numerical computations in

[1], and in fact, this has been essentially predicted there. In [1], we were not aware of the nature of the inner surface of the pot, since the boundary- and interior-layered solutions and the existence of singular limit point were not discovered yet.

Fig.5.3 shows the recovery of stability of the D_2-sheet. This picture shows the mechanism of a recovery of stability; this recovery is actually performed by a separation of a sub-branch which has both boundary- and interior layers.

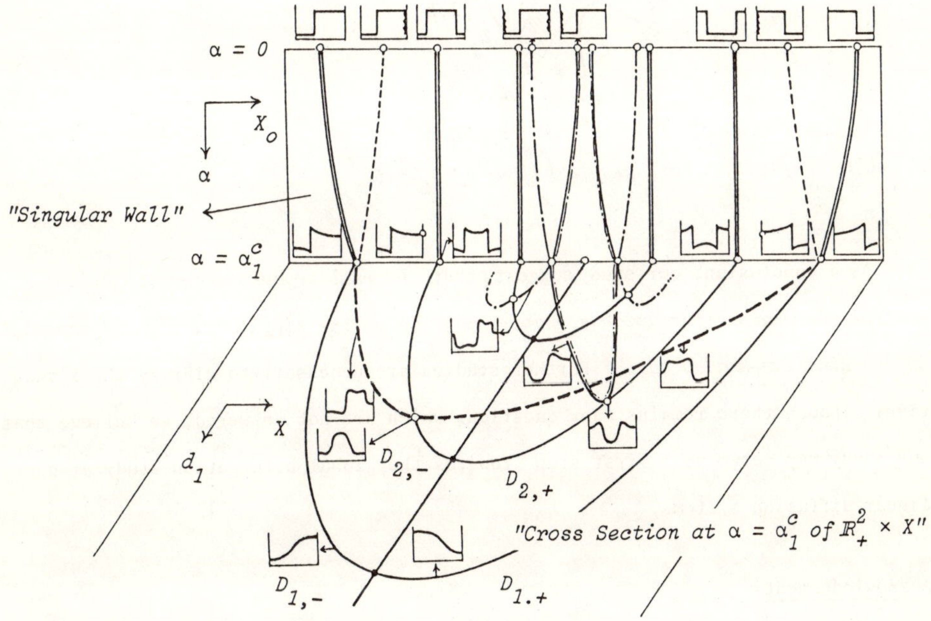

Fig. 5.3.

The Singular Limit Points of $D_{1,\pm}$, and the Recovery of $D_{2,\pm}$-branches.

It should be remarked that, as is mentioned in §1 (see, also [1]), the primary bifurcating D_2-sheet loses stability on the way to the singular wall $d_1 \downarrow 0$. One sees an outcrop of the stable region of D_2 between $\Gamma_{1,2}$ and $\Gamma_{2,3}$ in Fig.1.5. The above analyses suggest that the recovery line of stability of D_2, which starts $\Gamma_{1,2}$, enters into the singular-shadow edge $\alpha = d_1 = 0$, while the losing line which starts $\Gamma_{2,3}$, enters into the singular wall $d_1 = 0$ *at latest* at the singular limit point of D_2, namely, at $\alpha_2^c = 4\alpha_1^c$. As a result, the the stable region of the D_2-sheet occupies a band-shaped region of the D_2-pot. Similar statements hold in general to the D_n-sheets ($n \geqq 2$). See, Fig5.4 for the case of the D_2-sheet.

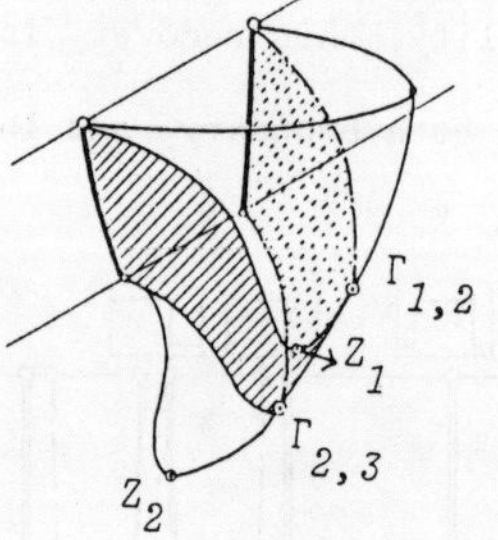

Fig.5.4. The Stable Region of D_2-pot.

As a conclusion, our study suggests that *the real organizing centre which control the whole bifurcation structure may lie on the singular wall, and especially in the singular-shadow edge.* Further studies are necessary to clarify the situation. Though there remains many questions which are not answered, we believe that our study may serve as a first step towards the global bifurcation study of non-linear diffusion systems.

Acknowledgements

We owe a special thank to our colleague Prof. Yuzo HOSONO for his discussion.

References

[1] H.Fujii, M.Mimura and Y.Nishiura, A Picture of the Global Bifurcation Diagram in Ecological Interacting and Diffusing Systems, *Physica D - Nonlinear Phenomena*, 5, No.1, 1-42 (1982).

[2] H.Fujii, Numerical Pattern Formation and Group Theory, *Computing Methods in Applied Sciences and Engineering*, Eds. R.Glowinski and J.L.Lions - Proc. of the Fourth International Symposium on Computing Methods in Applied Sciences and Engineering, North-Holland, 63-81 (1980).

[3] M.Mimura, Y.Nishiura and M.Yamaguti, Some Diffusive Prey and Predator Systems and Their Bifurcation Problems, *Annals of the New York Academy of Sciences*, 316, 490-510 (1979).

[4] M.Mimura, M.Tabata and Y.Hosono, Multiple Solutions of Two-Point Boundary Value Problems of Neumann Type with a Small Parameter, *SIAM J. Math. Anal.* 11, 613-631 (1980).

[5] Y.Nishiura, Global Structure of Bifurcating Solutions of Some Reaction-Diffusion Systems and their Stability Problem, *Computing Methods in Applied Sciences and Engineering*, *V*, Eds. R.Glowinski and J.L.Lions, North-Holland, 185-204 (1982).

[6] Y.Nishiura, Global Structure of Bifurcating Solutions of Some Reaction-Diffusion Systems, *SIAM J. Math. Anal.*, 13, 555-593 (1982).

[7] A.M.Turing, The Chemical Basis of Morphogenesis, *Phil. Trans. Roy. Soc.*, B237, 37-72 (1952).

Lecture Notes in Num. Appl. Anal., **5**, 37–54(1982)
Nonlinear PDE in Applied Science. U.S.-Japan Seminar, Tokyo, 1982

The Navier-Stokes Initial Value Problem in L^p and Related Problems

Yoshikazu Giga [1, 2]

Department of Mathematics
Faculty of Science
Nagoya University
Furo-cho, Chikusa-ku
Nagoya 464
JAPAN

We discuss the existence of a strong solution of the non-stationary Navier-Stokes system in L^p spaces. Our results generalize L^2 results of Kato and Fujita. To establish L^p theory we study the Stokes system and construct the resolvent of the Stokes operator.

Introduction and summary of results.

This is an introduction to the articles [3-7] which concern the Stokes and the Navier-Stokes equations.

Let D be a bounded domain in $\mathbb{R}^n$ $(n \geqq 2)$ with smooth boundary S. We consider the Navier-Stokes initial value problem concerning velocity $u = (u^1, \cdots, u^n)$ and pressure p:

$$\begin{aligned} \frac{\partial u}{\partial t} - \Delta u + (u,\mathrm{grad})u + \mathrm{grad}\, p &= f \quad \text{in } D \times (0,T), \\ \mathrm{div}\, u &= 0 \quad \text{in } D \times (0,T), \\ u &= 0 \quad \text{on } S \times (0,T), \\ u(x,0) &= a(x) \quad \text{in } D \end{aligned}$$

with given external force f and initial velocity a. Here

$$(u,\mathrm{grad}) = \sum_{j=1}^{n} u^j \frac{\partial}{\partial x_j}.$$

Many mathematicians, J. Leray, E. Hopf,··· have studied the solvability of this problem; see Ladyzhenskaya [9] and Temam [14] and papers cited there. On the existence of a regular (in time) solution there is a celebrated work established by Kato and Fujita [1,8].

Let us quickly review their theory. As is well known (see [9]) the space $(L_2(D))^n$ admits the orthogonal Helmholtz decomposition

$$(L_2(D))^n = X_2 \oplus G_2,$$

$$X_2 = \text{the closure in } (L_2(D))^n \text{ of } \{u \in (C_0^\infty(D))^n;\ \mathrm{div}\, u = 0\},$$

$$G_2 = \{\mathrm{grad}\, p;\ p \in W_2^1(D)\},$$

where $W_r^m(D)$ is the Sobolev space of order m such that $W_r^0(D) = L_r(D)$. Let P be the orthogonal projection from $(L_2(D))^n$ onto X_2. Using P, we can transform (I) to the evolution equation in X_2

$$\text{(II)} \qquad \frac{du}{dt} + Au = Fu + Pf \quad (t > 0), \quad u(0) = a,$$

where $Fu = -P(u,\mathrm{grad})u$. Here the operator $A = A_2 = -P\Delta$ is called *the Stokes operator* in X_2 with the domain

$$D(A) = \{u \in W_2^2(D);\ u = 0 \text{ on } S\} \cap X_2.$$

For simplicity we assume $Pf = 0$. Applying semigroup theory, Kato and Fujita have proved the existence of a unique global strong solution of (II) for every $a \in X_2$ when the space dimension n is two. While, when $n = 3$, they have proved the existence of a unique local strong solution of (II) for $a \in D(A^{1/4})$, where A^α denotes the fractional power of A.

Our aim is to show the existence of a unique strong solution *without* assuming that the initial velocity a is regular. To do this, we develop L_p theory $(1 < p < \infty)$ which extends the corresponding L_2 theory of Kato and Fujita. When $p = n$, our main result reads that $a \in L_n(D) \cap X_2$ *implies the existence of a unique strong solution.* With a particular choice of $p = 2$ this is just a result of [8]. On the other hand Serrin [12] raised a question how to show the existence of strong solutions in $n > 4$. Our L_p theory also answers to his problem.

To develop our theory the crucial step is to derive the following two properties of the Stokes operator A_p; here X_p and A_p are L_p-analogues of X_2 and A_2, respectively.

Theorem 1 ([3]). *The operator* $-A_p$ *generates a bounded analytic semigroup in* X_p. *Moreover, the estimate*

$$||(\lambda + A_p)^{-1}f|| \leqq \frac{C}{|\lambda|+1}||f||, \qquad f \in X_p, \qquad |\arg \lambda| \leqq \pi-\varepsilon \quad (\varepsilon > 0)$$

is valid with constant C, *where* $||f||$ *denotes the norm of* f *in* X_p.

Theorem 2 ([4]). *The space* $D(A_p^\alpha)$ *is the complex interpolation space* $[X_p, D(A_p)]_\alpha$, *where* $0 \leqq \alpha \leqq 1$.

Remark. When $p = 2$, we easily get the above properties by using the abstract functional analysis, because A_2 is a strictly positive self-adjoint operator.

Remark. Solonnikov [13] has proved the first part of Theorem 1 when $n = 3$, although his method is different from ours.

Remark. For the Laplace operator, the corresponding results are known by Fujiwara [2] and Seeley [11]. However, we cannot apply their results to our case since P does *not* commute with Δ.

To prove the theorems above we construct the resolvent of the Stokes operator, using pseudodifferential operators; see [3]. Since we use technique of pseudodifferential operators instead of integral kernels, our argument is more clear than the classical potential-theoretic discussion. Here our symbol class of pseudodifferential operators differs from that of Fujiwara [2] and Seeley [11] because the Stokes system is elliptic *not* in the sense of Petrowsky but in the sense of Agmon, Douglis and Nirenberg. Using L_p estimate for pseudodifferential operators, we can prove Theorems 1 and 2; for the detail, see [3,4].

Theorem 2 is useful in estimating the nonlinear term Fu in (II).

Lemma 1 ([5]). *Let* $0 \leqq \delta < \frac{1}{2} + n(1 - \frac{1}{p})\frac{1}{2}$. *We have*

$$\|A_p^{-\delta}P(u,\mathrm{grad})v\| \leqq M\|A_p^{\theta}u\|\,\|A_p^{\rho}v\|$$

with constant $M = M(\delta,\theta,\rho,p)$ *if* $\delta+\theta+\rho \geqq \frac{n}{2p} + \frac{1}{2}$, $\theta > 0$, $\rho > 0$, $\rho+\delta > \frac{1}{2}$.

In particular, if $p = n$ *the above estimate is valid for*

$$\delta = \frac{1}{4}, \quad \theta = \frac{1}{4}, \quad \rho = \frac{1}{2}.$$

We now consider (II) in X_p. The existence result follows from Theorem 1 and Lemma 1. Our method to prove is similar to that of Kato and Fujita.

Theorem 3 ([5]). *Fix* γ *such that* $n/2p - 1/2 \leqq \gamma < 1$. *Assume that* $a \in D(A_p^{\gamma})$. *Then there exists a unique local strong solution* u *of* (II) *with the following properties. For some* $T > 0$,

(i) u *is continuous from* $[0,T)$ *to* $D(A_p^{\gamma})$,

(ii) u *is continuous from* $(0,T)$ *to* $D(A_p^{\alpha})$ and

$$\|A^{\alpha}u(t)\| = o(t^{\gamma-\alpha}) \quad as \quad t \to 0 \quad for\ some \quad \alpha,\ \gamma < \alpha < 1.$$

Moreover, u *is smooth in* $\overline{D} \times (0,T)$. *If* $\|A_p^{\gamma}a\|$ *is small, then* u *can be extended to a global solution.*

One can easily see that Theorem 3 includes the results of Kato and Fujita [8] as a particular case $p = 2$, $n = 2, 3$.

Remark. When $p = n, \gamma$ can be taken to be zero and the assumption on initial data is $a \in X_n$. Kohn [15] pointed out this assumption is reasonable because the norm of a in X_n has zero dimension in "dimensional calculus".

Remark. Even when the zero-boundary condition in (I) is replaced by some first order boundary condition, we can prove similar results; see [6].

We next discuss the analyticity of the solution u of (II).

Theorem 4 ([7]). (i) *Let* u *be as in Theorem* 3. *Then* $u(t)$ *is analytic in* $(0,T)$ *with value in* $W_p^2(D)$.

(ii) *Suppose that* S *is analytic at* x_0. *Then,* $u(x,t)$ *is analytic in* (x,t) *at* (x_0,t), $t \in (0,T)$.

The first part implies the time-analyticity of the solution, while the second part implies the spatial analyticity up to the boundary S.

In [7] we extend the results of Masuda [10]. He discussed the time-analyticity in L_2 spaces and the interior spatial analyticity.

In the following sections we give heuristic arguments to prove the foregoing Theorems. We omit proof unless it is very short and understandable even to non-specialists. For the technical detail, see author's papers [3-7].

1. The resolvent of the Stokes operator.

We investigate the way of λ-dependence of the resolvent $(\lambda + A_p)^{-1}$. To do this we construct the resolvent. We begin with transforming the equation $(\lambda + A_p)u = f$ in X_p into the following Stokes equations

$$(\lambda - \Delta)u + \operatorname{grad} p = f \quad \text{in} \quad D,$$
$$\operatorname{div} u = 0 \quad \text{in} \quad D,$$
$$u = 0 \quad \text{on} \quad S,$$

where p is some scalar function. Since f determines u for $\lambda \in \mathbb{C} \setminus (-\infty, 0)$, we denote u by $u = G_\lambda f$. Our plan to construct G_λ is

1°. Reduce the problem to the Dirichlet problem with *tangential* boundary data.

2°. Solve an integral equation on the boundary for large λ.

To do Step 1° we recall the hydrodynamic potential. Set

$$k_\lambda^{ij}(\xi) = \left(\delta^{ij} - \frac{\xi_i \xi_j}{|\xi|^2}\right) \frac{1}{\lambda + |\xi|^2}, \quad \xi \in \mathbb{R}^n, \quad 1 \leqq i,j \leqq n,$$

where δ^{ij} is Kronecker's delta and $|\xi|^2 = \xi_1^2 + \cdots + \xi_n^2$. The hydrodynamic potential of f is

$$(K_\lambda f)(x) = (F^{-1} k_\lambda Ff)(x),$$

where F is the Fourier transformation with respect to x. The definition of $K_\lambda f$ implies that $u' = K_\lambda f$ satisfies the equations

$$(\lambda - \Delta)u' + \operatorname{grad} p' = f \quad \text{in} \quad \mathbb{R}^n,$$
$$\operatorname{div} u' = 0 \quad \text{in} \quad \mathbb{R}^n,$$

where p' is some scalar function on $\mathbb{R}^n$. Using $K_\lambda f$, we reduce the problem to the Dirichlet problem. More explicitly, $w = K_\lambda f - G_\lambda f$ satisfies

$$(\lambda - \Delta)w + \operatorname{grad} p'' = 0 \quad \text{in} \quad D,$$
$$\operatorname{div} w = 0 \quad \text{in} \quad D,$$
$$w = K_\lambda f \quad \text{on} \quad S,$$

where p'' is some scalar function on D.

We now reduce the problem to the Dirichlet problem with tangential boundary data. Let $z = N\varphi$ satisfy

$$\Delta z = 0 \quad \text{in} \quad D, \qquad \frac{\partial z}{\partial \nu} = \varphi \quad \text{on} \quad S, \qquad \int_D z(x)dx = 0,$$

where ν_x denotes the unit interior normal vector to S at $x \in S$. Let $< , >$ be the standard inner product in $\mathbb{R}^n$. Definition shows that

$$v = w - \text{grad}\, N<\nu, \gamma K_\lambda f>, \qquad q = p'' + \lambda N<\nu, \gamma K_\lambda f>$$

satisfy

$$\begin{aligned} (\lambda - \Delta)v + \text{grad}\, q &= 0 \quad \text{in} \quad D, \\ \text{div}\, v &= 0 \quad \text{in} \quad D, \\ v &= g \quad \text{on} \quad S, \\ <\gamma v, \nu> &= 0 \quad \text{on} \quad S, \end{aligned}$$

where γw denotes the trace of w on S. We call this problem *the Dirichlet problem with tangential boundary data* and denote v by $v = V_\lambda g$. If we notice that the projection P is defined by

$$Pf = f - \text{grad}\, N<\nu, \gamma f> \quad \text{for} \quad \text{div}\, f = 0,$$

we see that $v = PK_\lambda f - G_\lambda f$ and $g = \gamma PK_\lambda f$. We thus have

$$G_\lambda f = PK_\lambda f - V_\lambda M_\lambda f \quad \text{with} \quad M_\lambda f = \gamma PK_\lambda f.$$

Since P and K_λ are written explicitly, all we have to do is to construct V_λ.

Remark. We have reduced the original problem to the Dirichlet problem with tangential boundary data not with general boundary data. This is because the effect of the normal component differs from that of tangential component.

We now give a rough outline of Step 2°. By Step 1° our problem is reduced to construct V_λ. Let Y_λ be a pseudodifferential operator of order one on S. We consider the single layer potential $K_\lambda(\delta_S \otimes Y_\lambda h)$. We see

$$W_\lambda h = PK_\lambda(\delta_S \otimes Y_\lambda h)$$

satisfies the Dirichlet problem with tangential boundary data $\gamma W_\lambda h$; we denote $\gamma W_\lambda h$ by $S_\lambda h$. Our problem is now to solve the integral equation

$$g = S_\lambda h$$

for given tangential boundary data g. In [3] we construct Y_λ such that S has the inverse for large λ. We thus have $V_\lambda = W_\lambda S_\lambda^{-1}$. This yields

$$G_\lambda f = PK_\lambda f - W_\lambda S_\lambda^{-1} M_\lambda f.$$

We thus have constructed the resolvent.

To construct Y_λ the author introduced a symbol class $S^{m;k}$ (see below) of pseudodifferential operators.

Definition ([3]). Let m and k be real numbers. Then we denote by $S^{m;k}$ the set of all $p_\lambda \in C^\infty(\mathbb{R}^n \times \mathbb{R}^n)$ $(\lambda \in \mathbb{C} \setminus (-\infty,0])$ such that for all multi-indices α,β and positive numbers ε,ω the estimate

$$|\partial_\xi^\alpha \partial_x^\beta p_\lambda(x,\xi)| \leqq M\langle\xi\rangle^{m-|\alpha|}\langle\lambda;\xi\rangle^k, \quad (x,\xi) \in \mathbb{R}^n \times \mathbb{R}^n, \quad |\arg\lambda| \leqq \pi-\varepsilon, \quad |\lambda| > \omega$$

is valid with constant $M = M(\alpha,\beta,\varepsilon,\omega)$; here $\langle\lambda;\xi\rangle$ denotes $(|\lambda| + |\xi|^2 + 1)^{1/2}$ and $\langle\xi\rangle = \langle 0;\xi\rangle$.

Remark. Grubb [16] introduces a similar symbol class. In our situation her symbol class reads that the estimate above is replaced by

$$|\partial_\xi^\alpha \partial_x^\beta \partial_\lambda^j p_\lambda(x,\xi)| \leqq M((\langle\xi\rangle/\langle\lambda;\xi\rangle)^{m-|\alpha|} + 1)\langle\lambda;\xi\rangle^{k-|\alpha|-j}.$$

If we use this symbol class, our formulation will be more clear than that of [3,4].

2. Analyticity of the semigroup e^{-tA} and domains of A^α.

In section 1 we obtain an explicit form of G_λ, *i.e.*,

$$(\lambda + A)^{-1}f = G_\lambda f = PK_\lambda f - W_\lambda S_\lambda^{-1}M_\lambda f.$$

Since the spectrum of A is contained in $(-\infty,0)$, to prove Theorem 1 it suffices to prove the estimates

1. $\|PK_\lambda f\| \leqq \frac{C}{|\lambda|}\|f\|, \quad f \in (L_p(D))^n, \quad |\arg \lambda| \leqq \pi-\varepsilon$

2. $\|W_\lambda S_\lambda^{-1}M_\lambda f\| \leqq \frac{C}{|\lambda|}\|f\|, \quad f \in (L_p(D))^n, \quad |\arg \lambda| \leqq \pi-\varepsilon$

with constant $C = C(\varepsilon)$ for large λ. To show Step 1 is easy, so we give a proof.

Proof of Step 1. It is known that P is a bounded operator in $(L_p(D))^n$, so it suffices to prove

$$\|K_\lambda f\| \leqq \frac{C}{|\lambda|}\|f\| .$$

Since the symbol of K_λ satisfies the estimate

$$|\partial_\xi^\alpha k_\lambda(\xi)| \leqq \frac{C}{|\lambda|} |\xi|^{-|\alpha|},$$

this follows from Mihlin's L^p-boundedness theorem.

Remark. In principle L^p-boundedness of operator follows from estimates for its symbol. Classical results are the Calderón-Zygmund inequality and Mihlin's theorem for convolution type operators. In recent years L^p-boundedness is proved for more general class of operators, namely, pseudodifferential operators. We often use L^p-boundedness theorem to construct and estimate the resolvent; see [3,4].

We see Step 2 follows from the following three estimates:

$$\|W_\lambda g\| \leqq C|\lambda|^{-1/2p}|g|,$$

$$|M_\lambda f| \leqq C|\lambda|^{-1+1/2p}\|f\|,$$

$$|S_\lambda^{-1}h| \leqq C|h|, \quad \langle h,\nu\rangle = 0,$$

where $|h|$ denotes the norm of h in $(L_p(S))^n$. The first two estimates are easy to prove. The last one follows from the construction of Y_λ; for the detail see [3].

By Theorem 1 we can define fractional power A_p^z (Re $z < 0$), using the Dunford integral. To prove Theorem 2 it suffices to prove

$$\|A_p^z\| \leqq C_a e^{\varepsilon|\mathrm{Im}\ z|}$$

for all z such that $-a < \mathrm{Re}\ z < 0$, where $\|A_p^z\|$ is the operator norm of A_p^z. We prove this estimate, using the Dunford integral

$$A^z = \frac{1}{2\pi i}\int_\Gamma (-\lambda)^z\, G_\lambda d\lambda.$$

The expression of G_λ enable us to estimate the right side. The estimate

$$\left\|\int_\Gamma (-\lambda)^z\, PK_\lambda d\lambda\right\| \leqq Ce^{\varepsilon|\mathrm{Im}\ z|}$$

is easy to prove like the proof of Step 1. While

$$\left\|\int_\Gamma (-\lambda)^z\, V_\lambda M_\lambda d\lambda\right\| \leqq Ce^{\varepsilon|\mathrm{Im}\ z|}$$

is not easy. We have to study $V_\lambda M_\lambda = W_\lambda S_\lambda^{-1} M_\lambda$ more carefully. In [4] we decompose $V_\lambda M_\lambda$ into main term and error term to get the estimate.

Remark. Theorem 2 is very similar to the case of the Laplace operator B_p

with zero boundary condition. Fractional powers of both operators are closely related. Indeed, in [4] we show

$$D(A_p^\alpha) = D(B_p^\alpha) \cap X_p, \quad 0 < \alpha < 1.$$

3. Solutions to the evolution equation (II).

We now consider the nonlinear equation (II) and discuss only the existence results. In Theorem 3 the reader may have a canonical question.

What is T ? *Are there any estimate for* T *from below ?*

To answer this question we restrict ourselves in a simple case, for example, $p = n$ and give a rough outline of proof of Theorem 3; for the detail see [5].

We begin with estimates for nonlinear term Fu in (II). To avoid technical difficulties we only give a proof to the last part of Lemma 1. That is

$$\|A^{-1/4} P(u,\mathrm{grad})v\|_n \leqq M\|A^{1/4}u\|_n \|A^{1/2}v\|_n \quad (\|f\|_n : L_n\text{-norm of } f).$$

Proof. In a word this estimate follows from Sobolev's inequality and Theorem 2. Let $H_p^{2\alpha}(D)$ be the space of Bessel potentials, that is,

$$H_p^{2\alpha}(D) = [L_p(D), W_p^2(D)]_\alpha.$$

Theorem 2 now implies that the canonical injection $D(A_p^\alpha) \subset H_p^{2\alpha}(D)$ is continuous. Sobolev's inequality now implies that the injections

$$\text{(a)} \quad D(A_n^{1/4}) \subset L_{2n}(D)$$

$$\text{(b)} \quad D(A_{2n/3}^{1/4}) \subset L_n(D)$$

are continuous. The second inclusion (b) yields

$$\|A^{-1/4}\ P(u,\mathrm{grad})v\|_n \leqq C\|P(u,\mathrm{grad})v\|_{2n/3}.$$

Applying Hölder's inequality, we have

$$\|P(u,\mathrm{grad})v\|_{2n/3} \leqq C\|u\|_{2n}\ \|\mathrm{grad}\ v\|_n\ ,$$

since P is continuous from L_p to L_p. The estimates

$$\|u\|_{2n} \leqq C\|A^{1/4}u\|_n$$

$$\|\mathrm{grad}\ v\|_n \leqq C\|A^{1/2}u\|_n$$

follow from (a) and $D(A_p^{1/2}) \subset W_p^1(D)$. Combine the above to get the result.

To solve (II) we consider its integral form

(III) $$u(t) = e^{-tA}a + \int_0^t e^{-(t-s)A}\ Fu(s)\ ds,\quad t > 0.$$

We use Theorem 1 and Lemma 1 to prove existence theorem for the integral equation (III) in X_n. We construct approximate solutions by the iteration scheme

$$u_0(t) = e^{-tA}\ a,$$

$$u_{m+1}(t) = u_0(t) + \int_0^t e^{-(t-s)A}\ Fu_m(s)\ ds,\quad m \geqq 0.$$

We will estimate $\|A^\alpha u_m(t)\|$, where $\|f\|$ denotes the norm of f in X_n. Theorem 1 implies that

$$\|A^\alpha e^{-tA}\| \leqq C_\alpha,\quad \alpha \geqq 0.$$

This yields the estimate

$$\|A^\alpha u_0(t)\| \leqq K_{\alpha 0}\ t^{-\alpha},\quad \alpha \geqq 0$$

with

$$K_{\alpha 0} = \sup_{0<t\leqq T} t^{\alpha}\|A^{\alpha} e^{-tA} a\| \leqq C_{\alpha}\|a\| < \infty.$$

We consider the following problem:

If $K_{\alpha 0}$ *is small enough, is it possible to estimate* $\|A^{\alpha}u_m(t)\|$ *by* $K_{\alpha m} t^{-\alpha}$ *from above with constant* $K_{\alpha m} \leqq K < \infty$ *such that* K *is independent of* m *?*

The answer is yes and we will give a proof; see [1,5,8]. Suppose that for some $m \geqq 0$, $u_m(t)$ satisfies

$$\|A^{\alpha}u_m(t)\| \leqq K_{\alpha m} t^{-\alpha} \quad \text{for all} \quad \alpha \geqq 0.$$

Lemma 1 with $\delta = 1/4$, $\theta = 1/4$, $\rho = 1/2$ implies

$$\|A^{-\delta}Fu_m(s)\| \leqq M K_{\theta m} K_{\rho m} s^{\delta-1}.$$

We thus have

$$\|A^{\alpha}u_{m+1}(t)\| \leqq K_{\alpha 0} t^{-\alpha} + C_{\alpha+\delta}\int_0^t (t-s)^{-\alpha-\delta} \|A^{-\delta}Fu_m(s)\| ds$$

$$\leqq K_{\alpha,m+1} t^{-\alpha} \quad \text{for all} \quad \alpha, \quad 0 \leqq \alpha < 1 - \delta = 3/4$$

with

$$K_{\alpha,m+1} = K_{\alpha 0} + C_{\alpha+\delta} M\, B(1-\delta-\alpha,\delta)\, K_{\theta m} K_{\rho m},$$

where $B(a,b)$ is the beta function. This implies that $u_m(t)$ is well-defined for each $m \geqq 0$ as a element of $C([0,T], X_n) \cap C((0,T], D(A^{\alpha}))$ for all α, $0 \leqq \alpha < 3/4$ and that $u_m(t)$ satisfies

$$\|A^{\alpha}u_m(t)\| \leqq K_{\alpha m} t^{-\alpha}, \quad 0 \leqq \alpha < 3/4.$$

Put $k_m = \max\{K_{\theta m}, K_{\rho m}\}$ and note the definition of $K_{\alpha m}$ to get

$$k_{m+1} \leqq k_0 + C k_m^2$$

where C is a constant depending only on A. An elementary calculation shows that if

$$\text{(C)} \qquad k_0 < \frac{1}{4C},$$

then for each $m \geqq 1$ the estimates

$$k_m < K < 1/2C$$

$$K_{\alpha,m+1} \leqq K_{\alpha 0} + C_{\alpha+\delta} M\, B(1-\delta-\alpha,\delta)\, K^2 \equiv K_\alpha,$$

are valid for constant K. We thus have

$$\|A^\alpha u_{m+1}(t)\| \leqq K_\alpha\, t^{-\alpha}, \quad 0 \leqq \alpha < 3/4, \quad 0 \leqq t \leqq T.$$

Using again Theorem 1 and Lemma 1, we can prove

$$\|A^\alpha(u_{m+1}(t) - u_m(t))\| \leqq 2KC_{\alpha+\delta}(2CK)^{m-1}\, B(1-\delta-\alpha,\delta)t^{-\alpha}, \quad 0 \leqq \alpha < 3/4.$$

Since $2CK < 1$, this implies $u_m(t)$ converges a solution $u(t)$ of (III) and $u(t)$ is eventually a unique strong solution of (II); see [1,5,8].

We consider the meaning of (C). In (C) k_0 depends on T and a, so (C) is a condition for initial data and the length of time interval where the solution $u(t)$ of (III) exists. There are at least two types of sufficient conditions for (C). Conceptually speaking, these are

(i) T is fixed and a is taken so that $\|a\|$ is sufficiently small.

(ii) a is fixed and T is sufficiently small.

Let us explain (i). Suppose $\|a\|$ is small, say $\|a\| < 1/2C_\alpha C$ for $\alpha = 1/2, 1/4$. Then clearly $k_0 < 1/2C$ for *all* T. This implies that the solution $u(t)$ of (III) exists for all time if $\|a\| < 1/2C_\alpha C$ for $\alpha = 1/2, 1/4$. Namely, there exists a *global* solution of (III) if $\|a\|$ is *small* enough.

We next explain (ii). First, we prove that

$$t^{\alpha}\|A^{\alpha}e^{-tA}\, a\| \to 0 \quad (t \to 0) \quad \text{for all} \quad a \in X_n \quad (\alpha > 0).$$

If $a \in C^{\infty}_{0,\sigma} = (C^{\infty}_0(D))^n \cap X_2$, we see $b = A^{\alpha}a$ is in X_n. This yields

$$t^{\alpha}\|A^{\alpha}e^{-tA}\, a\| = t^{\alpha}\|e^{-tA}\, b\| \to 0 \quad (t \to 0).$$

Suppose now $a \in X_n$. Since $C^{\infty}_{0,\sigma}$ is dense in X_n, we can take a sequence $\{a_m\}$ in $C^{\infty}_{0,\sigma}$ such that a_m tends to a in X_n as $m \to \infty$. Using $\{a_m\}$, we have

$$\begin{aligned} t^{\alpha}\|A^{\alpha}e^{-tA}\, a\| &\leqq t^{\alpha}\|A^{\alpha}e^{-tA}(a - a_m)\| + t^{\alpha}\|A^{\alpha}e^{-tA}\, a_m\| \\ &\leqq C_{\alpha}\|a - a_m\| + t^{\alpha}\|A^{\alpha}e^{-tA}\, a_m\|. \end{aligned}$$

Since C_{α} is independent of m and since the result is valid for $a_m \in C^{\infty}_{0,\sigma}$, this estimate now implies

$$t^{\alpha}\|A^{\alpha}e^{-tA}\, a\| \to 0 \quad (t \to 0).$$

This result implies that we can take T sufficiently small so that $T^{\alpha}\|A^{\alpha}e^{-tA}\, a\| \leqq 1/2C$ $(\alpha = 1/2,\ 1/4)$ for fixed a. In other words for every $a \in X_n$ there is a solution on $[0,T]$ for small T. Note that this T heavily depends on a.

4. Analyticity of the solution of (II).

We can prove time-analyticity of the solution u of (III) as the proof of Theorem 3. Here we discuss spatial analyticity for u which is analytic in time.

To prove (ii) in Theorem 4 we consider the elliptic system of (u,v) of $n+2$ variables as in Masuda [10], where v is the vorticity, *i.e.*, $v = \text{curl}\ u$. Let us derive the elliptic system. For simplicity we assume $n = 3$. Since u is analytic in $(0,T]$, u is holomorphic in some complex neighborhood U of

[0,T]. This implies $u(x, \tau+i\sigma)$ is harmonic on $U \subset \mathbb{C} = \mathbb{R}^2$ for fixed x. That is,

$$\frac{\partial^2 u}{\partial\tau^2} + \frac{\partial^2 u}{\partial\sigma^2} = 0 \quad \text{in} \quad G = D \times U.$$

Since $\operatorname{div} u = 0$, we have

$$\Delta u - \operatorname{curl} v = 0 \quad \text{in} \quad G.$$

Sum up both sides to get

$$\Delta_G u - \operatorname{curl} v = 0 \quad \text{in} \quad G,$$

where Δ_G is the Laplacian on G.

We derive the equation for the vorticity v. For simplicity we assume $f = 0$. Apply curl both sides of (I) to get

$$\frac{\partial v}{\partial t} - \Delta v + \operatorname{curl}(u,\operatorname{grad})u = 0.$$

Since v is holomorphic in U, we have

$$-\left(\frac{\partial^2 v}{\partial\tau^2} + \frac{\partial^2 v}{\partial\sigma^2}\right) = 0.$$

Summing up both sides, we get

$$\Delta_G v - \frac{\partial v}{\partial\tau} = \operatorname{curl}(u,\operatorname{grad})u \quad \text{in} \quad G.$$

The foregoing argument shows that (v,u) satisfies

$$\Delta_G v - \frac{\partial v}{\partial\tau} = g \quad \text{in} \quad G,$$

$$\Delta_G u - \operatorname{curl} v = 0 \quad \text{in} \quad G,$$

$$u = 0 \quad \text{on} \quad S \times U,$$

$$v - \operatorname{curl} u = 0 \quad \text{on} \quad S \times U.$$

where $g = \text{curl}(u,\text{grad})u$.

If the right hand side is regarded as a given data, this system is a linear elliptic system with complementary boundary conditions on $S \times U$; see [7]. Moreover, this itself is a nonlinear elliptic boundary value problem for (v,u). Apply regularity theorems for elliptic system to get spatial analyticity of u.

Acknowledgments

I am grateful to Professor Robert Kohn for useful discussions during his stay in Japan. I am also grateful to Professor Gerd Grubb for discussions about symbol class of pseudodifferential operators.

Footnotes

1. *Current Address*: Courant Institute of Mathematical Sciences, New York University, 251 Mercer Street, New York, N.Y. 10012, U.S.A.

2. Author partially supported by THE SAKKOKAI FOUNDATION.

References

[1] H. Fujita and T. Kato, *On the Navier-Stokes initial value problem* I, Arch. Rational Mech. Anal. 16 (1964), 269-315.

[2] D. Fujiwara, *On the asymptotic behaviour of the Green operators for elliptic boundary problems and the pure imaginary powers of some second order operators*, J. Math. Soc. Japan 21 (1969), 481-521.

[3] Y. Giga, *Analyticity of the semigroup generated by the Stokes operator in* L_r *spaces*, Math. Z. 178 (1981), 297-329.

[4] Y. Giga, *Domains of fractional powers of the Stokes operator in* L_r *spaces,* Arch. Rational Mech. Anal. (to appear).

[5] Y. Giga and T. Miyakawa, *Solutions in* L_r *to the Navier-Stokes initial value problem,* ibid.

[6] Y. Giga, *The nonstationary Navier-Stokes system with some first order boundary condition,* Proc. Japan Acad., 58 A (1982), 101-104.

[7] Y. Giga, *Time and spatial analyticity of solutions of the Navier-Stokes equations,* Comm. in Partial Differential Equations. (to appear).

[8] T. Kato and H. Fujita, *On the nonstationary Navier-Stokes system,* Rend. Sem. Mat. Univ. Padova 32 (1962), 243-260.

[9] O. A. Ladyzhenskaya, *The Mathematical Theory of Viscous Incompressible Flow,* Revised English ed. New York-London: Gordon and Breach 1969.

[10] K. Masuda, *On the analyticity and the unique continuation theorem for solutions of the Navier-Stokes equation,* Proc. Japan Acad. 43 (1967), 827-832.

[11] R. Seeley, *Norms and domains of the complex powers* A_B^z, Amer. J. Math. 93 (1971), 299-309.

[12] J. Serrin, *The initial value problem for the Navier-Stokes equations,* Nonlinear problems, R.E. Langer ed. University of Wisconsin Press, Madison, 1963, 69-98.

[13] V. A. Solonnikov, *Estimates for solutions of nonstationary Navier-Stokes equations,* J. Sov. Math. 8 (1977), 467-529.

[14] R. Temam, *Navier-Stokes Equations,* Amsterdam-New York-Oxford, North Holland 1977.

[15] R. V. Kohn, *Partial regularity for the Navier-Stokes equations,* in this volume.

[16] G. Grubb, *Problèmes aux limites pseudo-différentiels dépendedant d'un paramètra,* C. R. Acad. Sci. Paris 292 (1981), 581-583.

Lecture Notes in Num. Appl. Anal., **5**, 55–60 (1982)
Nonlinear PDE in Applied Science. U.S.-Japan Seminar, Tokyo, 1982

Nash's Implicit Function Theorem and The Stefan Problem

Ei-Ichi Hanzawa

Department of Mathematics
Faculty of Science
Hokkaido University
SAPPORO 060, JAPAN

We sketch out a proof of the local existence of the classical solutions for the multidimensional Stefan problem and its relevance to Nash's implicit function theorem.

The Stefan problem is a mathematical model of the melting of a body of ice, where we suppose that a body of ice melts, at each point of the surface, with velocity in proportion to the normal gradient of the thermal distribution in water and that the thermal distribution satsfies the heat equation. We show the local existence of the classical solutios for the initial value problem of the Stefan problem, by using Nash's implicit function theorem, which enables us to reduce the solvability of a nonlinear problem to that of the linearized problem even if a loss of regularity of the solution for given data occurs.

Here we suppose that the initial surface Γ_0 of a body of ice is a closed C^∞ hypersurface in R^n, the exterior of Γ_0 is the ice part, a heater is in the interior of Γ_0 whose surface J is also a closed C^∞ hypersurface in R^n and the domain Ω_0 bounded by Γ_0 and J is the water part. (Except the conditions of regularity, these assumptions, e.g., that the water part is in the ice part, are not essential.) The locus of a surface of a body of ice, which is the free boundary to be determined, is denoted by $\Gamma_{\Phi,T} = \{(x,t) \in R^n \times [0, T];$

$\Phi(x, t) = 0\}$, where t is the time variable. We denote the free domain with water by $\Omega_{\Phi,T}$. See Figure 1.

Figure 1.

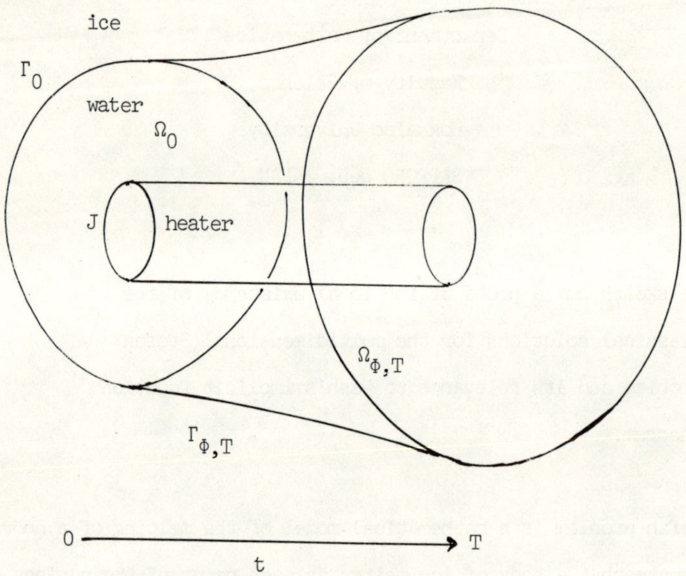

Our unknowns are the defining function Φ and the thermal distribution u in water in $\Omega_{\Phi,T}$. Our equations and result are as follows.

Equations.

(1) $\Phi|_{t=0} = \Phi_0$, $u|_{t=0} = a$, where $\{x; \Phi_0(x) = 0\} = \Gamma_0$.

(2) $(\partial_t - \Sigma_{i=1}^{n}\partial_{x_i}^2)u = 0$ in $\Omega_{\Phi,T}$.

(3) $u = b$ on $J \times [0, T]$.

(4) $u = 0$ on $\Gamma_{\Phi,T}$.

(5) $\partial_t \Phi - k\Sigma_{i=1}^{n}(\partial_{x_i} u)(\partial_{x_i} \Phi) = 0$ on $\Gamma_{\Phi,T}$, where k is a positive (from a physical reason) constant.

Theorem. Suppose that a and b are nonnegative C^{∞} functions and satisfy the compatibility conditions up to ∞ order on Γ_0 and J at $t = 0$ (which are necessary conditions of the existence of a C^{∞} solution for (1)-(5)). Then, for sufficiently small $T > 0$, there uniquely (in the essential sense) exist a C^{∞} function Φ and a C^{∞} function u on $\Omega_{\Phi,T}$ which satsfy (1)-(5).

Remark 1. For the one-dimensional Stefan problem, it is well known that the unique global classical solutions are obtained (see Rubinstein [5]). For the multidimensional problem, the unique global weak solutions are obtained by Kamenomostskaja [2].

Remark 2. The C^{∞}-ness of the solutions in Theorem is remarked by M. Tanigawa [6]. The author's original theorem is that for solutions with finite differentiability of any order. The reason for giving this limitation was purely technical. That is, the author did not know whether there are smoothing operators up to ∞ order on a scale of Banach spaces which is used in [1]. They are constracted by Tanigawa.

Remark 3. For the general existence theorem of the classical solutions for the multidimensional Stefan problem, it seems that Nash's implicit function theorem is necessary, i.e., we encounter an essential loss of regularity. It occurs because to solve the single first order equation (5) for Φ on $\Gamma_{\Phi,T}$ does not cover the loss of regularity of the normal derivative of the thermal distribution u, that is, the former gains the regularity only along the characteristic curves although the latter loses the regularity in every direction. See Figure 2. The situation of this phenomenon is clearly recognized when we linearize the problem (1)-(5). Note that this difficulty does not occur in the one-dimensional Stefan problem, because the free boundary is one-dimensional so that it is covered by the characteristic curve.

Figure 2: The reason why the loss of regularity occurs.

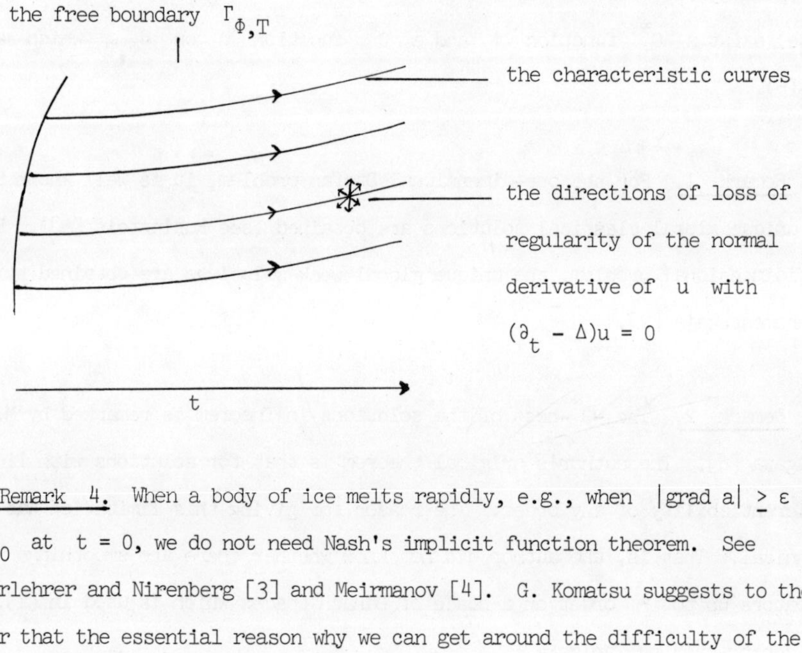

Remark 4. When a body of ice melts rapidly, e.g., when $|\text{grad } a| \geq \varepsilon > 0$ on Γ_0 at $t = 0$, we do not need Nash's implicit function theorem. See Kinderlehrer and Nirenberg [3] and Meirmanov [4]. G. Komatsu suggests to the author that the essential reason why we can get around the difficulty of the loss of regularity in this case is in the fact that a heat potential cover losses of regularity in the time direction when the melting is rapid.

Remark 5. The assumption that the initial data a and b are nonnegative (which is natural in physics) enables us to solve the linearized problem of (1)-(5). We resolve the linearized problem into a parabolic mixed problem and an initial value problem in $\Omega_{\Phi,T}$ for a first order operator which has the form

$\partial_t - k f^2(\partial_\nu u)\partial_\nu$ on $\Gamma_{\Phi,T}$, where ν is an outward unit normal to the surface $\{x; \Phi(x, t) = 0\}$ in R^n and u is the thermal distribution at which we liearize the problem. We can solve the latter problem if the characteristic curves starting from Ω_0 at $t = 0$ cover the domain $\Omega_{\Phi,T}$. This requirement is satisfied because $u = a \geq 0$ on Ω_0 at $t = 0$, $u = b \geq 0$ on $J \times [0, T]$, $u = 0$ on $\Gamma_{\Phi,T}$ and we have the maximum principle for the heat equation. See Figure 3. This fact is the core of the present work.

Figure 3: The reason why we can solve the linearized Stefan problem.

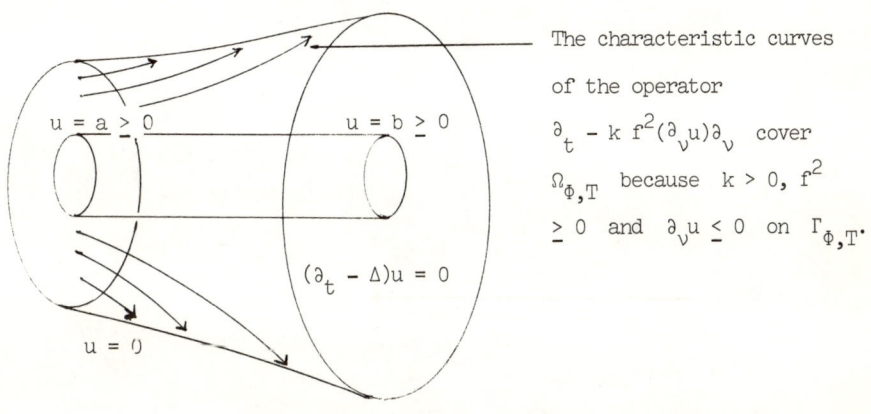

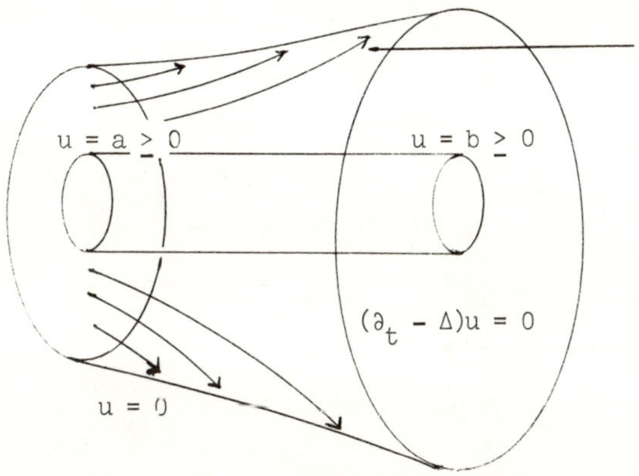

The characteristic curves of the operator $\partial_t - k f^2(\partial_\nu u)\partial_\nu$ cover $\Omega_{\Phi,T}$ because $k > 0$, $f^2 \geq 0$ and $\partial_\nu u \leq 0$ on $\Gamma_{\Phi,T}$.

References.

[1] E. Hanzawa, Classical solutions of the Stefan problem, Tôhoku Math. J. 33 (1981), 297-335.

[2] S. L. Kamenomostskaja, On Stefan's problem (in Russian), Naučn Dokl. Vysš. Školy 1 (1958), No.1, 60-62.

[3] D. Kinderlehrer and L. Nirenberg, The smoothness of the free boundary in the one phase Stefan problem, Comm. Pure Appl. Math. 31 (1978), 257-282.

[4] A. M. Meǐrmanov, On the classical solution of the multidimensional Stefan problem for quasilinear parabolic equations, Math. USSR Sbornik

40 (1981), 157-178.

[5] L. I. Rubinstein, The Stefan problem, Translations of Mathematical Monographs, Vol. 27, Amer. Math. Soc., Providence, R.I., 1972.

[6] M. Tanigawa, The C^∞-ness of solutions of the Stefan problem, to appear elsewhere.

Lecture Notes in Num. Appl. Anal., **5**, 61–76 (1982)
Nonlinear PDE in Applied Science. U.S.-Japan Seminar, Tokyo, 1982

Quasi-linear equations of evolution in nonreflexive Banach spaces

Tosio Kato[1]

Department of Mathematics

University of California, Berkeley

An existence-uniqueness theorem and a regularity theorem are given for the Cauchy problem for quasi-linear equations of evolution in nonreflexive Banach spaces. As an application, C^1-solutions are constructed for hyperbolic systems of partial differential equations in the "Schauder canonical form" (which include generic equations in two independent variables.)

1. Introduction

In previous publications [7,9,10], we considered the Cauchy problem for abstract quasi-linear equations of the form

$$\text{(Q)} \qquad du/dt + A(u)u = f(u), \qquad t \geq 0.$$

In this theory (at least) two Banach spaces X, Y are used, such that the solution $u(t)$ is in Y and $du(t)/dt$ is in X. The object of the present paper is to give another version of the theory which is more flexible and convenient in some applications.

The main differences of the new version from the previous ones are (a) elimination of the reflexivity assumptions on the basic Banach spaces X, Y, (b) elimination of the isomorphism S between Y and X, and (c) extension of the domain of the map $w \longmapsto A(w)$ to all of Y without imposing any restriction of the growth rate. In the simple form presented here, the new result will not completely cover the previous ones, but it will be applicable to problems that were not accessible to the latter.

In particular, we have in mind applications to first-order hyperbolic systems

$$\text{(HS)} \qquad du/dt + \sum_{j=1}^{m} a_j(u)du/dx_j = f(u), \quad t \geq 0, \quad x \in R^m,$$

in which the unknown $u = \{u_1(t,x),\ldots,u_N(t,x)\}$ is a real N-vector function, the $a_j(z)$ are real $N \times N$ matrices depending on $z \in R^N$ in the C^1-manner, and $f(z)$ is a similar N-vector function. It is assumed that the $a_j(z)$ are simultaneously diagonalizable by a common real matrix $q(z)$ depending on $z \in R^N$ in the C^1-manner. Our abstract results will be applicable to solve the Cauchy problem for (HS) in the class $u \in C([0,T];\dot{C}^1(R^m;R^N))$ for some $T > 0$, given an initial value $u(0) = \phi \in \dot{C}^1(R^m;R^N)$. [Here and in what follows $\phi \in \dot{C}(R^m;R^N)$ implies that $\phi(x) \longrightarrow 0$ as $|x| \longrightarrow \infty$, and similarly for $\dot{C}^1(R^m;R^N)$.]

Our old theory is not applicable to (HS) simply because $\dot{C}^1$ is not reflexive. It may be remarked that what is really required in the old theory is not necessarily the reflexivity of Y but rather that Y be locally closed in X : $Y^o = Y$ in the notation introduced below. Even this condition is not satisfied by the pair $Y = \dot{C}^1$, $X = \dot{C}$.

It will be recalled that for $m = 1$, classical C^1-theories for (HS) were given by Douglis [4] and Hartman and Wintner [6]. (For earlier works in this direction, see Schauder [14] and Friedrichs [5].) A generalization to the case $m > 1$ was given by Cinquini Cibrario [3]. All these proofs make essential use of the moduli of continuity of the first derivatives of the functions involved, indicating that there can be no cheap way to construct a C^1-theory for (HS). In fact our proof depends on the use of semigroups acting on spaces of functions with fixed moduli of continuity (see section 5).

In view of this example, it appears that the following is a natural and inevitable procedure. In the abstract theory, we shall be content with constructing only a weak solution $u(t)$ to (Q) which, however, will be uniquely determined by the initial value $u(0) = \phi \in Y^o$ and which will stay in Y^o, where Y^o is the local closure of Y in X. (For the definition of Y^o, see below.)

If we want to show that $u(t)$ is in fact a "strong solution" which stays in

Y provided $\emptyset \in Y$, we shall construct a weak solution in another pair $(\hat{X},\hat{Y})$ of spaces such that $\emptyset \in \hat{Y}^o \subset Y$. In view of the uniqueness, this will show that $u(t) \in Y$.

To ensure that the interval of existence $[0,T]$ of the solution is not diminished when we go over to the pair $(\hat{X},\hat{Y})$, we would need a <u>regularity theorem</u> to the effect that a weak solution in (X,Y) with $u(0) \in \hat{Y}^o$ is a weak solution in $(\hat{X},\hat{Y})$ on all of its interval of existence.

In the application to (HS), we choose $X = \dot{C}$, $Y = \dot{C}^1$ as above and then $\hat{X} = \dot{C}^{p+0}$ and $\hat{Y} = \dot{C}^{1+p+0}$. Here $\dot{C}^{p+0}$ is the closure of $\dot{C}^1$ in $\dot{C}^p$, the set of all functions in $\dot{C}$ with modulus of continuity dominated by a constant multiple of a fixed <u>modulus function</u> p, and $\dot{C}^{1+p+0}$ is the set of all $v \in \dot{C}^1$ with $d_j v \in \dot{C}^{p+0}$. p will be determined by the initial function $\emptyset \in \dot{C}^1$ and the continuity properties of the functions a_j, q and f. (see section 5 for details.)

2. The existence theorem

We start from a pair $Y \subset X$ of real Banach spaces, with the associated norms $|\ |_X$ $|\ |_Y$, with the inclusion continuous and dense.

We assume that for each $w \in Y$, a vector $f(w) \in Y$, a linear operator $A(w)$ in X and two norms $|\ |_{X_w}$, $|\ |_{Y_w}$, equivalent to $|\ |_X$, $|\ |_Y$, respectively, are given with the following properties. If $w, v \in Y$ with $|w|_Y \le r$, $|v|_Y \le r$, then

(N1) $\quad \mathrm{dist}(|\ |_{X_w}, |\ |_X) \le \lambda_1(r)$,

(N2) $\quad \mathrm{dist}(|\ |_{X_w}, |\ |_{X_v}) \le \mu_1(r)|w-v|_X$,

(N3) $\quad \mathrm{dist}(|\ |_{Y_w}, |\ |_Y) \le \lambda_2(r)$

(N4) $\quad \mathrm{dist}(|\ |_{Y_w}, |\ |_{Y_v}) \le \mu_2(r)|w-v|_X$,

(A1) $\quad A(w) \in G(X_w,1,\beta_1(r)) \cap G(Y_w,1,\beta_2(r))$.

(A2) $\quad |A(w)|_{Y,X} \le \lambda_3(r) \qquad (D(A(w)) \supset Y)$

(A3) $|A(w)-A(v)|_{Y,X} \le \mu_3(r)|w-v|_X$,

(f1) $|f(w)|_Y \le \lambda_4(r)$,

(f2) $|f(w)-f(v)|_X \le \mu_4(r)|w-v|_X$.

Here $\lambda_1,\ldots, \beta_1,\ldots, \mu_4$ are monotone increasing functions on R_+ to R_+ , and will be called the <u>parameters</u> of the system (X,Y,A,f).

REMARK 2.1. (a) The distance $\mathrm{dist}(|\ |,|\ |')$ between two equivalent norms is defined by $\log \sup_x\{|x|/|x|',|x|'/|x|\}$. $A \in G(X,1,\beta)$ means that $-A$ generates a C_0-semigroup $\{e^{-tA};\ t \ge 0\}$ on X such that $|e^{-tA}|_X \le e^{\beta t}$. We denote by $|A|_{Y,X}$ the operator-norm of $A \in B(Y,X)$, and $|A|_X = |A|_{X,X}$. X_w is the Banach space X with the norm $|\ |_{X_w}$. (b) In (N1) --- (f2), the argument r in the parameters $\lambda_1(r),\ldots,\mu_4(r)$ are (for simplicity and generality) assumed to be $r = |w|_Y$ or $r = |w|_Y \vee |v|_Y$. (We write $a \vee b$ for $\sup\{a,b\}$.) In some problems, however, (N1) holds with $\lambda_1(|w|_Y)$ replaced by $\lambda_1(|w|_X)$, thereby strengthening some of the results. Similar remarks apply to other parameters.

DEFINITION 2.2 A sequence $\{u_n\}$ is called a null sequence (of approximate solutions) to (Q) on $[0,T]$ if the u_n are bounded in $C([0,T];Y \cap \mathrm{Lip}_*([0,T];X)$ and

$$du_n/dt + A(u_n)u_n - f(u_n) \longrightarrow 0 \quad \text{in} \quad L^\infty([0,T];X) \ . \tag{2.1}$$

Here $u_n \in \mathrm{Lip}_*([0,T];X)$ means that u_n is an indefinite Bochner integral of a function $\dot{u}_n \in L^\infty([0,T];X)$ so that $du_n/dt = \dot{u}_n$. Note that $A(u_n)u_n \in C([0,T];X)$ by virtue of (A2) and (A3). [In the existence proof, approximate solutions u_n will be mostly piecewise $C^1([0,T];X)$ and no measure theory will be required.]

DEFINITION 2.3. $u \in C([0,T];X)$ is called a weak solution to (Q) if there is a partition of $[0,T]$ into a finite number of subintervals such that on each closed subinterval I, u is the limit in $C(I;X)$ of a null sequence. [Hence

$u \in Lip([0,T];X)$.]

DEFINITION 2.4 Let $B_Y(r)$ be the ball in Y with center 0 and radius $r > 0$, and $cl_X(B_Y(r))$ its closure in X . We denote by Y^o the union of $cl_X(B_Y(r))$ for all $r > 0$. Y^o will be called the local closure of Y in X .

REMARK 2.5 As is easily seen, Y^o is a Banach space with the norm $|x|_{Y^o}$ defined as the infimum of $r > 0$ such that $x \in cl_X(B_Y(r))$. Obviously we have $Y \subset Y^o \subset X$, with the inclusions continuous. If Y is reflexive one has $Y^o = Y$, including the norm. In general Y need not be dense in Y^o. For example, let $X = C[0,1]$, $Y = C^1[0,1]$; then $Y^o = Lip[0,1]$, and Y is a closed subspace of Y^o.

THEOREM I. (existence) Given $\emptyset \in Y^o$, there is $T > 0$, depending only on $|\emptyset|_{Y^o}$ (and the parameters of the system), and a unique weak solution u to (Q) on $[0,T]$ with $u(0) = \emptyset$. The map $\emptyset \longmapsto u$ is bounded on a bounded subset of Y^o to $B([0,T];Y^o)$, and is continuous from the X-topology to $C([0,T];X)$ within a bounded set of Y^o. [Here $B(I;Y^o)$ denotes the set of bounded functions on I to Y^o. We cannot replace it with $L^\infty(I;Y^o)$ since the functions considered may not be strongly measurable.]

REMARK 2.6 After introducing the space Y^o, one might try to extend the map $w \longmapsto A(w)$ to all $w \in Y^o$, to be able to work in the space pair $Y^o \subset X$ instead of $Y \subset X$. There are two difficulties in this attempt. First, there is no general method to extend A in this manner so as to make $A(w) \in B(Y^o,X)$. Second, even if this is possible, $A(w)$ may not become a generator in Y^o. This may be expected from the typical example (HS) in which Y^o = Lip (see section 5). In fact there are no reasonable C_0-semigroups on the space Lip.

REMARK 2.7. If Y is reflexive, we have $Y^o = Y$ and the solution u in Theorem I belongs to $C_w([0,T];Y)$, where C_w indicates weak continuity, and u is a solution to (Q) with $du/dt \in C_w([0,T];X)$. Thus u is <u>almost</u> a strong solution to (Q). In favorable cases one may be able to show that $u \in C([0,T];Y)$

(strong solution) by auxiliary considerations such as those given in [8;Remark 5.3]

3. Sketch of the proof of Theorem I

For simplicity we assume $f = 0$.

As in previous works [7,9,10], we use successive approximation based on the theory of linear evolution equations given in [8].

(a) First we assume that $\phi \in Y$, and find a ball in Y in which we can expect to confine the values of the approximate solutions u_n for a fixed interval $[0,T]$. To this end, fix an $R > |\phi|_Y$. Then we can determine R', R'' such that

$$\phi \in B_Y(R) \subset B_{Y_\phi}(R') \subset B_Y(R'') . \tag{3.1}$$

Indeed, in view of (N3) it suffices to set $R' = R \exp[\lambda_2(R)]$ and $R'' = R'\exp[\lambda_2(R)]$. All approximate solutions u_n and related functions we introduce below will take values in $B_Y(R'')$, so that we shall be able to set $r = R''$ in all the parameters $\lambda_1(r),\dots,\mu_4(r)$.

REMARK 3.1. R, R', R'' and L, T (introduced below) are determined by $|\phi|_Y$ only. This is a great advantage over the situation in [7,10], where T depended only on ϕ but not necessarily on $|\phi|_Y$ only.

(b) Let E be the set of all functions $v \in C([0,T];Y)$ such that

$$v(0) = \phi, \quad v(t) \in B_{Y_\phi}(R'), \quad |v(t)-v(s)|_X \le L|t-s|, \tag{3.2}$$

where T and L are constants to be determined.

For each $v \in E$, let $\tilde{v}$ be a step-function approximation for v (by which it is implied that the values of $\tilde{v}$ are a subset of the values of v). It follows from (3.2) and (N2), (N4), (A1) that $A^v(t) = A(v(t))$ and $A^{\tilde{v}}(t) = A(\tilde{v}(t))$ form stable families of generators in X as well as in Y (see [8]), with uniform stability constants. Therefore, there is an evolution operator $\{U^{\tilde{v}}(t,s)\}$ associated with $\{A^{\tilde{v}}(t)\}$ (see [8]; here we may disregard finitely many discontinuities for the derivatives of $U^{\tilde{v}}(t,s)$).

It follows from the uniform estimates for the stability constants that $u \equiv U^{\tilde{v}}(\cdot,0)\emptyset \in E$ if L and T are chosen appropriately, for any $v \in E$ and any step function approximation $\tilde{v}$ of v. Moreover, the map $\tilde{v} \longmapsto u \equiv \Phi\tilde{v}$ can be shown to be a contraction in the metric of $L^{\infty}([0,T];X)$, by reducing the size of T if necessary.

(c) We can now construct a null sequence $\{u_n\}$ to (Q) on $[0,T]$ such that $u_n(0) = \emptyset$. Assuming that $u_n \in E$ has been constructed, we choose a step function approximation $\tilde{u}_n$ to u_n such that $\|\tilde{u}_n - u_n\|_X \leq 2^{-n}$, where $\| \ \|_X$ denotes the $L^{\infty}([0,T];X)$-norm. Then $u_{n+1} = \Phi\tilde{u}_n$ will be the next element, and $\{u_n\}$ is shown to be a null sequence.

Finally we show that $\lim u_n = u$ exists in $L^{\infty}([0,T];X)$, so that u is a weak solution to (Q) on $[0,T]$ with $u(0) = \emptyset$. The proof is based on the following lemma.

LEMMA 3.2. If $\{u_n\}$ is a null sequence to (Q) on $[0,T]$ and if $\lim u_n(0)$ exists in X, then $u = \lim u_n$ exists in $C([0,T];X)$ (so that u is a weak solution).

(d) In the general case in which $\emptyset \in Y^o$, we choose a sequence $\emptyset_j \in Y$ such that $|\emptyset_j|_Y = |\emptyset|_{Y^o}$ and $|\emptyset_j - \emptyset|_X \longrightarrow 0$. Let u_j be a weak solution to (Q) with $u_j(0) = \emptyset_j$, which exists on an interval $[0,T]$ independent of j by the previous result.

Each u_j is constructed as the limit in $L^{\infty}([0,T];X)$ of a null sequence $\{u_{jn}\}$. Thus we can find a sequence $\{u_{j,n_j}\} = \{v_j\}$ such that $\|u_j - v_j\|_X \leq j^{-1}$ and $\|Qv_j\|_X \leq j^{-1}$, where $Qw = dw/dt + A(w)w$. Thus $\{v_j\}$ is a null sequence for (Q) on $[0,T]$ such that $v_j(0) = \emptyset_j \longrightarrow \emptyset$ in X. According to Lemma 3.2, it follows that $\lim v_j = u$ exists and defines a weak solution to (Q) on $[0,T]$ with $u(0) = \emptyset$.

(e) The uniqueness of the weak solution is also a direct consequence of Lemma 3.2.

4. The regularity theorem

In general different choices of the spaces X, Y are possible for a given equation (Q). In other words, there are many systems (X,Y,A,f) satisfying conditions (N1) --- (f2) with different pairs (X,Y) but with the same A and f. [To explain the last expression, we say that the two systems (X,Y,A,f) and (X',Y',A',f') have the same A and f if $A(w)y = A'(w)y \in X \cap X'$ and $f(w) = f'(w) \in Y \cap Y'$ whenever $w, y \in Y \cap Y'$.]

For simplicity, suppose that we have two systems (X,Y,A,f) and $(\hat{X},\hat{Y},A,f)$ such that $\hat{X} \subset X$, $\hat{Y} \subset Y$ with the inclusions continuous. Then we have obviously $\hat{Y}^o \subset Y^o$. If we choose the initial value $\phi \in \hat{Y}^o$, Theorem I gives a weak solution $u \in B([0,T];Y^o)$ to (Q) and another solution $\hat{u} \in B([0,\hat{T}];\hat{Y}^o)$, both satisfying the initial condition $u(0) = \hat{u}(0) = \phi$. In view of the uniqueness result, we may assume that $\hat{T} \leq T$ and $\hat{u} = u$ on $[0,\hat{T}]$. The question arises whether or not we can take $\hat{T} = T$.

More generally, one may ask whether every weak solution in the system (X,Y,A,f) with $u(0) \in \hat{Y}^o$ is automatically a weak solution in $(\hat{X},\hat{Y},A,f)$ with the same interval of existence. This is the problem of regularity for (Q).

To answer this question, we would need further assumptions. To formulate such assumptions, we find it convenient to introduce the notions of <u>norm-compression</u> and <u>compressible systems.</u>

DEFINITION 4.1. Given two Banach spaces $\hat{X} \subset X$ with the inclusion continuous, we may introduce in $\hat{X}$ new equivalent norms. A family of equivalent norms $|\ |_{\varepsilon,\hat{X}}$ in $\hat{X}$, depending on a parameter $\varepsilon > 0$, will be called a compressible norm in $\hat{X}$ (relative to the X-norm) if

$$\limsup_{\varepsilon\to 0} |x|_{\varepsilon,\hat{X}} \leq |x|_X \quad \text{for each } x \in \hat{X}. \tag{4.1}$$

EXAMPLE 4.2. Set $|x|_{\varepsilon,\hat{X}} = |x|_X \vee \varepsilon|x|_{\hat{X}}$. In this case we have $|x|_\varepsilon = |x|_X$ for sufficiently small ε (depending on x).

Consider now the two systems (X,Y,A,f) and $(\hat{X},\hat{Y},A,f)$ for (Q) mentioned above, with $\hat{X} \subset X$ and $\hat{Y} \subset Y$. Suppose that we introduce compressible norms in various spaces: $|\ |_{\varepsilon,\hat{X}}$ relative to X-norm, $|\ |_{\varepsilon,\hat{Y}}$ relative to Y-norm, $|\ |_{\varepsilon,\hat{X}_w}$ relative to X_w-norm, and $|\ |_{\varepsilon,\hat{Y}_w}$ relative to Y_w-norm, where $w \in \hat{Y} \subset Y$. Assume that with these new norms, condition (N1) --- (f2) remain satisfied with a set of parameters $\lambda_{1,\varepsilon},\ldots,\mu_{4,\varepsilon}$ depending on ε .

DEFINITION 4.3. If these parameters $\lambda_{1,\varepsilon},\ldots,\mu_{4,\varepsilon}$ stay bounded as $\varepsilon \longrightarrow 0$, we say that the system $(\hat{X},\hat{Y},A,f)$ is compressible to the system (X,Y,A,f). [The parameter $\lambda_{1,\varepsilon}$ is bounded as $\varepsilon \longrightarrow 0$ if $\lambda_{1,\varepsilon}(r) \leq \lambda_{1,0}(r)$ for $r > 0$, for some monotone increasing function $\lambda_{1,0}$.]

REMARK 4.4 Definition 4.3 is admittedly rather implicit and complicated. But it is not unreasonable, as is seen from the example (HS) discussed in the next section. As a matter of fact, compressibility holds in most well-posed systems as soon as (X,Y,A,f) is a reasonably good system, although the definition would require a generalization to systems (X,Y,Z,A,f) involving three Banach spaces if wider applications are desired. Among simple systems in which two Banach spaces X, Y suffice, we may mention the KdV equation, for which $X = H^{-1}(R)$, $Y = H^2(R)$ will give a "good" system, and $\hat{X} = H^{s-1}(R)$, $\hat{Y} = H^{s+2}(R)$ with $s > 0$ will give a compressible system. For relevant results see [11,12] (though compressibility was not formally introduced there).

THEOREM II. (regularity) Suppose that the system $(\hat{X},\hat{Y},A,f)$ for (Q) is compressible to (X,Y,A,f). If u is a weak solution to (Q) on $[0,T]$ in the system (X,Y,A,f) (so that $u(t) \in Y^o$) and if $u(0) \in \hat{Y}^o$, then u is a weak solution to (Q) on $[0,T]$ in the system $(\hat{X},\hat{Y},A,f)$ (so that $u(t) \in \hat{Y}^o$).

Proof (sketch). Since $u(0) \in \hat{Y}^o$, there is by Theorem I a weak solution in the system $(\hat{X},\hat{Y},A,f)$ on an interval $[0,\hat{T}]$ with the initial value $u(0)$. By the uniqueness result, this solution coincides with u on $[0,\hat{T}]$. If the assertion were not true, $|u(t)|_{\hat{Y}^o}$ must blow up at some $t = s < T$. If we take $s' < s$ sufficiently close to s, and choose the compressible norms with sufficiently small ε , we may achieve that $|u(s')|_{\varepsilon,\hat{Y}^o} \le |u(s')|_{Y^o} + 1 \le K$. Since the parameters stay bounded with compression, the weak solution in the system $(\hat{X},\hat{Y})$ will exist on an interval $[s',s'+T_1]$ with $T_1 > 0$ independent of s'. Thus we have a contradiction by choosing s' sufficiently close to s.

REMARK 4.5. Since norm-compression is used in the proof, we have no simple estimate for the growth rate of $|u(t)|_{\hat{Y}^o}$. But we do have sufficient control over the interval of existence $[0,\hat{T}]$ to show that $\hat{T} = T$.

5. Application to (HS)

For simplicity we assume $f = 0$ in (HS).

It is assumed that the a_j are simultaneously diagolizable:

$$a_j(z) = q(z)^{-1}a_j^0(z)q(z) \qquad (j = 1,\dots,m;\ z \in R^N), \tag{5.1}$$

where the $a_j^0(z)$ are real diagonal matrices and $q(z)$ is a real nonsingular matrix, such that

$$a_j^0\ ,\ q \in C^1(R^m;R^{N^2}). \tag{5.2}$$

We may assume, if necessary, that $\det q(z) = 1$.

(a) Let

$$X = \dot{C}(R^m;R^N), \qquad Y = \dot{C}^1(R^m;R^N). \tag{5.3}$$

X is the set of all vector-valued continuous functions u such that $u(x) \longrightarrow 0$ as $|x| \longrightarrow \infty$. Y is the set of all $u \in C^1$ such that $u(x)$ and all $d_j u(x) \longrightarrow 0$ as $|x| \longrightarrow \infty$. ($d_j = d/dx_j$).

The norms $|\ |_X$ and $|\ |_Y$ are given by

$$|u|_X = |u_1|_{L^\infty} \vee \cdots \vee |u_N|_{L^\infty}\ , \tag{5.4}$$

$$|u|_Y = |u|_X \vee |d_1 u|_X \vee \cdots \vee |d_m u|_X\ ,$$

where $u = (u_1,\dots,u_N)$. In general we agree to use the norm

$$(5.5) \qquad |z| = |z_1| \vee \cdots \vee |z_N| \quad \text{for} \quad z = (z_1,\ldots,z_N) \in R^N,$$

while we use as usual the euclidean norm $|x|$ for $x \in R^m$.

For each $w \in Y$, the norms $|\ |_{X_w}$ and $|\ |_{Y_w}$ are given by

$$(5.6) \qquad |u|_{X_w} = |q(w)u|_X ,$$

$$|u|_{Y_w} = |u|_{X_w} \vee |d_1 u|_{X_w} \vee \cdots \vee |d_m u|_{X_w} .$$

With these norms, conditions (N1) to (N4) are easily verified. Here we may take $r = |w|_X$ instead of $|w|_Y$, etc.

Next we define the operator $A(w)$ formally by

$$(5.7) \qquad A(w) = \sum_{j=1}^{m} a_j(w(x))d_j = \sum_{j=1}^{m} (q^{-1}a_j^0 q)(w(x))d_j$$

$$= q(w)^{-1}A^0(w)q(w) - B(w),$$

where

$$(5.8) \qquad A^0(w) = \sum_{j=1}^{m} a_j^0(w)d_j ,$$

$$B(w) = q(w)^{-1} \sum_{j=1}^{m} a_j^0(w)(d_j q(w)) \qquad (\in B(X)).$$

To be precise, these operators must be defined by carefully specifying their domains. In any case it is obvious that $A^0(w)$ is a first-order differential operator acting separately on each component. Using the well-known results for the first-order operators in one unknown, it is possible, with some efforts, to determine these domains and verify conditions (A1) to (A3).

Thus we are able to apply Theorem I to construct a unique weak solution u for (HS) in the system (X,Y,A). Since $Y^o = \mathrm{Lip}(R^m;R^N)$ (with appropriate behavior at infinity), $u(t)$ is Lipschitzian for each t but as yet unknown to be in C^1, even when $\emptyset = u(0) \in C^1$. This corresponds to the results proved by Cesari [1] and Cinquini Cibrario [2].

(b) To prove the sharper result that $u(t) \in Y$ if $\phi \in Y$, we introduce a new system $(\hat{X},\hat{Y},A)$ with spaces $\hat{Y} \subset \hat{X}$ related to the moduli of continuity.

If $\phi \in Y = \dot{C}^1$, the $d_j\phi$ have uniform modulus of continuity. Since the a_j^0 and q are C^1, their first derivatives have uniform moduli of continuity on any compact subset of R^N. Since we have already found a weak solution u to (HS) staying in a bounded set in Y^o, we may assume that the a_j^0 and q have first derivatives with a common modulus of continuity on all of R^N, modifying these functions if necessary for large $|z|$.

Thus we are able to find a "modulus function" p such that the $d_j\phi$, $d_k a_j^0$, and $d_k q$ have moduli of continuity dominated by constant multiples of p.

Then we define the space $\hat{X} \subset X = \dot{C}(R^m;R^N)$ with the norm

(5.9) $$|u|_{\hat{X}} = |u|_X \vee |u|_{[p]},$$

$$|u|_{[p]} = \sup_{x,y} |u(x)-u(y)|_X/p(|x-y|).$$

Similarly we define the space $\hat{Y} \subset Y = \dot{C}^1(R^m;R^N)$ with the norm

(5.10) $$|u|_{\hat{Y}} = |u|_{\hat{X}} \vee |d_1u|_{\hat{X}} \vee \cdots \vee |d_mu|_{\hat{X}}.$$

For each $w \in \hat{Y}$, we now introduce equivalent norms:

(5.11) $$|u|_{\hat{X}_w} = |q(w)u|_{\hat{X}},$$

(5.12) $$|u|_{\hat{Y}_w} = |u|_{\hat{X}_w} \vee |d_1u|_{\hat{X}_w} \vee \cdots \vee |d_mu|_{\hat{X}_w}.$$

It can be shown, with some computations, that conditions (N1) --- (N4) are satisfied with X, Y replaced by $\hat{X}$, $\hat{Y}$, respectively. Here again, r may be taken to be $|w|_{\hat{X}}$ rather than $|w|_{\hat{Y}}$, etc.

Finally, conditions (A1) to (A3) can be verified for the system $(\hat{X},\hat{Y},A)$ by making use of the explicit formulas known for the operator $A^0(w)$. Before doing so, however, we have to make a small correction to the previous definitions by replacing the space $\hat{X}$ with its subspace spanned by $\dot{C}^1$, and accordingly modifying the space $\hat{Y}$. (We denote these spaces by $\hat{X} = \dot{C}^{p+0}$, $\hat{Y} = \dot{C}^{1+p+0}$.)

This is necessary because otherwise the operator $-A(w)$ would not be densely defined and therefore not generate a C_0-semigroup on $\hat{X}$ or $\hat{Y}$. Note that $\emptyset$ may be assumed to be in the modified $\hat{Y}$, by weakening p slightly if necessary.

Theorem I can now be applied to the system $(\hat{X},\hat{Y},A)$, with the result that $u(t) \in \hat{Y}^o$ at least for a short time. Using the special properties of the space Y, then, it is not difficult to show that $u \in C([0,T];Y)$. Thus u is a strong solution in the system (X,Y,A) on $[0,\hat{T}]$.

Actually we can take $\hat{T} = T$. To prove this, we apply Theorem II by showing that the system $(\hat{X},\hat{Y},A)$ is compressible to (X,Y,A). In the present problem, however, compressibility is almost a built-in property. Indeed, there is nothing that distinguishes a modulus function p from its constant multiple $\varepsilon^{-1}p$. If we choose ε sufficiently small, the associated seminorm $|u|_{[p]}$ in (5.9) becomes small so that we have $|u|_{\hat{X}} = |u|_X$. This can be done simultaneously for any finite number of functions u. Thus it is not surprising that the parameters $\hat{\lambda}_1,\dots,\hat{\mu}_3$ in the system $(\hat{X},\hat{Y},A)$ can be made equal to $\lambda_1,\dots,\mu_3$ by choosing ε sufficiently small, although the proof is by no means trivial. (For details cf. Nakata [13], where norm-compression is systematically used.)

(c) Thus we have shown that (HS) has a unique strong solution $u \in C([0,T];Y)$ for any $u(0) = \emptyset \in Y = \dot{C}^1(R^m;R^N)$, with $T > 0$ depending only on $|\emptyset|_Y$ (for the a_j fixed). Moreover, we shall show that the dependence $\emptyset \longmapsto u$ is continuous from Y to $C([0,T];Y)$.

To this end let $\emptyset_j \in Y$, $j = 1,2,\dots$, such that $\emptyset_j \longrightarrow \emptyset$ in Y. Let $u_j \in C([0,T];Y)$ be the strong solution with $u_j(0) = \emptyset_j$, where T can be taken common to all u_j and u. We have to show that $u_j \longrightarrow u$ in $C([0,T];Y)$.

Since $\emptyset_j \longrightarrow \emptyset$ in $Y = \dot{C}^1$, it can be shown that there is a modulus function p such that all $\emptyset_j$ and $\emptyset$ are in $\hat{Y} = \dot{C}^{1+p+0}$ with the norms bounded. we may also assume, by weakening p if necessary, that the a_j^0 and q are in C^{1+p+0}. Then Theorem I applied to the system $(\hat{X},\hat{Y},A)$ (where $\hat{X} = \dot{c}^{p+0})$ shows that $u_j \longrightarrow u$ in $C([0,T];\hat{X})$. Since the u_j are bounded in $B([0,T];\hat{Y})$ (because the $\emptyset_j$ are bounded in $\hat{Y}$), it follows that $u_j \longrightarrow u$ in $C([0,T];Y)$.

6. An example of compressible system

Let us illustrate the notion of compressible systems by a simple example.

EXAMPLE 6.1. Consider the first-order scalar equation

$$u_t + uu_x = 0 , \quad x \in R , \quad t \geq 0. \tag{6.1}$$

choose

$$X = \hat{X} = H^0(R), \quad Y = H^2(R), \quad \hat{Y} = H^3(R), \quad A(w) = wd_x . \tag{6.2}$$

It is known (see [9]) that (X,Y,A) is a "good" system. We shall show that $(\hat{X},\hat{Y},A)$ is compressible to (X,Y,A). Since $\hat{X} = X$, we may choose the norms

$$|u|_{\varepsilon,\hat{X}} = |u|_X = |u| \quad (L^2\text{-norm}), \tag{6.3}$$

$$|u|_Y^2 = |u|^2 + |u_{xx}|^2 , \quad |u|^2_{\varepsilon,\hat{Y}} = |u|_Y^2 + \varepsilon^2|u_{xxx}|^2$$

In this problem we do not need variable norms $|\ |_{X^w}$, etc. Thus the only parameters we have to consider are $\hat{\beta}_{1,\varepsilon}$, $\hat{\beta}_{2,\varepsilon}$, $\hat{\lambda}_{3,\varepsilon}$, and $\hat{\mu}_{3,\varepsilon}$. Among them, only $\hat{\beta}_{2,\varepsilon}$ is nontrivial, since it is easy to see that $\hat{\beta}_{1,\varepsilon}(r) \leq \beta_1(r)$, etc. due to $\hat{X} = X$.

To estimate $\hat{\beta}_{2,\varepsilon}$, we compute

$$\begin{aligned} |(A(w)u,u)_{\varepsilon,\hat{Y}}| &\leq c|w_{xx}|(|u|^2 + |u_{xx}|^2 + \varepsilon^2|u_{xxx}|^2) \\ &\quad + c\varepsilon^2|w_{xxx}||u_{xx}||u_{xxx}| \leq c|w|_{\varepsilon,\hat{Y}}|u|^2_{\varepsilon,\hat{Y}} . \end{aligned} \tag{6.4}$$

It follows that $|(A(w)+\lambda)u|_{\varepsilon,\hat{Y}} \geq (\lambda - c|w|_{\varepsilon,\hat{Y}})|u|_{\varepsilon,\hat{Y}}$. Hence we can take $\hat{\beta}_{2,\varepsilon}(r) = cr$, which is independent of ε . This shows that $(\hat{X},\hat{Y},A)$ is compressible to (X,Y,A).

REMARK 6.2. It is instructive to see what happens if in the above example we replace Y by $H^1(R)$ and $\hat{Y}$ by $H^2(R)$, with $|u|_Y^2 = |u|^2 + |u_x|^2$ and $|u|^2_{\varepsilon,\hat{Y}} = |u|_Y^2 + \varepsilon^2|u_{xx}|^2$. In this case (X,Y,A) is not a "good" system, so that $\hat{\beta}_{2,\varepsilon}$ could not stay bounded as $\varepsilon \longrightarrow 0$. Indeed, the best estimate one can expect of the sort of (6.4) will be

$$|(A(w)u,u)_{\varepsilon,\hat{Y}}| \leq c|w_{xx}|(|u|^2 + |u_x|^2 + \varepsilon^2|u_{xx}|^2)$$

$$\leq c\varepsilon^{-1}|w|_{\varepsilon,\hat{Y}}|u|^2_{\varepsilon,\hat{Y}} .$$

This gives $\hat{\beta}_{2,\varepsilon}(r) = c\varepsilon^{-1}r$, which blows up as $\varepsilon \longrightarrow 0$.

Footnotes

1. This work was partially supported by NSF Grant MCS 79-02578.

References

[1] Cesari, L., A boundary value problem for quasilinear hyperbolic systems in the Schauder canonic form, Ann. Scuola Norm. Sup. Pisa (4) 1 (1974), 311-358.

[2] Cinquini Cibrario, M. and Cinquini, S., Equazioni a derivate parziali di tipo iperbolico (Edizioni Cremonese, Roma 1964).

[3] Cinquini Cibrario, M., Ulteriori resultati per systemi di equazioni quasi lineari a derivate parziali in piu variabili independenti, Ist. Lombardo Accad. Sci. Lett. Rend. A 103 (1969), 373-407.

[4] Douglis, A., Some existence theorems for hyperbolic systems of partial differential equations in two independent variables, Comm. Pure Appl. Math. 5 (1952), 119-154.

[5] Friedrichs, K. O., Nonlinear hyperbolic differential equations of two independent variables, Amer. J. Math. 70 (1948), 555-589.

[6] Hartman, P. and Wintner, A., On the hyperbolic partial differential equations, Amer. J. Math. 74 (1952), 834-864.

[7] Hughes, T. J. R., Kato, T., and Marsden, J. R., Well-posed quasi-linear second-order hyperbolic systems with applications to nonlinear elastodynamics and general relativity, Arch. Rational Mech. Anal. 63 (1977), 273-294.

[8] Kato, T., Linear evolution equations of "hyperbolic" type, J. Fac. Sci. Univ. Tokyo, Sec. I, 17 (1970), 241-258.

[9] Kato, T., Quasi-linear equations of evolution, with applications to partial differential equations, Spectral Theory and Differential Equations, Lecture Notes in Math., 448 (Springer 1975, pp. 25-70).

[10] Kato, T., Linear and quasi-linear equations of evolution of hyperbolic type, C. I. M. E., II CICLO, 1976, Hyperbolicity, pp. 125-191.

[11] Kato, T., The Cauchy problem for the Korteweg-de Vries equation, Nonlinear partial differential equations and their applications, in: Brézis, H. and Lions J. L. (eds.), Collège de France Seminar, VOL. I (Pitman 1980, pp. 293-307).

[12] Kato, T., On the Cauchy problem for the (generalized) Korteweg-de Vries equation, Advances in Mathematics Supplementary Studies (Academic Press, to appear).

[13] Nakata, M., Quasi-linear evolution equation in nonreflexive Banach spaces, with applications to hyperbolic systems, Dissertation, University of California, Berkeley, 1983.

[14] Schauder, J., Cauchy'sches Problem für partielle Differentialgleichungen erster Ordnung. Anwendungen einiger sich auf die absolutbeträge der Lösungen beziehenden Abschätzungen, Commentarii Math. Helv. 9 (1937), 263-283.

Lecture Notes in Num. Appl. Anal., **5**, 77–100 (1982)
Nonlinear PDE in Applied Science. U.S.-Japan Seminar, Tokyo, 1982

Asymptotic Behaviors of the Solution of an Elliptic Equation with Penalty Terms

Hideo Kawarada* and Takao Hanada**

* Department of Applied Physics,
Faculty of Engineering, University of Tokyo
Bunkyo-ku, Tokyo 113, JAPAN
** Department of Information Mathematics,
The University of Electro-Communications
1-5-1, Chofugaoka, Chofu-shi, Tokyo 182, JAPAN

1. Introduction

Let Ω_0 be connected domain in R^2 with smooth boundary Γ. Take Ω so as to satisfy (i) $\Omega \supset \overline{\Omega}_0$; (ii) $\Omega_1 = \Omega - \Omega_0$ is a connected domain; (iii) the measure of $\partial\Omega_1^{(1)} \cap \partial\Omega$ is positive or Ω_1 is unbounded; (iv) $\partial\Omega$ is smooth (see Fig.1).

We shall consider the boundary value problem defined in Ω for every $\varepsilon > 0$ and $\alpha, \beta \in R$.

Find $\psi^\varepsilon = \begin{cases} \psi_0^\varepsilon & \text{in } \Omega_0, \\ \psi_1^\varepsilon & \text{in } \Omega_1 \end{cases}$ such that

(1.1) $$-\Delta \psi_0^\varepsilon + \lambda_0 \psi_0^\varepsilon = f \quad \text{in } \Omega_0$$

(1.2) $$-\varepsilon^{2\alpha} \cdot \Delta\psi_1^\varepsilon + \varepsilon^{-2\beta} \cdot \psi_1^\varepsilon = 0 \quad \text{in } \Omega_1$$

(1.3) $$\psi_0^\varepsilon = \psi_1^\varepsilon \quad \text{on } \Gamma$$

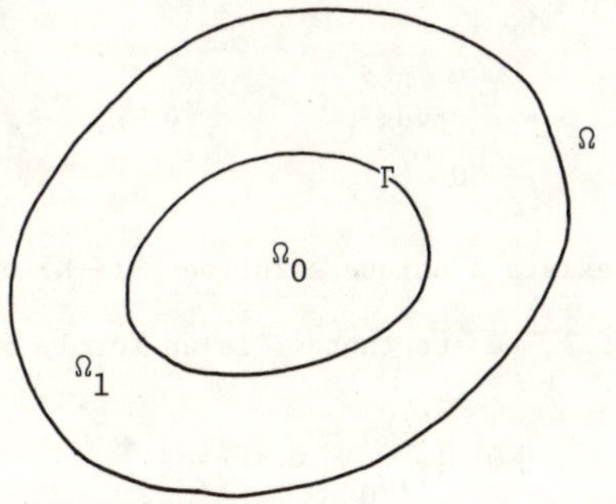

Fig.1

(1.4) $$\frac{\partial \psi_0^\varepsilon}{\partial n} = \varepsilon^{2\alpha} \cdot \frac{\partial \psi_1^\varepsilon}{\partial n} \qquad \text{on } \Gamma$$

(1.5) $$\psi_1^\varepsilon = 0 \qquad \text{on } \partial\Omega$$

$$\text{and} \quad \psi_1^\varepsilon \to 0 \quad (|x| = \sqrt{x_1^2 + x_2^2} \to \infty).$$

Here n is the outward normal on Γ to Ω_0 and λ_0 is a positive constant. It is found in Lions ([3], Chapter 1, p.80) that the boundary condition which the limit function of ψ_0^ε as $\varepsilon \to 0$ satisfies on Γ is classified into three types, which depend upon the relative value of α and β.

In this paper, we study an asymptotic behavior of ψ^ε on Γ when ε is small enough. We now summarize the contents of this paper. Section 2 includes four Theorems. In section 3, we prepare some Lemmas for the proofs of Theorems. Sections from 4 to 7 are devoted to the proofs of Theorems.

2. Theorems

2.1 We put

(2.1) $$K = \{\psi \in H^1(\Omega) \mid \psi = 0 \ \text{ on } \partial\Omega \ \text{ and } \ \psi \to 0 \ (|x| \to \infty)\}.$$

Then (1.1)-(1.5) are reformulated as follows:

Find $\psi^\varepsilon \in K$ such that

(2.2) $$\int_{\Omega_0} \nabla\psi^\varepsilon \nabla v dx + \lambda_0 \int_{\Omega_0} \psi^\varepsilon v dx + \varepsilon^{2\alpha} \cdot \int_{\Omega_1} \nabla\psi^\varepsilon \nabla v dx + \varepsilon^{-2\beta} \cdot \int_{\Omega_1} \psi^\varepsilon v dx$$

$$= \int_{\Omega_0} f v dx , \qquad \forall v \in K.$$

There exists a unique solution $\psi^\varepsilon (\in K)$ of (2.2) for $\forall f \in H^{-1}(\Omega_0)$. Putting $v = \psi^\varepsilon$ in (2.2), we see that ψ_0^ε is uniformly bounded in ε:

(2.3) $$\| \psi^\varepsilon \|_{1,\Omega_0}^{(2)} \leq C < +\infty$$

where C depends upon only the data f.

When ε tends to zero, we can extract a sequence ε_n $(n = 1, 2, \dots)$ such that

$$(2.4) \qquad \psi_0^{\varepsilon_n} \to \psi_0^0 \qquad \text{weakly in } H^1(\Omega_0).$$

Then

$$(2.5) \qquad \psi_0^{\varepsilon_n} \to \psi_0^0 \qquad \text{strongly in } H^s(\Omega_0) \qquad (\forall s < 1, [5]).$$

Let $v \in H_0^1(\Omega_0)$ and $\tilde{v}$ be the zero extension of v to Ω. Passing to the limit in (2.2) for $\tilde{v} \in H^1(\Omega)$ yields

$$(2.6) \qquad \int_{\Omega_0} \nabla\psi_0^0 \nabla v \, dx + \lambda_0 \int_{\Omega_0} \psi_0^0 v \, dx = \int_{\Omega_0} f v \, dx, \qquad \forall v \in H_0^1(\Omega_0).$$

from which, we have

$$(2.7) \qquad -\Delta \psi_0^0 + \lambda_0 \psi_0^0 = f \qquad \text{in } H^{-1}(\Omega_0).$$

If we assume $f \in H^{m-1}(\Omega_0)$ $(m \geqq 0)$, then we have

$$(2.8) \qquad \psi_0^0 \in H^{m+1}(\Omega_0).$$

By the trace theorem (Nečas [5]),

$$(2.9) \qquad \psi_0^0\Big|_\Gamma \in H^{m+\frac{1}{2}}(\Gamma) \quad \text{and} \quad \frac{\partial \psi_0^0}{\partial n}\Big|_\Gamma \in H^{m-\frac{1}{2}}(\Gamma).$$

Moreover, ψ_0^0 satisfies on Γ:

<u>Theorem 1</u>(3) Suppose $f \in H^{m-1}(\Omega_0)$ $(m \geqq 0)$.

(a) If $\beta > |\alpha|$, then

$$(2.10) \qquad \psi_0^0\Big|_\Gamma = 0 \qquad \text{in } H^{m+\frac{1}{2}}(\Gamma).$$

(b) If $\beta = \alpha > 0$, then

(2.11) $$\left(\frac{\partial\psi_0^0}{\partial n} + \psi_0^0\right)\Big|_\Gamma = 0 \qquad \text{in } H^{m-\frac{1}{2}}(\Gamma).$$

(c) If $\beta < \alpha$ and $\alpha > 0$, then

(2.12) $$\frac{\partial\psi_0^0}{\partial n}\Big|_\Gamma = 0 \qquad \text{in } H^{m-\frac{1}{2}}(\Gamma).$$

Remark 1 There also holds $\psi_0^0|_\Gamma = 0$ in the case $\alpha + \beta \leqq 0$ and $\alpha < 0$ under the same assumption.

2.2 We now state our main result as follows.

Theorem 2 Suppose $f \in H^m(\Omega_0)$ $(m \geqq 0)$ and let ε be small enough.

(a) If $\beta > |\alpha|$, then

(2.13) $$\psi^\varepsilon|_\Gamma = -\varepsilon^{\beta-\alpha} \cdot \frac{\partial\psi_0^0}{\partial n}\Big|_\Gamma + O(\varepsilon^{2(\beta-\alpha)} + \varepsilon^{\beta-3\alpha} + \varepsilon^{2(\alpha+\beta)}) \qquad \text{in } H^{m-\frac{1}{2}}(\Gamma)$$

where ψ_0^0 satisfies (2.10).

(b) If $\beta = \alpha > 0$, then

(2.14) $$\psi^\varepsilon|_\Gamma = \psi_0^0|_\Gamma + O(\varepsilon^{4\alpha}) \qquad \text{in } H^{m+\frac{1}{2}}(\Gamma)$$

where ψ_0^0 satisfies (2.11).

(c) If $|\beta| < \alpha$, then

(2.15) $$\psi^\varepsilon|_\Gamma = \psi_0^0|_\Gamma + O(\varepsilon^{\alpha-\beta}) \qquad \text{in } H^{m+\frac{1}{2}}(\Gamma)$$

where ψ_0^0 satisfies (2.12).

2.3 By using (a) of Theorem 2, we have the regularity results about ψ^{ε}.

Theorem 3 Suppose $f \in H^{k}(\Omega_0)$ $(k \geqq 5)$ and let ε be small enough.

If $\beta > |\alpha|$, then

$$\| \psi^{\varepsilon} \|_{W^{0,\infty}(\Gamma)} \leqq O(\varepsilon^{\beta-\alpha}), \tag{2.16}$$

$$\| \psi^{\varepsilon} \|_{W^{1,\infty}(\Omega)} \leqq O(1+\varepsilon^{-2\alpha}), \tag{2.17}$$

$$\| \psi^{\varepsilon} \|_{W^{2,\infty}(\Omega_1)} \leqq O(\varepsilon^{-(3\alpha+\beta)}). \tag{2.18}$$

2.4 The motivation of this paper consists in the integrated penalty method presented by one of the author [2]. The mathematical justification of this method was done in the sense of distribution. If we use (a) of Theorem 2, we can prove the key-point of this method in the framework of the Soborev space.

Theorem 4 Suppose $f \in H^{m}(\Omega_0)$ $(m \geqq 0)$ and let ε be small enough.

If $\beta > |\alpha|$, then

$$\| \varepsilon^{-2\beta} \cdot \int_{\Gamma^{\perp}(s)} \psi_1^{\varepsilon} d\Gamma^{\perp} + \frac{\partial \psi_0^{0}}{\partial n}(s) \|_{m-\frac{1}{2},\Gamma} = O(\varepsilon^{2(\alpha+\beta)}). \tag{2.19}$$

Here s stands for the length of the arc along Γ.

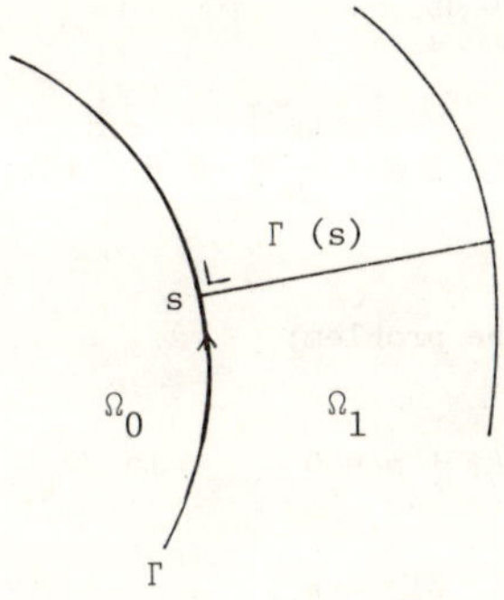

Fig.2

3. Preliminaries

The aim of this section is to give some preparatory lemmas which will be needed in the proofs of Theorems.

3.1 We first introduce some operators defined between traces on Γ.

(i) Define the mapping

$$T_f : H^{\frac{1}{2}}(\Gamma) \ni a \to \left.\frac{\partial \psi_a}{\partial n}\right|_{\Gamma} \in H^{-\frac{1}{2}}(\Gamma): \tag{3.1}$$

ψ_a is the solution of the problem;

$$-\Delta \psi + \lambda_0 \psi = f \quad \text{in } \Omega_0 \tag{3.2}$$

$$\psi|_{\Gamma} = a \tag{3.3}$$

where $f \in H^{m-1}(\Omega_0)$ $(m \geq 0)$.

(ii) Define the mapping

$$R^{\varepsilon} : H^{-\frac{1}{2}}(\Gamma) \ni b \mapsto \left.\psi_b^{\varepsilon}\right|_{\Gamma} \in H^{\frac{1}{2}}(\Gamma); \tag{3.4}$$

ψ_b^{ε} is the solution of (3.2) with $f \equiv 0$ and the boundary condition

$$\left.\left(\varepsilon^{\beta-\alpha} \cdot \frac{\partial \psi}{\partial n} + \psi\right)\right|_{\Gamma} = b. \tag{3.5}$$

(iii) Define the mapping

$$S^{\varepsilon} : H^{\frac{1}{2}}(\Gamma) \ni a \mapsto \left.\frac{\partial \psi_a^{\varepsilon}}{\partial n}\right|_{\Gamma} \in H^{-\frac{1}{2}}(\Gamma); \tag{3.6}$$

ψ_a^{ε} is the solution of the problem;

$$-\varepsilon^{\alpha+\beta} \cdot \Delta\psi + \psi = 0 \quad \text{in } \Omega_1 \tag{3.7}$$

$$\psi|_{\Gamma} = a \tag{3.8}$$

(3.9) $\psi|_{\partial\Omega} = 0$ and $\psi \to 0$ $(|x| \to \infty)$.

We denote T_f^m, S_m^ε and R_m^ε by the restriction of T_f, S^ε and R^ε to $H^{m+\frac{1}{2}}(\Gamma)$. But we abbreviate the suffix m hereafter.

3.2

Lemma 1 Let a, b be arbitrary in $H^{m+\frac{1}{2}}(\Gamma)$. Then

(3.10) $T_f(a) - T_f(b) = T_0(a-b)$

where T_0 implies $T_{f=0}$.

Proof Let ψ_γ $(\gamma = a,b)$ be the solution of (3.2) under the boundary condition

(3.11) $\psi|_\Gamma = \gamma$.

Put $\Psi = \psi_a - \psi_b$. Ψ satisfies

(3.12) $-\Delta\Psi + \lambda_0\Psi = 0$ in Ω_0,

(3.13) $\Psi|_\Gamma = a - b$.

Then

(3.14) $\left.\dfrac{\partial\Psi}{\partial n}\right|_\Gamma = T_0(a-b)$.

On the other hand,

(3.15) $\left.\dfrac{\partial\Psi}{\partial n}\right|_\Gamma = \left.\dfrac{\partial\psi_a}{\partial n}\right|_\Gamma - \left.\dfrac{\partial\psi_b}{\partial n}\right|_\Gamma = T_f(a) - T_f(b)$.

From (3.14) and (3.15) follows (3.10). Here we should note that T_0 is linear and T_f is non-linear. ∎

Lemma 2

T_f and S are homeomorphic from $H^{m-\frac{1}{2}}(\Gamma)$ to $H^{m+\frac{1}{2}}(\Gamma)$ and R is homeomorphic from $H^{m+\frac{1}{2}}(\Gamma)$ to $H^{m-\frac{1}{2}}(\Gamma)$ for any $m \geqq 0$.

Proof

1° T_f is injective from $H^{m+\frac{1}{2}}(\Gamma)$ into $H^{m-\frac{1}{2}}(\Gamma)$. In fact, let $a,b \in H^{m+\frac{1}{2}}(\Gamma)$ $(a \neq b)$. Suppose $T_f(a) = T_f(b)$. Then, by (3.10)

$$0 = T_f(a) - T_f(b) = T_0(a-b) \neq 0$$

because of the strong maximum principle under the assumption $\lambda_0 > 0$. This is a contradiction.

2° T_f is surjective from $H^{m+\frac{1}{2}}(\Gamma)$ onto $H^{m-\frac{1}{2}}(\Gamma)$. In fact, we choose any $b \in H^{m+\frac{1}{2}}(\Gamma)$. Then, the following problem:

(3.16) $$-\Delta\psi + \lambda_0\psi = f \qquad \text{in } \Omega_0$$

(3.17) $$\left.\frac{\partial\psi}{\partial n}\right|_\Gamma = b$$

has a unique solution $\psi_b \in H^{m+1}(\Omega_0)$ if $\lambda_0 > 0$, which satisfies

(3.18) $$\psi_b\big|_\Gamma \in H^{m+\frac{1}{2}}(\Gamma) \qquad \text{and} \qquad b = T_f(\psi_b\big|_\Gamma).$$

3° It is checked that T_f and $(T_f)^{-1}$ are continuous between $H^{m+\frac{1}{2}}(\Gamma)$ and $H^{m-\frac{1}{2}}(\Gamma)$ (see [1]).

4° Summing up 1°, 2° and 3°, we see that T_f is a homeomorphism from $H^{m+\frac{1}{2}}(\Gamma)$ onto $H^{m-\frac{1}{2}}(\Gamma)$. The repeated use of the above arguments gives that $(R^\varepsilon)^{-1}$ and S^ε are also homeomorphic between $H^{m+\frac{1}{2}}(\Gamma)$ and $H^{m-\frac{1}{2}}(\Gamma)$. ∎

3.3 Here we give the estimates of the norm of R^ε and S^ε, which are crucial for the proof of Theorems 1 and 2.

Lemma 3 Let ε be small enough and suppose $\beta \geq \alpha$ and $m \geq 0$. Then

$$\| R^{\varepsilon}(a) \|_{m+\frac{1}{2},\Gamma} = O(\varepsilon^{\alpha-\beta}) \| a \|_{m-\frac{1}{2},\Gamma}, \qquad \text{for } \forall a \in H^{m-\frac{1}{2}}(\Gamma) \tag{3.19}$$

$$\| R^{\varepsilon}(a) \|_{m-\frac{1}{2},\Gamma} = O(1) \| a \|_{m-\frac{1}{2},\Gamma}, \qquad \text{for } \forall a \in H^{m-\frac{1}{2}}(\Gamma) \tag{3.20}$$

and

$$\| R^{\varepsilon}(a) - a \|_{m-\frac{1}{2},\Gamma} = O(\varepsilon^{\beta-\alpha}) \| a \|_{m+\frac{1}{2},\Gamma} \qquad \text{for } \forall a \in H^{m+\frac{1}{2}}(\Gamma). \tag{3.21}$$

<u>Proof</u> Using Green's formula in the problem defining R^{ε}, we have

$$\varepsilon^{\beta-\alpha}\cdot\int_{\Omega_0} (|\nabla\psi|^2 + \lambda_0|\psi|^2)dx + \int_{\Gamma} |\psi|^2 ds = \int_{\Gamma} a\psi ds. \tag{3.22}$$

From (3.22) it follows

$$\|\psi\|_{0,\Gamma} \leqq \|a\|_{0,\Gamma} \tag{3.23}$$

and

$$\varepsilon^{\beta-\alpha} \|\psi\|_{\frac{1}{2},\Gamma} \leqq \|a\|_{-\frac{1}{2},\Gamma} . \tag{3.24}$$

Using the standard technique to raise up the regularity property of the solution of partial differential equations, we obtain (3.19) and (3.20).

Rewriting (3.5) with an aid of T_0 and R^{ε}, we have for $\forall a \in H^{m+\frac{1}{2}}(\Gamma)$

$$\begin{aligned} \| R^{\varepsilon}(a) - a \|_{m-\frac{1}{2},\Gamma} &= \varepsilon^{\beta-\alpha} \| T_0 R^{\varepsilon}(a) \|_{m-\frac{1}{2},\Gamma} \\ &= O(\varepsilon^{\beta-\alpha}) \| R^{\varepsilon}(a) \|_{m+\frac{1}{2},\Gamma} \\ &= O(\varepsilon^{\beta-\alpha}) \|a\|_{m+\frac{1}{2},\Gamma} \qquad \text{(by 3.20).} \end{aligned} \tag{3.25}$$

Here we have used the continuity of T_0 from $H^{m+\frac{1}{2}}(\Gamma)$ to $H^{m-\frac{1}{2}}(\Gamma)$. ∎

<u>Lemma 4</u> Let ε be small enough and $m \geqq 0$.

(a) If $\sigma = \alpha+\beta+\rho(\alpha-\beta) > 0$ $(\rho \in R)$, then

$$\| \varepsilon^{\sigma} S^{\varepsilon}(a) + \varepsilon^{\rho(\alpha-\beta)} a \|_{m-\frac{1}{2},\Gamma} = O(\varepsilon^{\sigma}) \| a \|_{m+\frac{1}{2},\Gamma}, \quad \forall a \in H^{m+\frac{1}{2}}(\Gamma). \tag{3.26}$$

(b) If $\alpha+\beta > 0$, then

$$\| S^{\varepsilon}(a) + \varepsilon^{-(\alpha+\beta)} a \|_{m-\frac{1}{2},\Gamma} = O(\varepsilon^{\alpha+\beta}) \| a \|_{m+\frac{1}{2},\Gamma}, \quad \forall a \in H^{m+\frac{1}{2}}(\Gamma) \tag{3.27}$$

$$\| (S^{\varepsilon})^{-1}(b) \|_{m+\frac{1}{2},\Gamma} = O(1) \| b \|_{m-\frac{1}{2},\Gamma}, \quad \forall b \in H^{m-\frac{1}{2}}(\Gamma) \tag{3.28}$$

$$\| (S^{\varepsilon})^{-1}(b) + \varepsilon^{\alpha+\beta} b \|_{m-\frac{1}{2},\Gamma} = O(\varepsilon^{3(\alpha+\beta)}) \| b \|_{m+\frac{1}{2},\Gamma}, \tag{3.29}$$

$$\forall b \in H^{m+\frac{1}{2}}(\Gamma).$$

<u>Proof</u> We prove this lemma in the two cases. In the first case, we prove the special case $\Omega_1 = R^2_+ = \{(x_1,x_2) \mid 0 < x_1, -\infty < x_2 < +\infty\}$ $(\Omega_0 = R^2_-)$ by using the fourier transformation. Subsequently, we give the plan of the proof in the general geometry.

1° Let

$$\hat{\psi}^{\varepsilon}(x_1,\xi) = \int_{-\infty}^{\infty} e^{2\pi i \xi x_2} \cdot \psi^{\varepsilon}(x_1,x_2)\,dx_2$$

and

$$\hat{a}(\xi) = \int_{-\infty}^{\infty} e^{2\pi i \xi x_2} \cdot a(x_2)\,dx_2.$$

Here $\psi^{\varepsilon}(x_1,x_2)$ is the solution of the problem (3.7)-(3.9). Then $\hat{\psi}^{\varepsilon}$ satisfies

$$(3.30)\qquad -\frac{\partial^2\hat{\psi}^{\varepsilon}}{\partial x_1^2} + \varepsilon^{-2(\alpha+\beta)}\cdot(1+4\pi^2|\xi|^2\varepsilon^{2(\alpha+\beta)})\hat{\psi}^{\varepsilon} = 0 \qquad \text{in } R_+^2,$$

$$(3.31)\qquad \hat{\psi}^{\varepsilon}\Big|_{x_1=0} = \hat{a}.$$

Solving (3.30) and (3.31), we have

$$(3.32)\qquad \hat{\psi}^{\varepsilon} = \hat{a}\cdot\exp\{-\varepsilon^{-(\alpha+\beta)}\cdot(1+4\pi^2\cdot|\xi|^2\varepsilon^{2(\alpha+\beta)})^{\frac{1}{2}}\, x_1\}.$$

From (3.32)

$$(3.33)\qquad \frac{\partial\hat{\psi}^{\varepsilon}}{\partial x_1}\Bigg|_{x_1=0} = \widehat{S^{\varepsilon}(a)} = -\varepsilon^{-(\alpha+\beta)}\cdot(1+4\pi^2\cdot|\xi|^2\cdot\varepsilon^{2(\alpha+\beta)})^{\frac{1}{2}}\cdot\hat{a}.$$

We compute

$$(3.34)\qquad \varepsilon^{\sigma}\cdot\widehat{S^{\varepsilon}(a)} + \varepsilon^{\rho(\alpha-\beta)}\cdot\hat{a}$$

$$= \varepsilon^{\rho(\alpha+\beta)}\{1-(1+4\pi^2\cdot|\xi|^2\cdot\varepsilon^{2(\alpha+\beta)})^{\frac{1}{2}}\}\,\hat{a}$$

$$= -4\pi^2\cdot\varepsilon^{\sigma}\cdot\frac{|\xi|^2\cdot\varepsilon^{\alpha+\beta}\hat{a}}{1+(1+4\pi^2\cdot|\xi|^2\cdot\varepsilon^{2(\alpha+\beta)})^{\frac{1}{2}}}\,.$$

From (3.34), it follows

$$(3.35)\qquad \int_{-\infty}^{\infty}(1+4\pi^2\cdot|\xi|^2)^{m-\frac{1}{2}}\left|\varepsilon^{\sigma}\cdot\widehat{S^{\varepsilon}(a)} + \varepsilon^{\rho(\alpha-\beta)}\cdot\hat{a}\right|^2 d\xi$$

$$= 16\pi^4\cdot\varepsilon^{2\sigma}\int_{-\infty}^{\infty}\frac{(1+4\pi^2|\xi|^2)^{m-\frac{1}{2}}\cdot|\xi|^2\cdot\varepsilon^{2(\alpha+\beta)}\cdot|\xi|^2\cdot|\hat{a}|^2}{\{1+(1+4\pi^2\cdot|\xi|^2\varepsilon^{2(\alpha+\beta)})^{\frac{1}{2}}\}^2}\,d\xi$$

$$\leq O(\varepsilon^{2\sigma})\int_{-\infty}^{\infty}(1+4\pi^2\cdot|\xi|^2)^{m+\frac{1}{2}}\cdot|\hat{a}|^2\,d\xi.$$

Hence we obtain

$$(3.36)\qquad \|\varepsilon^{\sigma}\cdot S^{\varepsilon}(a)+\varepsilon^{\rho(\alpha-\beta)}\cdot a\|_{m-\frac{1}{2},\Gamma}\leq O(\varepsilon^{\sigma})\|a\|_{m+\frac{1}{2},\Gamma}.$$

Repeating the simular arguments as above, we conclude (3.27)-(3.29).

2° Let us now deal with the general case. The domain Ω_1 is a regular simply connected domain; then there exists a (fixed) regular conformal mapping $w=f(z)=u_1+iu_2$ $(z=x_1+ix_2)$ which maps Ω_1 into R^2_+. As a matter of fact, Γ is mapped into the u_2-axis of w-plane. Then the transformed solution $\Psi^{\varepsilon}=\psi^{\varepsilon}(f^{-1}(w))$ satisfies

$$(3.37)\qquad -\varepsilon^{2(\alpha+\beta)}\cdot\Delta\Psi^{\varepsilon}+\left|\frac{dz}{dw}\right|^2\Psi^{\varepsilon}=0 \qquad \text{in } R^2_+,$$

$$(3.38)\qquad \Psi^{\varepsilon}|_{u_1=0}=A(u_2)=a(f^{-1}(w)).$$

By means of the iterative method proposed in the theory of singular perturbation (see [3]), Ψ^{ε} is asymptotically developed in the following way:

$$(3.39)\qquad \psi^{\varepsilon}=\psi^0_{\varepsilon}+\varepsilon^{\alpha+\beta}\psi^1_{\varepsilon}+\varepsilon^{2(\alpha+\beta)}\psi^2_{\varepsilon}+\ldots+\varepsilon^{n(\alpha+\beta)}\psi^n_{\varepsilon}+w_{\varepsilon}.$$

Using (3.39), we obtain (3.26)-(3.29) (see the appendix). ∎

3.4 Define

$$(3.40)\qquad \varphi^{\varepsilon}=\psi^{\varepsilon}|_{\Gamma}\in H^{m+\frac{1}{2}}(\Gamma),$$

$$(3.41)\qquad \varphi^{0}=\psi^{0}_{0}\Big|_{\Gamma}\in H^{m+\frac{1}{2}}(\Gamma).$$

Then we have

Lemma 5 Let $\varepsilon_n\to 0$ $(n=1, 2, \ldots)$. Then

(3.42) $$\varphi^{\varepsilon_n} \to \varphi^0 \qquad \text{weakly in } H^{\frac{1}{2}}(\Gamma),$$

(3.43) $$T_f(\varphi^{\varepsilon_n}) \to T_f(\varphi^0) \qquad \text{weakly in } H^{-\frac{1}{2}}(\Gamma).$$

<u>Proof</u> Recalling (2.4), (3.42) is obvious. Let a, b be arbitrary in $H^{\frac{1}{2}}(\Gamma)$. Then, we denote by ψ_γ $(\gamma = a,b)$ the solution of the problem:

(3.44) $$-\Delta\psi + \lambda_0\psi = 0 \qquad \text{in } \Omega_0,$$

(3.45) $$\psi|_\Gamma = \gamma.$$

By using Green's formula, we have

(3.46) $$\int_\Gamma b\, T_0(a)ds - \int_\Gamma a\, T_0(b)ds = \int_{\Omega_0} (\psi_b\Delta\psi_a - \psi_a\Delta\psi_b)dx = 0.$$

Using (3.10) and taking $a = \varphi^{\varepsilon_n} - \varphi^0$ in (3.46),

(3.47) $$\int_\Gamma \{T_f(\varphi^{\varepsilon_n}) - T_f(\varphi^0)\}\, b\, ds = \int_\Gamma T_0(\varphi^{\varepsilon_n} - \varphi^0)\, b\, ds$$

$$= \int_\Gamma (\varphi^{\varepsilon_n} - \varphi^0)T_0(b)ds \to 0 \qquad (\varepsilon_n \to 0).$$ ∎

<u>Lemma 6</u> Suppose $\sigma = \alpha + \beta + \rho(\alpha-\beta) > 0$ $(\rho \in R)$. Then

(3.48) $$\varepsilon_n^{(\rho-1)(\alpha-\beta)} \cdot T_f(\varphi^{\varepsilon_n}) + \varepsilon_n^{\rho(\alpha-\beta)} \cdot \varphi^{\varepsilon_n} \to 0$$

strongly in $H^{-\frac{1}{2}}(\Gamma)$ as $\varepsilon_n \to 0$.

<u>Proof</u> By using the definition of T_f and S^ε, (1.4) is rewritten as follows:

(3.49) $$T_f(\varphi^\varepsilon) = \varepsilon^{2\alpha} \cdot S^\varepsilon(\varphi^\varepsilon).$$

Taking $a = \varphi^{\varepsilon_n}$ and $\varepsilon = \varepsilon_n$ in (3.26) for $m = 0$ and substituting (3.49), we have

(3.50) $$\| \varepsilon_n^{(\rho-1)(\alpha-\beta)} \cdot T_f(\varphi^{\varepsilon_n}) + \varepsilon_n^{\rho(\alpha-\beta)} \cdot \varphi^{\varepsilon_n} \|_{-\frac{1}{2},\Gamma} = O(\varepsilon_n^\sigma)\|\varphi^{\varepsilon_n}\|_{\frac{1}{2},\Gamma} \cdot$$

Let $\varepsilon_n \to 0$. Then we conclude (3.48) with an aid of (2.3). ∎

4. Proof of Theorem 1

(a) Let $\rho = 0$ and $\beta > |\alpha|$ in the assumption of Lemma 6. Then $\sigma = \alpha + \beta > 0$ and (3.48) becomes

$$(4.1) \qquad \varepsilon_n^{\beta-\alpha} \cdot T_f(\varphi^{\varepsilon_n}) + \varphi^{\varepsilon_n} \to 0 \qquad \text{strongly in } H^{-\frac{1}{2}}(\Gamma).$$

By (3.42) and (3.43),

$$(4.2) \qquad \varphi^0 \quad (= \psi_0^0\big|_\Gamma) = 0.$$

(b) Let $\alpha = \beta > 0$. Then $\sigma = 2\alpha > 0$ and (3.48) becomes

$$(4.3) \qquad T_f(\varphi^{\varepsilon_n}) + \varphi^{\varepsilon_n} \to 0 \qquad \text{strongly in } H^{-\frac{1}{2}}(\Gamma)$$

which implies

$$(4.4) \qquad T_f(\varphi^0) + \varphi^0 = \left(\frac{\partial \psi_0^0}{\partial n} + \psi_0^0\right)\Big|_\Gamma = 0$$

by the definition of T_f and φ^0.

(c) Let $\rho = 1$, $\alpha > \beta$ and $\alpha > 0$. Then $\sigma = 2\alpha > 0$ and (3.48) becomes

$$(4.5) \qquad T_f(\varphi^{\varepsilon_n}) + \varepsilon_n^{\alpha-\beta} \cdot \varphi^{\varepsilon_n} \to 0 \qquad \text{strongly in } H^{-\frac{1}{2}}(\Gamma)$$

from which

$$(4.6) \qquad T_f(\varphi^0) \ \left(= \frac{\partial \psi_0^0}{\partial n}\Big|_\Gamma\right) = 0 \qquad \text{in } H^{-\frac{1}{2}}(\Gamma).$$

Combining (2.8) with the results obtained above, we conclude (2.10)-(2.12). ∎

5. Proof of Theorem 2

5.1 Using (3.49), the problem (1.1)-(1.5) is transformed into the following one:

Find $a \in H^{m+\frac{1}{2}}(\Gamma)$ such that

$$T_f(a) = \varepsilon^{2\alpha} \cdot S^{\varepsilon}(a). \tag{5.1}$$

Hereafter we call (5.1) the transmission equation. As a matter of fact, the solution of (5.1) is equal to the trace $\psi^{\varepsilon}|_{\Gamma}$ of the solution of the problem (1.1)-(1.5).

5.2 Let b be arbitrary in $H^{m+\frac{1}{2}}(\Gamma)$. Then, combining (5.1) and (3.10), we have

$$T_0(a-b) - \varepsilon^{2\alpha} \cdot S^{\varepsilon}(a) = -T_f(b) \in H^{m-\frac{1}{2}}(\Gamma). \tag{5.2}$$

Let us begin to prove (a), in which $\beta > |\alpha|$ is assumed. By (a) of Theorem 1, we have $\psi_0^0\big|_{\Gamma} = 0$. Therefore we choose $b = 0$ in (5.2). On substituting (3.27) into (5.2), we get

$$T_0(a) + \varepsilon^{\alpha-\beta} \cdot a - \varepsilon^{2\alpha} \cdot S_1(a) = -T_f(0), \qquad \text{or} \tag{5.3}$$

$$\varepsilon^{\beta-\alpha} \cdot T_0(a) + a = \varepsilon^{\alpha+\beta} \cdot S_1^{\varepsilon}(a) - \varepsilon^{\beta-\alpha} \cdot T_f(0)$$

where $\quad S_1^{\varepsilon}(a) = S^{\varepsilon}(a) + \varepsilon^{-(\alpha+\beta)} \cdot a.$

The definition of R^{ε} allows us to rewrite (5.3) by

$$a = \varepsilon^{\alpha+\beta} \cdot R^{\varepsilon} S_1^{\varepsilon}(a) - \varepsilon^{\beta-\alpha} \cdot R^{\varepsilon}(T_f(0)). \tag{5.4}$$

Let $\varepsilon\,(>0)$ be small enough in (5.4).

Using Lemmas 2, 3 and 4, we see that the mapping $\varepsilon^{\alpha+\beta} \cdot R^{\varepsilon} S_1^{\varepsilon}$ becomes the contraction mapping from $H^{m+\frac{1}{2}}(\Gamma)$ onto itself if ε is small enough and $\beta > \alpha > -\frac{1}{3}$.

Indeed,

$$(5.5)\qquad \varepsilon^{\alpha+\beta}\| R^{\varepsilon} S_1^{\varepsilon}(a) \|_{m+\frac{1}{2},\Gamma} \leqq O(\varepsilon^{2\alpha})\| S_1^{\varepsilon}(a) \|_{m-\frac{1}{2},\Gamma} \qquad \text{(by 3.19)}$$

$$\leqq O(\varepsilon^{3\alpha+\beta})\, \| a \|_{m+\frac{1}{2},\Gamma} . \qquad \text{(by 3.27)}$$

On the other hand, by (3.21)

$$(5.6)\qquad R^{\varepsilon}(T_f(0)) = T_f(0) + O(\varepsilon^{\beta-\alpha}) \quad \text{in } H^{m-\frac{1}{2}}(\Gamma).$$

Here we note that $T_f(0)$ should be included in $H^{m+\frac{1}{2}}(\Gamma)$. Therefore, we have to assume $f \in H^m(\Omega_0)$.

Summing up (5.4), (5.5) and (5.6), we have

$$(5.7)\qquad a = \psi^{\varepsilon}|_{\Gamma} = (I - \varepsilon^{\alpha+\beta}\cdot R^{\varepsilon} S_1^{\varepsilon})^{-1}\{-\varepsilon^{\beta-\alpha}\cdot T_f(0) + O(\varepsilon^{2(\beta-\alpha)})\}$$

$$= -\varepsilon^{\beta-\alpha}\cdot T_f(0) + O(\varepsilon^{2(\beta-\alpha)} + \varepsilon^{2(\alpha+\beta)}) \qquad \text{in } H^{m-\frac{1}{2}}(\Gamma)$$

if $\beta > \alpha > -\frac{1}{3}\beta$.

We remove into the case $-\alpha < \beta \leqq -3$. Operating $\varepsilon^{-2\alpha}\cdot(S^{\varepsilon})^{-1}$ on both sides of (5.2), we have

$$(5.8)\qquad a = \varepsilon^{-2\alpha}(S^{\varepsilon})^{-1}T_0(a) + \varepsilon^{-2\alpha}\cdot(S^{\varepsilon})^{-1}(T_f(0)).$$

Let ε (> 0) be small enough. Then $\varepsilon^{-2\alpha}\cdot(S^{\varepsilon})^{-1}T_0$ becomes the contraction mapping from $H^{m+\frac{1}{2}}(\Gamma)$ onto itself if $\alpha < 0$ and $\alpha + \beta > 0$. In fact, the boundedness of T_0 and (3.28) yields

$$(5.9)\qquad \varepsilon^{-2\alpha}\, \| (S^{\varepsilon})^{-1}T_0(a) \|_{m+\frac{1}{2},\Gamma} \leqq O(\varepsilon^{-2\alpha})\, \| a \|_{m+\frac{1}{2},\Gamma} .$$

Therefore, by using (5.8), (5.9) and (3.29), we have

$$(5.10)\qquad a = \psi^\varepsilon|_\Gamma = \{I - \varepsilon^{-2\alpha}(S^\varepsilon)^{-1}T_0\}^{-1}\{-\varepsilon^{\beta-\alpha}T_f(0) + O(\varepsilon^{\alpha+3\beta})\}$$

$$= -\varepsilon^{\beta-\alpha}T_f(0) + O(\varepsilon^{\beta-3\alpha} + \varepsilon^{\alpha+3\beta}) \qquad \text{in } H^{m-\frac{1}{2}}(\Gamma),$$

if $\alpha + \beta > 0$ and $\alpha > 0$.

Combining (5.7) and (5.10), we obtain (2.13). ∎

5.3 We shall prove (b) of Theorem 2, in which $\alpha = \beta > 0$ is assumed. From (b) of Theorem 1 follows $T_f(\psi_0^0) + \psi_0^0 = 0$ on Γ. Choose $b = \psi_0^0|_\Gamma$ in (5.2). Then we have

$$(5.11)\qquad T_0(a - \psi_0^0) - \varepsilon^{2\alpha}S^\varepsilon(a) = -T_f(\psi_0^0) = \psi_0^0 .$$

By (3.27), we have

$$(5.12)\qquad T_0(a - \psi_0^0) + a - \psi_0^0 = \varepsilon^{2\alpha}\cdot S_1^\varepsilon(a)$$

where $S_1^\varepsilon(a) = S^\varepsilon(a) + \varepsilon^{-2\alpha}a$. By use of R^ε with $\alpha = \beta$,

$$(5.13)\qquad a - \psi_0^0 = \varepsilon^{2\alpha}\cdot R^\varepsilon S_1^\varepsilon(a).$$

Then $\varepsilon^{2\alpha}\cdot R^\varepsilon S_1^\varepsilon$ becomes the contraction mapping from $H^{m+\frac{1}{2}}(\Gamma)$ onto itself if $\alpha > 0$ and ε is small enough. In fact, by (3.19) and (3.27), we have

$$(5.14)\qquad \varepsilon^{2\alpha}\,\|R^\varepsilon S_1^\varepsilon(a)\|_{m+\frac{1}{2},\Gamma} = O(\varepsilon^{4\alpha})\,\|a\|_{m+\frac{1}{2},\Gamma} .$$

Therefore we have

$$(5.15)\qquad a = \psi^\varepsilon|_\Gamma = (I - \varepsilon^{2\alpha}R^\varepsilon S_1^\varepsilon)^{-1}\psi_0^0 = \psi_0^0 + O(\varepsilon^{4\alpha}) \qquad \text{in } H^{m+\frac{1}{2}}(\Gamma).$$ ∎

5.4 Now we are in the final step to prove (c). In this case, $\alpha > \beta$ and $\alpha > 0$ are assumed. (c) of Theorem 1 gives us $T_f(\psi_0^0) = 0$. Put $b = \psi_0^0|_\Gamma$ in (5.2). Then we have

$$(5.16)\qquad T_0(a - \psi_0^0) - \varepsilon^{2\alpha}\cdot S^\varepsilon(a) = -T_f(\psi_0^0) = 0.$$

Operating $(T_0)^{-1}$ on both sides of (5.16), we have

$$a - \psi_0^0 = \varepsilon^{2\alpha}(T_0)^{-1}S^\varepsilon(a - \psi_0^0) + \varepsilon^{2\alpha}(T_0)^{-1}S^\varepsilon(\psi_0^0). \tag{5.17}$$

Repeating the similar arguments as in the proofs of (a) and (b), $\varepsilon^{2\alpha}\cdot(T_0)^{-1}S^\varepsilon$ becomes the contraction mapping from $H^{m+\frac{1}{2}}(\Gamma)$ onto itself if $\alpha > \beta$ and ε is small enough. Then we have

$$\begin{aligned} a = \psi^\varepsilon|_\Gamma &= \psi_0^0 + \{I - \varepsilon^{2\alpha}(T_0)S^\varepsilon\}^{-1}\{\varepsilon^{2\alpha}(T_0)^{-1}S^\varepsilon(\psi_0^0)\} \\ &= \psi_0^0 + O(\varepsilon^{\alpha-\beta}) \qquad \text{in } H^{m+\frac{1}{2}}(\Gamma). \end{aligned} \tag{5.18}$$

∎

6. Proof of Theorem 3

6.1 Assume $f \in H^k(\Omega_0)$ $(k \geq \frac{5}{2})$ and $\beta > |\alpha|$. Then, from (2.7) and (2.10), we have

$$\psi_0^0 \in H_0^1(\Omega_0) \cap H^{k+2}(\Omega_0), \tag{6.1}$$

$$\left.\frac{\partial\psi_0^0}{\partial n}\right|_\Gamma \in H^{k+\frac{1}{2}}(\Gamma) \cap C^{k-1,\delta}(\Gamma) \qquad (0 < \delta < 1). \tag{6.2}$$

By using (2.13) and (6.2), we have

$$\psi^\varepsilon|_\Gamma = O(\varepsilon^{\beta-\alpha}) \qquad \text{in } H^{k+\frac{1}{2}}(\Gamma) \cap C^{k-1,\delta}(\Gamma). \tag{6.3}$$

By applying the maximum principle to the problem (1.2), (1.3) and (1.5) and using (6.3), we obtain

$$\| \psi_1^\varepsilon \|_{C(\Omega_1)} \leq O(\varepsilon^{\beta-\alpha}). \tag{6.4}$$

We compute on Γ;

$$\| \psi_1^\varepsilon \|_{k-\frac{1}{2},\Gamma} \leq \left\| \frac{\partial\psi_1^\varepsilon}{\partial n} \right\|_{k-\frac{1}{2},\Gamma} + \left\| \frac{\partial\psi_1^\varepsilon}{\partial s} \right\|_{k-\frac{1}{2},\Gamma} \tag{6.5}$$

where s is the arc length of Γ.

By (6.3), we have

(6.6) $$\left\| \frac{\partial \psi_1^\varepsilon}{\partial s} \right\|_{k-\frac{1}{2},\Gamma} \leqq O(\varepsilon^{\beta-\alpha}).$$

From the definition of S^ε, we have

(6.7) $$\left.\frac{\partial \psi_1^\varepsilon}{\partial n}\right|_\Gamma = S^\varepsilon(\psi^\varepsilon|_\Gamma) \in H^{k-\frac{1}{2}}(\Gamma).$$

By (3.27) and (2.13),

(6.8) $$\left\| \frac{\partial \psi_1^\varepsilon}{\partial n} \right\|_{k-\frac{1}{2},\Gamma} \leqq \varepsilon^{-(\alpha+\beta)} \cdot \| \psi_1^\varepsilon \|_{k+\frac{1}{2},\Gamma} \leqq O(\varepsilon^{-2\alpha}).$$

Combining (6.6) and (6.8),

(6.9) $$\| \nabla \psi_1^\varepsilon \|_{k-\frac{1}{2},\Gamma} \leqq O(\varepsilon^{-2\alpha}).$$

Similarly, we have

(6.10) $$\| \nabla \psi_1^\varepsilon \|_{k-\frac{1}{2},\partial\Omega} \leqq \left\| \frac{\partial \psi^\varepsilon}{\partial n} \right\|_{k-\frac{1}{2},\partial\Omega} \leqq O(1)$$

because of (6.4), $\psi_1^\varepsilon|_{\partial\Omega} = 0$ and $\left.\frac{\partial \psi_1^\varepsilon}{\partial n}\right|_{\partial\Omega} \in H^{k-\frac{1}{2}}(\partial\Omega) \cap C^{k-2,\delta}(\partial\Omega)$.
Put $\Psi^\varepsilon = \nabla\psi_1^\varepsilon$. Then Ψ^ε satisfies

(6.11) $$-\varepsilon^{2(\alpha+\beta)} \Delta\Psi^\varepsilon + \Psi^\varepsilon = 0 \qquad \text{in } \Omega_1.$$

From the maximum principle together with (6.9) and (6.10), it follows

(6.12) $$\| \nabla \psi^\varepsilon \|_{C(\bar{\Omega}_1)} \leqq O(1+\varepsilon^{-2\alpha}).$$

Here we have to assume $k \geqq 4$ to obtain the good regularity of ψ^ε. Repeating the similar argument, we have

(6.13) $$\left\| \frac{\partial^2 \psi^\varepsilon}{\partial x_i \partial x_j} \right\|_{C(\overline{\Omega}_1)} \leqq O(\varepsilon^{-(3\alpha+\beta)}) \qquad (k \geqq 5).$$

7. Proof of Theorem 4

In the final section we give the proof of Theorem 4 under the drastic assumption. Suppose $\Omega_1 = R^2_+$.

In the same way as in 1° of the proof of Lemma 4, we transform ψ^ε_1 into $\widehat{\psi^\varepsilon}$. Then $\widehat{\psi^\varepsilon}$ satisfies

(7.1) $$\widehat{\psi^\varepsilon}(x_1, \xi) = \widehat{\psi^\varepsilon}|_{x_1=0} \cdot \exp\{-\varepsilon^{-(\alpha+\beta)}(1 + 4\pi^2 \cdot |\xi|^2 \cdot \varepsilon^{2(\alpha+\beta)})^{\frac{1}{2}} x_1\} \quad \text{in } R^2_+.$$

By (2.13) of Theorem 2, we have

(7.2) $$\widehat{\psi^\varepsilon}\Big|_{x_1=0} = -\varepsilon^{\beta-\alpha} \cdot \widehat{\frac{\partial \psi^0_0}{\partial x_1}}\Big|_{x_1=0} + \cdots \qquad \text{in } H^{m-\frac{1}{2}}(\Gamma).$$

By substituting (7.2) into (7.1), we have

(7.3) $$I(\xi) = \frac{1}{\varepsilon^{2\beta}} \cdot \int_0^\infty \widehat{\psi^\varepsilon}(x_1, \xi)\,dx_1$$

$$= -\widehat{\frac{\partial \psi^0_0}{\partial x_1}}\Big|_{x_1=0} \cdot \frac{1}{(1+4\pi^2 \cdot |\xi|^2 \cdot \varepsilon^{2(\alpha+\beta)})^{\frac{1}{2}}} + \cdots .$$

We compute

(7.4) $$I(\xi) + \widehat{\frac{\partial \psi^0_0}{\partial x_1}}\Big|_{x_1=0}$$

$$= 4\pi^2 \cdot \varepsilon^{2(\alpha+\beta)} \cdot \frac{|\xi|^2}{(1+4\pi^2 \cdot |\xi|^2 \cdot \varepsilon^{2(\alpha+\beta)})} \cdot \frac{1}{\{1+(1+4\pi^2 \cdot |\xi|^2 \cdot \varepsilon^{2(\alpha+\beta)})\}}$$

From (7.4), we have

$$\left\| \frac{1}{\varepsilon^{2\beta}} \int_0^\infty \psi_1^\varepsilon \, dx_1 + \frac{\partial \psi_0^0}{\partial x_1} \right\|_{m-\frac{1}{2}, \Gamma} = O(\varepsilon^{2(\alpha+\beta)}). \tag{7.5}$$

Appendix

For simplicity, we assume $A(u_2) \in C_0^\infty(\hat{\Gamma})$ and rewrite $|\frac{dz}{dw}| = a(u_1, u_2)$.

1° Here we state how to construct ψ_ε^n $(n = 0, 1, 2, \ldots)$ in (3.39). Let ψ_ε^0 be the solution of the following ordinary differential equation:

$$-\varepsilon^{2(\alpha+\beta)} \cdot \frac{\partial^2 \psi_\varepsilon^0}{\partial u_1^2} + a(0, u_2)^2 \psi_\varepsilon^0 = 0 \qquad \text{in } R_+^2, \tag{A.1}$$

$$\psi_\varepsilon^0 \Big|_{u_1=0} = A(u_2) \qquad \text{and} \qquad \psi_\varepsilon^0 \to 0 \quad (u_2 \to +\infty). \tag{A.2}$$

Solving (A.1,2), we get

$$\psi_\varepsilon^0 = A(u_2) \cdot \exp\{-\varepsilon^{-(\alpha+\beta)} \cdot a(0, u_2) u_1\} \tag{A.3}$$

We compute

$$\begin{aligned} &-\varepsilon^{2(\alpha+\beta)} \cdot \Delta \psi_\varepsilon^0 + a(u_1, u_2)^2 \psi_\varepsilon^0 \\ &= -\varepsilon^{2(\alpha+\beta)} \cdot \frac{d^2 A}{du_2^2} + \varepsilon^{-2(\alpha+\beta)} \cdot \{a(0, u_2)^2 \\ &\qquad\qquad - a(u_1, u_2)^2\} \cdot \exp\{-\varepsilon^{-(\alpha+\beta)} \cdot a(0, u_2) u_1\} \\ &\equiv \varepsilon^{2(\alpha+\beta)} \cdot g_\varepsilon^0(u_1, u_2). \end{aligned} \tag{A.4}$$

Let ψ_ε^1 be the solution of the problem:

$$(A.5)\qquad -\varepsilon^{2(\alpha+\beta)}\cdot\frac{\partial^2\psi_\varepsilon^1}{\partial u_1^2} + a(0,u_2)^2\psi_\varepsilon^1 = -g_\varepsilon^0(u_1,u_2) \qquad \text{in } R_+^2$$

$$(A.6)\qquad \psi_\varepsilon^1\Big|_{u_1=0} = 0 \quad \text{and} \quad \psi_\varepsilon^1 \to 0 \quad (u_2 \to +\infty).$$

Solving (A.5,6) and computing

$$(A.7)\qquad -\varepsilon^{2(\alpha+\beta)}\cdot\Delta\psi_\varepsilon^1 + a(u_1,u_2)^2\cdot\psi_\varepsilon^1 = \varepsilon^{2(\alpha+\beta)}\cdot g_\varepsilon^1(u_1,u_2),$$

we can construct the equation which ψ_ε^2 satisfies in the following way:

$$(A.8)\qquad -\varepsilon^{2(\alpha+\beta)}\frac{\partial^2\psi_\varepsilon^2}{\partial u_1^2} + a(0,u_2)^2\psi_\varepsilon^2 = g_\varepsilon^1(u_1,u_2),$$

$$(A.9)\qquad \psi_\varepsilon^2\Big|_{u_1=0} = 0 \quad \text{and} \quad \psi_\varepsilon^2 \to 0 \quad (u_2 \to +\infty).$$

Using the cascade system defined above, we can obtain ψ_ε^n $(n = 0, 1, 2, \ldots)$.

2° We put

$$(A.10)\qquad \Theta^\varepsilon = \psi_\varepsilon^0 + \varepsilon^{2(\alpha+\beta)}\cdot\psi_\varepsilon^1 + \cdots + \varepsilon^{2n(\alpha+\beta)}\cdot\psi_\varepsilon^n$$

and

$$(A.11)\qquad w_\varepsilon = \psi^\varepsilon - \Theta^\varepsilon.$$

Then w_ε satisfies

$$(A.12)\qquad -\varepsilon^{2(\alpha+\beta)}\cdot\Delta w_\varepsilon + a(u_1,u_2)^2 w_\varepsilon = O(\varepsilon^{(n+2)(\alpha+\beta)}) \qquad \text{in } R_+^2,$$

$$(A.13)\qquad w_\varepsilon\big|_{u_1=0} = 0 \quad \text{and} \quad w_\varepsilon \to 0 \quad (u_2 \to +\infty).$$

From (A.12,13), we have

(A.14) $$\| w_\varepsilon \|_{L^2(R^2_+)} \leqq O(\varepsilon^{(n+2)(\alpha+\beta)}),$$

(A.15) $$\| w_\varepsilon \|_{H^1(R^2_+)} \leqq O(\varepsilon^{(n+1)(\alpha+\beta)})$$

and moreover

(A.16) $$\| w_\varepsilon \|_{H^{n+2}(R^2_+)} \leqq O(1).$$

3° We compute

(A.17) $$\left.\frac{\partial\psi^\varepsilon}{\partial u_1}\right|_{u_1=0} = \left.\frac{\partial\Theta^\varepsilon}{\partial u_1}\right|_{u_1=0} - \left.\frac{\partial w^\varepsilon}{\partial u_1}\right|_{u_1=0}$$

where

(A.18) $$\left.\frac{\partial\Theta^\varepsilon}{\partial u_1}\right|_{u_1=0} = \sum_{k=0}^{n} \varepsilon^{2k(\alpha+\beta)} \cdot \left.\frac{\partial\psi^k_\varepsilon}{\partial u_1}\right|_{u_1=0} .$$

On the other hand, from (A.16)

(A.19) $$\left\| \left.\frac{\partial w_\varepsilon}{\partial u_1}\right|_{\hat\Gamma} \right\|_{H^{n+\frac{1}{2}}(\hat\Gamma)} \leqq O(1) \qquad (\hat\Gamma\text{: the } u_2 \text{ axis of w plane}).$$

If we choose $n+\frac{1}{2} \geqq m-\frac{1}{2}$ (or $n \geqq m-1$), then we have

(A.20) $$\left.\frac{\partial\psi^\varepsilon}{\partial u_1}\right|_{\hat\Gamma} = a(0\,,\,u_2)\{-\varepsilon^{-(\alpha+\beta)} + S_1^\varepsilon\}A(u_2) \in H^{m-\frac{1}{2}}(\hat\Gamma)$$

for any $A \in C_0^\infty(\hat\Gamma)$.

By noting $\left.\frac{\partial\psi^\varepsilon}{\partial n}\right|_\Gamma = \frac{1}{a(0,u_2)} \cdot \left.\frac{\partial\psi^\varepsilon}{\partial u_1}\right|_{\hat\Gamma}$ and using the density argument, we conclude (3.26)-(3.29).

Footnotes

1. $\partial\Omega$ stands for the boundary of Ω.
2. $\|v\|_{m,E}$ stands for the norm of v in $H^m(E)$.
3. This theorem was proved in [3] for the case $\alpha > 0$ and $\beta > 0$. In this paper, we give another proof, which is simpler than in [3].

References

[1] S. Agmon, A. Douglis and L. Nirenberg, *Estimates near the boundary for solutions of elliptic partial differential equations satisfying general boundary conditions* I, Comm. Pure Appl. Math. 12 (1959), 623-727.

[2] H. Kawarada, *Numerical methods for free surface problems by means of penalty*, Lecture Notes in Mathematics, 704, Springer-Verlag, 1979..

[3] J.L. Lions, *Perturbations singulières dans les problèmes aux limites et en contrôle optimal*, Springer-Verlag, 1973.

[4] J.L. Lions and E. Magenes, *Nonhomogeneous boundary value problems and Applications*, Springer-Verlag, Berlin, New York, 1972.

[5] J. Nečas, *Les methodes directes en théorie des équations elliptiques*, Masson, Paris, 1967.

Lecture Notes in Num. Appl. Anal., **5**, 101–118(1982)
Nonlinear PDE in Applied Science. U.S.-Japan Seminar, Tokyo, 1982

PARTIAL REGULARITY AND THE NAVIER-STOKES EQUATIONS

Robert V. Kohn

Courant Institute of Mathematical Sciences

It is a pleasure and an honor to participate in this U.S.-Japan Seminar. My talk concerns recent joint work with L. Nirenberg and L. Caffarelli, in which we prove

Theorem 1: *The singular set of a "suitable weak solution" of the Navier-Stokes equations has "parabolic one-dimensional measure zero" in spacetime.*

I shall explain what we mean by a "suitable weak solution," and by the phrase "parabolic one-dimensional measure zero"; and I shall describe the structure of the proof, avoiding the more technical parts. A fully complete discussion can be found in [1].

Theorem 1 extends and strengthens results of V. Scheffer [15-19], and our arguments draw extensively from his ideas. Scheffer has recently proved a result on "partial regularity at the boundary" [19]; here and in [1] we consider only the interior problem.

Section 1. Remarks on existence and regularity.

Let Ω be a smoothly bounded domain in R^3, and consider the initial-boundary value problem

$$(1.1) \qquad u_t + u\cdot\nabla u - \Delta u + \nabla p = f$$
$$\Delta\cdot u = 0 \quad \text{on} \quad \Omega x(0,T)$$

(1.2) $u(x,0) = u_0(x)$ on Ω, $u(x,t) = 0$ on $\partial\Omega x(0,T)$

where

$$u_0\big|_{\partial\Omega} = 0, \ \nabla\cdot u_0 = 0, \ \text{and } \nabla\cdot f = 0 \ .$$

The function $u = (u^1,u^2,u^3)$ represents the velocity of an incompressible fluid with unit viscosity; p is the pressure; and f is a nonconservative force.

It is well-known that if u_0 and f are C^∞ then (1.1), (1.2) has a unique C^∞ solution on $\Omega x(0,T)$ for some $T > 0$ [7]. There is also an extensive theory of strong solutions with less regular data [9,11,20]. If, for example, u_0 has "one-half derivative in L^2" or if $u_0 \in L^3$, one can still show the existence of a unique strong solution locally in time [2,3,5,8]. One might conjecture that the strong solution exists for all time; but this has been proved up to now only when the data u_0, f are sufficiently small.

The concept of a weak solution of (1.1), (1.2) was introduced by J. Leray, in order to obtain an existence theorem that is global in time. Pioneering work of Leray [10] and Hopf [6] showed the existence of a function u and a distribution p such that

(1.3a) $u \in L^\infty(0,T;L^2(\Omega)) \cap L^2(0,T;H^1(\Omega))$ for each $T < \infty$;

(1.3b) equations (1.1), (1.2) hold weakly;

(1.3c) $$\int_{\Omega x\{t\}} |u|^2 dx + 2 \int_0^t\!\!\int_\Omega |\nabla u|^2 dxdt \le 2 \int_0^t\!\!\int_\Omega u\cdot f dsdt + \int_\Omega |u_0|^2 dx.$$

In relation (1.3c), the "energy inequality", we write

$$|\nabla u|^2 = \sum_{i,j=1}^{3} (\nabla_i u^j)^2$$

A sufficiently regular "strong solution" is known to be unique in the class of Leray-Hopf weak solutions [13]. However, weak solutions are not known to be unique.

The fundamental regularity problem for the Navier-Stokes equations in three space dimensions remains open: even if $f = 0$, one does not know whether weak solutions of (1.1) remain smooth for all time. The work presented here achieves a much more limited goal: we show that a "suitable weak solution" can be singular only on a rather small set. Results of this type, called partial regularity theorems, are well-known in the theory of minimal surfaces and quasi-linear elliptic systems. It was Scheffer's remarkable idea to study the Navier-Stokes equations from this point of view.

Section 2. Basic tools.

The proof of Theorem 1 makes extensive use of the following four tools: (a) Interpolation inequalities; (b) Solving for p in terms of u; (c) Dimensional analysis; and (d) The generalized energy inequality. Of these, (a)-(c) are quite standard, while (d) was introduced by Scheffer in [16]. We review each briefly.

Interpolation inequalities

The energy (or generalized energy) inequality gives information about $|u|^2$ and $|\nabla u|^2$. To draw conclusions about other L^p norms one uses the well-known relation

$$(2.1)\qquad \int_{B_r} |u|^q < C\Big(\int_{B_r} |\nabla u|^2\Big)^a\Big(\int_{B_r} |u|^2\Big)^{q/2-a} + \frac{C}{r^{2a}}\Big(\int_{B_r} |u|^2\Big)^{q/2} ,$$

where

$$2 \le q \le 6,\ a = \frac{3}{4}(q-2) ,$$

B_r is a ball of radius r in R^3, and C does not depend on r [4,12].

A typical use of (2.1) is this: if $u : \Omega \times (0,T) \to R^3$, $u|_{\partial\Omega} = 0$, and

(2.2a) $$\int_{\Omega \times \{t\}} |u|^2 dx \le M \quad \text{a.e. } t$$

(2.2b) $$\int_0^T \int_\Omega |\nabla u|^2 dx dt \le M$$

Then

(2.3) $$\int_0^T \int_\Omega |u|^{\frac{10}{3}} dx dt \le CM^{\frac{5}{3}} .$$

To prove (2.3), extend u by zero off Ω and apply (2.1) with $q = \frac{10}{3}$, $a = 1$, $r \to \infty$ to see that

$$\int_{\Omega \times \{t\}} |u|^{\frac{10}{3}} dx \le CM^{\frac{2}{3}} \int_{\Omega \times \{t\}} |\nabla u|^2 dx ;$$

(2.3) follows by integration in time.

<u>Solving for the pressure.</u>

The generalized energy inequality gives information about u, but not about p. To draw conclusions about the pressure, one uses the relation

(2.4) $$\Delta p = - \sum_{i,j=1}^{3} \nabla_i \nabla_j (u^i u^j) ,$$

which follows from (1.1) by differentiation. If $\Omega = R^3$ then (2.4) determines p explicitly as a sum of singular integral operators acting on $u^i u^j$.

For partial regularity theory, it is more important to represent p locally, using only local information about u. Let $\phi(x)$ be C^∞, with supp $\phi \subset B_r$ and $\phi = 1$ on $B_{\frac{r}{2}}$. Then

(2.5a) $$\phi(x)p(x) = -\frac{3}{4\pi}\int \frac{1}{|x-y|}\Delta(\phi p)(y)dy ,$$

and

(2.5b) $$\Delta(\phi p) = p\Delta\phi + 2 < \nabla\phi, \nabla p > - \phi \sum \nabla_i\nabla_j(u^i u^j) .$$

Substituting (2.5b) into (2.5a), one obtains a formula for p on $B_{\frac{r}{2}}$ as a sum of harmonic functions and integral transforms of u.

Dimensional analysis.

Though elementary, the scaling properties of the equations are of fundamental importance. If (u,p) solve (1.1), then so do

(2.6) $$u_\lambda(x,t) = u(\lambda x,\lambda^2 t)$$
$$p_\lambda(x,t) = \lambda^2 p(\lambda x,\lambda^2 t)$$

for any λ . We encode this information by assigning a scaling dimension to each quantity:

(2.7)
x_i has dimension 1
t has dimension 2
u^i has dimension -1
p has dimension -2

so that each term in (1.1) has dimension -3. The fact that u has dimension -1 is consistent with its interpretation as a velocity, i.e. dimensionally as space/time.

In view of (2.6), it is natural to use parabolic cylinders in space-time instead of Euclidean balls; we therefore define

(2.8) $$Q_r(x,t) = \{(y,\tau) : |y-x| < r, \quad |t-\tau| < r^2\} .$$

Using (2.7), one can assign a dimension to any integral

involving u and p. For example, the estimates

$$(2.9)\qquad \iint |\nabla u|^2 dxdt < \infty \quad \text{and} \quad \int |u|^2 dx < \infty$$

have dimension one; this corresponds to the fact that

$$\iint_{Q_r} |\nabla u_\lambda|^2 dxdt = \lambda^{-1} \iint_{Q_{\lambda r}} |\nabla u|^2 dxdt$$

$$\int_{B_r} |u_\lambda|^2 dx = \lambda^{-1} \int_{B_{\lambda r}} |u|^2 dx .$$

The relevance of this "scaling dimension" will become clearer as we proceed. We note here, however, that various estimates of dimension ≤ 0 imply regularity. For example, every known existence theorem for a strong solution requires a hypothesis of dimension ≤ 0 on the initial data. Moreover, Serrin has proved that if f is C^∞ and

$$(2.10)\qquad \int \left(\int |u|^q dx \right)^{\frac{s}{q}} dt < \infty, \quad \frac{3}{q} + \frac{2}{s} < 1 ,$$

then u is C^∞ in space; the estimate (2.10) has dimension < 0 [14].

It is thus not surprising that the estimates (2.9), which have scaling dimension one, lead to an estimate of the one-dimensional measure of the singular set.

Generalized energy inequality.

We shall work only with weak solutions that satisfy the following generalization of the standard energy inequality: if $\phi(x,t)$ is C^∞, compactly supported, and $\phi \geq 0$ then

$$(2.11) \qquad \int_{\Omega\times\{t\}} |u|^2\phi + 2\iint_{0\,\Omega}^{t} |\nabla u|^2\phi \le \iint_{0\,\Omega}^{t} |u|^2(\phi_t+\Delta\phi) +$$

$$+ \iint_{0\,\Omega}^{t} (|u|^2+2p)u\cdot\nabla\phi + 2\iint_{0\,\Omega}^{t}(u\cdot f)\phi \;.$$

One obtains this relation formally, with $\le$ replaced by =, by multiplying (1.1) with $u\phi$ and integrating by parts. That procedure is, of course, not admissible for a weak solution. There is, however, at least one weak solution of (1.1) - (1.2) which satisfies the inequality (2.11) [1].

The advantage of (2.11) over (1.3c) should be clear: by choosing ϕ appropriately in (2.11), one can obtain <u>local</u> or <u>weighted</u> estimates of u.

We close this section with some definitions.

<u>Definition 1</u>: We call (u,p) a <u>suitable weak solution</u> of the Navier-Stokes equations with force f if

i) u, p, and f are defined on a space-time cylinder $D = B_r\times(a,b)$

$$f \in L^q(D) \quad \text{for some } q > \frac{5}{2}$$

$$\int_D |\nabla u|^2 dxdt < C$$

$$\int_{B_r\times\{t\}} |u|^2 dx < C \quad \text{a.e. } t \in (a,b)$$

$$p \in L^{\frac{5}{4}}(D)$$

ii) $\nabla\cdot f = \nabla\cdot u = 0$, and

$$u_t^i + \sum_j \nabla_j(u^iu^j) - \Delta u^i + \nabla_i p = f^i$$

in the sense of distributions.

iii) The generalized energy inequality (2.11) holds for every $\phi \geq 0$, C^∞ and compactly supported in D.

Definition 2: If (u,p) is a suitable weak solution, we call a point (x,t) regular if u is L^∞_{loc} in a neighborhood of (x,t). Other points are called singular. Notice that

$$S = \{\text{singular points of } u\} \tag{2.12}$$

is by definition a closed set (relative to the domain of u).

Definition 3: A set X in $R^3 \times R$ has parabolic k-dimensional measure zero if for every $\varepsilon > 0$ there exists a finite family of parabolic cylinders, $\{Q_i = Q_{r_i}(x_i,t_i)\}_{i=1}^N$, with

$$X \subset \bigcup_i Q_i$$

and

$$\sum_{i=1}^N r_i^k \leq \varepsilon$$

How big can X be and still have parabolic one-dimensional measure zero? Certainly it cannot contain a smooth arc; indeed, its one-dimensional Hausdorff measure must be zero. Moreover, it is easily shown that the projection onto the t-axis

$$\pi(X) = \{t : \exists x, (x,t) \in X\}$$

must have one-half dimensional Hausdorff measure zero.

Section 3. Sufficiently small solutions are regular.

We have already noted that "sufficiently small" solutions of the initial-boundary value problem are regular. The key to proving a partial regularity theorem lies in showing an analogous local result:

Proposition 1: There are absolute constants ε_1, $C_1 > 0$, and a constant $\varepsilon_2(q)$ depending only on q, with the following property. If (u,p) is a suitable weak solution of the Navier-Stokes equations on Q_1 with force $f \in L^q (q > \frac{5}{2})$, and if

$$(3.1) \qquad \iint_{Q_1} (|u|^3 + |u||p|) + \int_{-1}^{1} \Big(\int_{|x|<1} |p|dx \Big)^{\frac{5}{4}} dt \le \varepsilon_1$$

and

$$(3.2) \qquad \iint_{Q_1} |f|^q \le \varepsilon_2$$

then

$$(3.3) \qquad |u(x,t)| \le C_1 \text{ a.e. on } Q_{1/2} .$$

Proposition 1 is a local version of the main lemma in [16]. To understand it, one should ignore f and p; heuristically, it says that if $|u|$ is small enough in the L^3-norm on a unit-sized cylinder Q_1, then u is regular on $Q_{\frac{1}{2}}$.

One proves Proposition 1 by a variant of Scheffer's clever inductive procedure. As motivation, consider trying to bound $|u(0,0)|^2$ by using a fundamental solution of the backward heat equation: suppose ϕ^* is defined for $t < 0$,

$$\phi^*(\cdot,0) = \delta(\cdot)$$

$$\phi^*_t + \Delta\phi^* = 0 \qquad \text{near } (x,t) = (0,0) .$$

$$\phi^* = 0 \qquad \text{near } \{|x|=1\} \cup \{t=-1\}.$$

Substitution of ϕ^* into (2.11) leads formally to a bound for $|u(0,0)|^2$; but one lacks sufficient information to estimate the right hand side.

Next, notice that the worst term, $\iint (|u|^2+2p)u\cdot\nabla\phi$, is cubic in u, while the left side of (2.11) is quadratic (p is "like $|u|^2$" by (2.5)). Moreover, using (2.1)

$$r^{-5}\iint_{Q_r} |u|^3 \leq C\alpha(r)^{\frac{3}{2}} , \tag{3.4}$$

for any $r > 0$, where

$$\alpha(r) = (\sup_{|t|<r^2} r^{-3} \int_{B_r} |u|^2) + r^{-3} \iint_{Q_r} |\nabla u|^2$$

is roughly the information available on the left of (2.11).

The inductive argument, then, uses not ϕ^* but a sequence of test functions $\{\phi_n\}$, ϕ_n being essentially a mollification of ϕ^* of order 2^{-n}. As one enters the nth stage, one knows

$$r^{-5}\iint_{Q_r} |u|^3 \leq \varepsilon_1^{\frac{3}{2}} , \quad 2^{-n} \leq r \leq 1, \tag{3.5}$$

and an analogous bound for the pressure. The function ϕ_n satisfies

$$\left.\begin{array}{l} \phi_n \sim 2^{-3n} \\ |\nabla\phi_n| \leq C2^{-4n} \end{array}\right\} \text{ on } Q_{2^{-n-1}}$$

$$\left.\begin{array}{l} \phi_n \leq C2^{-3k} \\ |\nabla\phi_n| \leq C2^{-4k} \end{array}\right\} \text{ on } Q_{2^{-k}}\backslash Q_{2^{-k-1}}, \ k \leq n .$$

$$|(\phi_n)_t+\Delta\phi_n| \leq C \qquad \text{on } Q_1 .$$

Therefore, using (3.5),

$$\iint |u|^3|\nabla\phi_n| \leq C\varepsilon_1^{\frac{2}{3}} .$$

One bounds the other terms in (2.11) similarly, to obtain

$$\alpha(2^{-n-1}) \leq C\varepsilon_1^{\frac{2}{3}} .$$

Using (3.4), and assuming that ε_1 is small, it follows that

$$(3.6) \qquad r^{-5} \iint_{Q_r} |u|^3 \leq C\varepsilon_1 \leq \varepsilon_1^{\frac{2}{3}}, \qquad 2^{-n-1} \leq r \leq 2^{-n} .$$

One proves a corresponding estimate for p using (2.5) (this is the most technical part), and the induction continues. The key is the different homogeneity on the left and the right in (2.11), which allows the smallness hypothesis to be useful in (3.6).

One can rescale Proposition 1 to obtain a result on $Q_r = Q_r(x,t)$ for any r, using (2.6):

<u>Corollary 1</u>: <u>For</u> <u>any</u> $r > 0$, <u>if</u>

$$(3.7) \qquad r^{-2} \iint_{Q_r} (|u|^3 + |u||p|) + r^{-\frac{13}{4}} \int_{Q_r} \left(\int |p| dy\right)^{\frac{5}{4}} d\tau \leq \varepsilon_1$$

<u>and</u>

$$(3.8) \qquad r^{3q-5} \iint_{Q_r} |f|^q \leq \varepsilon_2$$

<u>then</u>

$$|u| \leq C_1 r^{-1} \quad \text{a.e. on } Q_{\frac{r}{2}} .$$

Again, the way to understand Corollary 1 is to ignore f and p. It says, in essence, that if the dimensionaless quantity

$$(3.9) \qquad R(r;x,t) = \frac{1}{\text{meas }(Q_r)} \iint_{Q_{r(x,t)}} (r|u|)^3 dxdt$$

is small enough, then u is regular on $Q_{\frac{r}{2}}$. One may view R(r) as a <u>local</u> <u>Reynolds</u> <u>number</u> for the flow on the cylinder Q_r.

Section 4. A dimension $\frac{5}{3}$ result.

If (u,p) is a suitable weak solution on all of R^3, then by (2.3) and (2.4)

$$(4.1) \qquad \int_0^\infty \int_{R^3} |u|^{\frac{10}{3}} + |p|^{\frac{5}{3}} < \infty .$$

As Scheffer observed in [16], Corollary 1 and (4.1) imply an estimate for the parabolic $\frac{5}{3}$-dimensional measure of the singular set S. The idea is simply this:

if $u \in L^{\frac{10}{3}}$, then "at most points" the average of u on $Q_r(x,t)$ will not be too large; at such points $R(r;x,t) \to 0$ as $r \to 0$. To quantify this, one uses the following Vitali-type covering lemma.

Lemma 1: Let J be any set of parabolic cylinders $Q_r(x,t)$ contained in a bounded subset of $R^3 \times R$. There exists a finite or denumerable family $J' = \{Q_{r_i}(x_i,t_i)\}$ such that

(4.2) the elements of J' are disjoint;

(4.3) for each $Q \in J$ there exists

$Q_i = Q_{r_i}(x_i,t_i) \in J'$ such that

$Q \subset Q_{5r_i}(x_i,t_i)$.

Given Lemma 1, we argue as follows. Fix $\delta > 0$; since $f \in L^q$ and $q > \frac{5}{2}$, we may assume that (3.8) holds whenever $r < \delta$, by choosing δ small enough. By (3.7) and Holder's inequality, there exists $\varepsilon_1' > 0$ such that

$$(4.4) \qquad r^{-\frac{5}{3}} \iint_{Q_r} |u|^{\frac{10}{3}} + |p|^{\frac{5}{3}} \leq \varepsilon_1' \Rightarrow \text{ u is regular on } Q_{\frac{r}{2}}$$

whenever $r < \delta$.

Let V be any open, bounded subset of $R^3 x(0,\infty)$, and let J consist of all cylinders $Q_r(x,t)$ such that

$$Q_r \subset V \quad , \quad r < \delta \ ,$$

(4.5) $$\iint_{Q_r} |u|^{\frac{10}{3}} + |p|^{\frac{5}{3}} > \varepsilon_1' r^{\frac{5}{3}} \quad .$$

By (4.4), J covers $S \cap V$. If J' is as in Lemma 1 then

$$S \cap V \subset \bigcup_i Q_{5r_i}(x_i,t_i)$$

by (4.3), and

(4.6) $$\sum r_i^{\frac{5}{3}} < (\varepsilon_1')^{-1} \iint_{\cup Q_{r_i}} |u|^{\frac{10}{3}} + |p|^{\frac{5}{3}} < \infty$$

by (4.1), (4.2), and (4.5). As $\delta \to 0$, we conclude that $S \cap V$ has Lebesque measure zero. Since $\cup Q_{r_i}$ is contained in a δ-neighborhood of S, the right side of (4.6) actually tends to zero as $\delta \to 0$, so the $\frac{5}{3}$ - dimensional measure of S is zero.

Section 5. The dimension 1 result.

The argument in section 4 gives a $\frac{5}{3}$ - dimensional estimate for S because it uses the global estimate (4.1), which has scaling dimension $\frac{5}{3}$. To prove a dimension-one result by this method, one must use the dimension-one estimates (2.2) instead of (4.1).

Returning to Corollary 1, suppose that the point (x_0,t_0) is singular. Then (3.7) must fail for <u>every</u> sufficiently small $r > 0$. Heuristically, this means that $R(r;x_0,t_0)$ is bounded away from zero, i.e. that

(5.1) $$|u| > \frac{C}{r} \ , \ r = |x-x_0| + |t-t_0|^2 \to 0,$$

"in the L^3-mean". Thus Corollary 1 specifies a minimum rate at which singularities can develop. If $|u|$ grows as $\frac{1}{r}$, it is natural to guess that $|\nabla u|$ should grow as $\frac{1}{r^2}$. These considerations motivate

Proposition 2: *There is an absolute constant* $\varepsilon_3 > 0$ *with the following property. If* (u,p) *is a suitable weak solution of the Navier-Stokes equations near* (x,t) *and if*

$$\limsup_{r\to 0} r^{-1} \iint_{Q_r(x,t)} |\nabla u|^2 \, dxdt \le \varepsilon_3 \tag{5.2}$$

then (x,t) *is a regular point.*

Proposition 2 implies Theorem 1 by the covering argument of section 4, using (5.2) in place of (4.4).

The essential idea in the proof of Proposition 2 is contained in the following calculus lemma

Lemma 2: *Let* $w(x,t)$ *be a* $W^{1,2}$ *function defined near* $(0,0) \in R^3 x R$. *For* $r > 0$, *let*

$$R(r) = r^{-2} \iint_{Q_r} |w|^3$$

$$\beta(r) = r^{-\frac{1}{2}} \iint_{Q_r} |w_t|^{\frac{3}{2}}$$

$$\gamma(r) = r^{-1} \iint_{Q_r} |\nabla w|^2 .$$

Then for any $\rho < r$,

$$R(\rho) \le C\{(\tfrac{\rho}{r})^3 [R(r)+\beta^2(r)] + (\tfrac{r}{\rho})^3 [R^{\frac{1}{2}}(r)+\beta(r)]\gamma^{\frac{3}{4}}(r)\} . \tag{5.3}$$

Notice that $R(r)$, $\beta(r)$ and $\gamma(r)$ are dimensionless in the sense of (2.7). Proving the lemma is an amusing exercise, using the interpolation inequality (2.1), Hölder's inequality, and the fundamental theorem of calculus. The conclusion (5.3) is a sort of decay estimate for $R(\rho)$:

<u>Corollary to Lemma 2</u>: <u>For any</u> $\varepsilon > 0$ <u>there exists</u> $\delta > 0$ <u>such that</u>

$$(5.4) \qquad \limsup_{r\to 0} \beta(r) + \gamma(r) < \delta \Rightarrow \liminf_{r\to 0} R(r) < \varepsilon .$$

Indeed, (5.3) implies

$$R(\rho) \le C_2 R(r)\cdot[(\tfrac{\rho}{r})^3 + \varepsilon^{-\frac{1}{2}}(\tfrac{r}{\rho})^3 \gamma^{\frac{3}{4}}(r)]$$

$$+ C_3[\beta^2(r) + (\tfrac{r}{\rho})^3 \beta(r)\gamma^{\frac{3}{4}}(r)]$$

whenever $R(r) \ge \varepsilon$. Choosing $0 < \Theta < 1$ so that $C_2\Theta^3 < \frac{1}{6}$, then $\delta > 0$ so that

$$\varepsilon^{-\frac{1}{2}} C_2 \Theta^{-3} \delta^{\frac{3}{4}} < \frac{1}{6} \quad \text{and} \quad C_3[\delta^2 + \Theta^{-3}\delta^{\frac{7}{4}}] \le \frac{\varepsilon}{6} ,$$

we conclude

$$R(\Theta r) \le \tfrac{1}{3}R(r) + \tfrac{\varepsilon}{6} \le \tfrac{1}{2}R(r)$$

whenever $R(r) \ge \varepsilon$ and $\beta(r) + \gamma(r) < \delta$. The assertion (5.4) follows, with this choice of δ.

The proof of Proposition 2 is roughly parallel to the above argument. For a weak solution u, one has no bound on $\iint |u_t|^{\frac{3}{2}}$, but the generalized energy inequality lets one bound the oscillation in time of $\int_{B_r} |u|^2$. One proves a "decay estimate" like (5.3), for the entire left side of (3.7) instead of for $R(\rho)$.

Section 6. Concluding remarks.

One reason for studying partial regularity is the hope of settling, by this method, certain classical open questions about weak solutions. Might one prove uniqueness or strong continuity, for example, without actually proving regularity? Theorem 1 alone does not suffice; one appears to need information about the maximum rate at which singularities can develop. We note in this context a qualitative difference between Corollary 1 and Proposition 2: the hypothesis of the former concerns a fixed Q_r, and the conclusion asserts a bound for u ; the hypothesis of the latter concerns all Q_r, and the conclusion gives no explicit estimate.

Might similar methods be used to prove an estimate of the singular set of dimension less than one? This would require a global estimate with scaling dimension less than one. Proving such an estimate would take, it seems, a fundamental new idea.

It may be, of course, that weak solutions are not regular. An attractive scheme for constructing a solution with a self-similar singularity is proposed in [10].

Finally, I note that the generalized energy inequality may have uses other than for partial regularity theory. In [1], for example, it is used to prove weighted norm estimates for the Cauchy problem, in case the initial velocity satisfies $\int_{R^3} |u_0|^2|x| < \infty$, or $\int_{R^3} |u_0|^2|x|^{-1}$ sufficiently small.

REFERENCES

[1] Caffarelli, L.; Kohn, R.; Nirenberg, L; "Partial regularity of suitable weak solutions of the Navier-Stokes equations," Comm. Pure Applied Math., to appear.

[2] Fabes, E.; Lewis, J,; Riviere, N.; "Boundary value problems for the Navier-Stokes equations," Amer. J. Math., vol. 99, 1977, pp. 626-668.

[3] Fujita, H.; Kato, T.; "On the Navier-Stokes initial value problem I," Arch. Rat. Mech. Anal., vol. 16, 1964, pp. 269-315.

[4] Gagliardo, E., "Ulteriori proprietà di alcune classi di funzioni in più variabili," Richerche di Mat. Napoli, vol. 8, 1959, pp. 24-51.

[5] Giga, Y.; Miyakawa, T.; "Solutions in L_r to the Navier-Stokes initial value problem." to appear.

[6] Hopf , E.; "Uber die Aufangswertaufgabe für die hydrodynamischen Grundgleichungen," Math. Nachr., vol. 4, 1950-51, pp. 213-231.

[7] Kaniel, S. and Shinbrot, M., "Smoothness of weak solutions of of the Navier-Stokes equations," Arch. Rat. Mech. Anal., vol. 24, 1967, pp. 302-324.

[8] Kato, T.; Fujita, H.; "On the nonstationary Navier-Stokes system," Rend. Sem. Mat. Padova, vol. 32, 1962, pp. 243-260.

[9] Ladyzhenskaya, O.A.; The Mathematical Theory of Viscous Incompressible Flow. Gordon and Breach, New York, 1969.

[10] Leray, J.; "Sur le mouvement d'un liquide visquex emplissant l'espace," Acta Math., vol. 63, 1934, pp. 193-248.

[11] Lions, J.-L.; Quelques Methodes de Résolution des Problèmes aux Limitex non Linéaires. Dunod-Gauthiers-Villars, Paris, 1969.

[12] Nirenberg, L.; "On elliptic partial differential equations," Ann. di Pisa, vol. 13, 1959, pp. 116-162.

[13] Serrin, J., "The initial value problem for the Navier-Stokes equations," in Nonlinear Problems, R.F. Langer, ed., Univ. of Wisconsin Press. 1963, pp. 69-98.

[14] Serrin, J., "On the interior regularity of weak solutions of the Navier-Stokes equations," Arch. Rat. Mech. Anal. vol. 9, 1962, pp. 187-195.

[15] Scheffer, V., "Partial regularity of solutions to the Navier-Stokes equations," Pac. J. Math., vol. 66, 1976, pp. 535-552.

[16] Scheffer, V., "Hausdorff measure and the Navier-Stokes equations," Comm. Math. Phys., vol. 55, 1977, pp. 97-112.

[17] Scheffer, V., "The Navier-Stokes equations in space dimension four," Comm. Math. Phys., vol. 61, 1978, pp. 41-68.

[18] Scheffer, V., "The Navier-Stokes equations on a bounded domain," Comm. Math. Phys., vol. 73, 1980, pp. 1-42.

[19] Scheffer, V., to appear in Comm. Math. Phys.

[20] Temam, R., Navier-Stokes Equations. Theory and Numerical Analysis. North-Holland, Amsterdam and New York, 1977.

Lecture Notes in Num. Appl. Anal., **5**, 119–131 (1982)
Nonlinear PDE in Applied Science. U.S.-Japan Seminar, Tokyo, 1982

Blow-Up of Solutions of Some Nonlinear Diffusion Equations

Kyûya Masuda

Mathematical Institute
Tohoku University
Sendai, Japan 980

Abstract

We prove that any solution of $u_t = \Delta u + u^2$ can be analytically continued (in t) through the upper (and lower) half- complex plane into some infinite interval (t^*, ∞), $t^* > 0$.

1. Introduction Consider the nonlinear diffusion equation of the form

(1) $$u_t = \Delta u + u^2, \quad x \in \Omega , \quad t > 0$$

with the homogeneous Neumann boundary condition:

(2) $$\partial u/\partial n_{|\Gamma} = 0$$

and the initial condition:

(3) $$u_{|t=0} = u_0$$

where Ω is a bounded domain (in R^n) with smooth boundary Γ , and u_0 is a given smooth function satisfying the homogeneous Neumann boundary condition; $\partial/\partial n$ denotes differentiation in the direction of the exterior normal to Γ .

We know that the local (in t) solution of (1), (2), (3) exists and unique. On the other hand, it is easy to see that any solution of (1),(2),(3) does not exist globally, and blows up in a finite time if Pu_0 is positive, where

$$Pf = \frac{1}{|\Omega|} \int_\Omega f(x)dx \tag{4}$$

($|\Omega|$: the volume of Ω) ; see H. Fujita [1]. In particular, if u_0 is a positive constant function, then the solution $u = u(t)$ is explicitly given by : $u(t) = u_0 /(1- u_0 t)$, and surely blows up at $t = 1/u_0$; note $Pu_0 = u_0$. Moreover, this local solution can be analytically continued into the infinite interval $(1/u_0, \infty)$. More generally we shall show that the local solution of (1),(2),(3) can be analytically continued through the upper (and lower) half-(complex)plane into some infinite interval (t^*, ∞), $t^* > 0$.

Fix $p > n$. We suppose that u_0 is a real-valued function in $L^p(\Omega)$ with $Pu_0 \neq 0$. Put $\alpha = 1/Pu_0$, and let δ be such that $0 < \delta < |\alpha|$. Let D_0 be the angular domain in the complex-plane : $|\arg t| < \pi/4$. D_1 is the angular domain:

$$5\pi/4 < \arg(t-\alpha-\delta i) < 7\pi/4$$

($i = \sqrt{-1}$). Set

$$D_+ = D_0 - \mathrm{cl.}(D_1) \ ;$$

$$D_- = \{\bar{z} \ ; \ z \in D_+ \}$$

(cl. means the closure) ; note that $D_+ = D_- = D_0$ if $\alpha < 0$. See Fig. 1.

We define the operator Q in $L^p(\Omega)$ by $Q = I - P$ (I ; the identity operator in $L^p(\Omega)$).

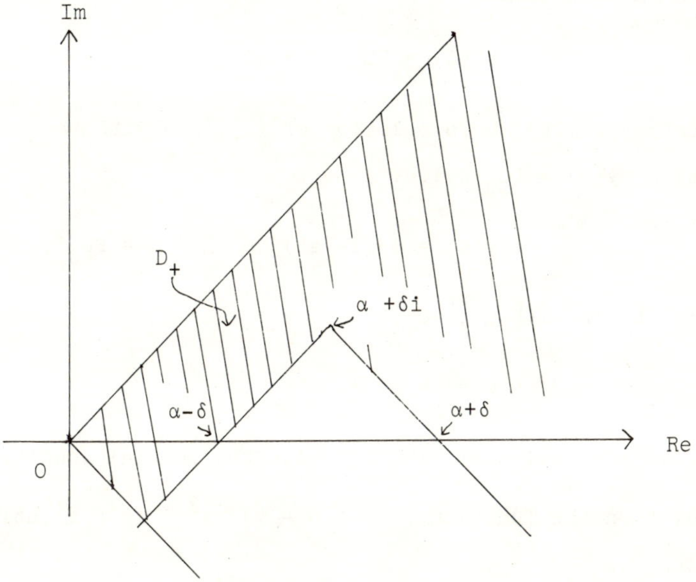

Fig. 1

Our results now read as follows.

Theorem 1 Let $p > n$. Let $u_0 \varepsilon L^p(\Omega)$, real. Suppose that $\|Qu_0\|_{L^p}/|Pu_0|^2$ is sufficiently small. Then the initial-boundary value problem (1),(2),(3) has one and only one solution $u^+ = u^+(t)$ (resp. $u^- = u^-(t)$) that is analytic for t in D_+ (resp. D_-) as a $W^{1,p}(\Omega)$-valued function ($W^{1,p}(\Omega)$; the usual Sobolev space).

Theorem 2 Let u_0, $u_\pm$ be as in Theorem 1. Suppose that Pu_0 is positive, and that $\|Qu_0\|_{L^p}/|Pu_0|^2$ is sufficiently small. Then if $u^+(t) = u^-(t)$ for some real $t : t > \alpha + \delta$, then u_0 is real constant. Conversely, if u_0 is a real constant, then $u^+(t) = u^-(t)$ for all $t : t > \alpha + \delta$.

2. Proof of Theorem 1 We have decomposition

$$L^p(\Omega) = PL^p(\Omega) + QL^p(\Omega) .$$

Let $y = y(t)$ be a solution of (1),(2) with the initial condition $y(0) = Pu_0$, i.e.

$$y(t) = a/(1- at) , \quad (a = Pu_0). \tag{5}$$

For any b in $QL^p(\Omega)$, we set

$$\phi(b) = a + a^2 b . \tag{6}$$

If u is a solution of (1),(2) with the initial condition $u_{|t=0} = \phi(b)$, then the function $v = (u - y)/ y^2$ is a solution of the equation

$$v_t = \Delta v + c(t)v^2 , \quad (c(t) = y(t)^2), \tag{7}$$

satisfying the conditions:

$$\partial v/\partial n_{|\Gamma} = 0 ; \tag{8}$$

$$u_{|t=0} = b \tag{9}$$

and vice versa. Using the theory of equations of evolution, we shall show that the initial-boundary value problem (7),(8),(9) has one and only one solution that is analytic in D_+ (and D_-) as a $W^{1,p}(\Omega)$-valued function. To this end, we define the operator A in $L^p(\Omega)$ by $Au = - \Delta u$ with domain $D(A)$;

$$D(A) = \{ u \in W^{2,p}(\Omega) ; \partial u/\partial n_{|\Gamma} = 0 \} .$$

($W^{2,p}(\Omega)$ is the usual Sobolev space). The following is then well known.

Proposition 3 The spectrum of $-A$ consists only of non-negative and discrete eigenvalues $\{\lambda_j\}_{j=1}^{\infty}$, tending to the infinity. Moreover, the first eigenvalue is zero, which is a simple one; the projection corresponding to the eigenvalue zero is given by (4).

Setting

$$\Sigma_\varepsilon = \{ t \in C ; |\arg t| < \pi/2 - \varepsilon \}, \quad \varepsilon > 0 ,$$

we have

Propostion 4 The operator $-A$ generates the holomorphic semi-groups $\{e^{-tA}\}_{t>0}$ in $L^p(\Omega)$ such that for each $\varepsilon > 0$ e^{-tA} is holomorphic in Σ_ε. Moreover, it satisfies the estimates:

(10) $$\|e^{-tA}\| \le M_\varepsilon ;$$

(11) $$\|Qe^{-tA}\| \le M_\varepsilon e^{-\beta \mathrm{Re}\, t};$$

(12) $$\|Ae^{-tA}\| \le M_\varepsilon e^{-\beta \mathrm{Re}\, t} /|t|$$

for t in Σ_ε, M_ε and β being positive constants.

(For the proof, we refer the reader to H. Tanabe [3])

In what follows we fix $\varepsilon = \pi/4$ and $\Sigma \equiv \Sigma_{\pi/4}$.

Using a priori estimate: (M denotes various constants)

$$\|\nabla u\|_{L^p} \le M \|u - Pu\|_{L^p} + M \|\Delta u\|_{L^p}, \quad u \in D(A)$$

(see [3]), we get

(13) $$\|\nabla e^{-tA}\| \le M |t|^{-1/2} e^{-\beta \mathrm{Re}\, t}, \quad t \in \Sigma$$

by the proposition 3 and (12), and by noting

$$\| u - Pu \|_{L^p} \leq M \| Au \|_{L^p}$$

Let γ be such that $0 < \gamma < \beta$, β being as in Proposition 4, and we introduce function spaces Y_0, Y_1. Y_0 is the set of all holomorphic functions u in D_+ with values in $L^p(\Omega)$ such that

$$|||u|||_0 \equiv \sup_{t \varepsilon D_+} (\| Pu(t) \|_{L^p} + e^{\gamma \mathrm{Re}\, t} \| Qu(t) \|_{L^p}) < \infty ;$$

and Y_1 is the set of all holomorphic functions u in D_+ with values in $W^{1,p}(\Omega)$ such that

$$|||u|||_1 \equiv |||u|||_0 + \sup_{t \varepsilon D_+} (e^{\gamma \mathrm{Re}\, t} \left|\frac{t}{1+t}\right|^{1/2} \| \nabla u(t) \|_{L^p}) < \infty$$

Equipped with the norm $|||u|||_0$ (resp. $|||u|||_1$), Y_0 (resp. Y_1) is a Banach space.

We define the map F of $QL^p(\Omega) \times Y_1$ into Y_1 by:

(14) $$F(b,w) = w(t) - e^{-tA} b - \int_0^t e^{-(t-s)A} c(s) w(s)^2 ds$$

$(t \varepsilon D_+)$ where the path of integration is taken in the interior of D_+. Then:

<u>Lemma 5</u> The F is an analytic map : $QL^p(\Omega) \times Y_1 \to Y_1$, satisfying

(15) $$F(0,0) = 0 ;$$

(16) $$F_w(0,0) = I$$

($F_w(0,0)$ denotes the Frechet derivative with respect to w at $b = 0$, $w = 0$).

<u>Proof</u> : The map I_1 defined by :

$$I_1[b] = e^{-tA} b , \qquad b \in QL^p(\Omega)$$

is analytic map of $QL^p(\Omega)$ into Y_1 by Proposition 4 and (13). Next the map I_2 defined by :

$$I_2[w] = \int_0^t e^{-(t-s)A} g(s) w(s)ds$$

is an analytic map of Y_0 into Y_1, where $g(s) = ((1+s)/s)^{1/2} c(s)$. In fact, we have the inequalities:

$$\| PI_2[w] \|_{L^p} \leq M \int_0^t |g(s)| |ds| \ ||| w |||_0 \ ;$$

$$e^{\gamma \mathrm{Re}\, t} \| QI_2[w] \|_{L^p} \leq M \int_0^t |g(s)| |ds| ||| w |||_0 \ ;$$

$$e^{\gamma \mathrm{Re}\, t} \| \nabla I_2[w] \|_{L^p} \leq M \int_0^t |t-s|^{1/2} |g(s)| |ds| ||| w |||_0$$

$(t \in D_+)$ by (10), (11) and (13) from which it follows that I_2 is an analytic map : $Y_0 \to Y_1$.

We finally define the map I_3 by:

$$I_3[w] = (t/(1+t))^{1/2} w^2.$$

Then the I_3 is an analytic map of Y_1 into Y_0 since we have the following inequalities by the Sobolev inequality

$$\| Qw \|_{L^\infty} \leq M (\| Qw \|_{L^p} + \| w \|_{L^p})$$

and by the decomposition

$$Qw^2 = Q((Qw)^2) + 2 Pw \cdot Qw$$

;

$$|PI_3[w]| \leq M \| w \|_{L^p} \leq M ||| w |||_0$$

$$e^{\gamma \mathrm{Re}\ t} \| QI_3[w] \|_{L^p} \le M e^{\gamma \mathrm{Re}\ t} |t/(1+t)|^{1/2} (\|Qw\|_{L^\infty} + $$
$$+ |Pw|) \|Qw\|_{L^p}$$
$$\le M e^{\gamma \mathrm{Re}\ t} |t/(1+t)|^{1/2} (\|\nabla w\|_{L^p} +$$
$$+ \|Qw\|_{L^p} + |Pw|) \| Qw\|_{L^p}$$
$$\le M ||| w |||_1^2 .$$

Since

$$F(b,w) = I_1[b] + I_2 o I_3[w]$$

($I_2 o I_3$ is the composition of I_2 and I_3), it follows that F is an analytic map of $QL^p(\Omega)\times Y_1$ into Y_1. It is easy to see (15) and (16). Thus the proof of the lemma is completed.

Now we are in a position to apply the implicit function theorem, and can see that there is one and only one solution $w = v^+(t,b)$ of $F(b,v^+(\cdot,b)) = 0$ near $(0,0)$ that is an analytic function with values in Y_1 of b near 0. The v^+ satisfies

$$(17) \qquad v^+(t,b) = e^{-tA} b + \int_0^t e^{-(t-s)A} c(s)(v^+(s,b))^2 ds$$

($t \in D_+$). By the standard argument in the theory of equations of evolution, it follows that $v^+(t,b)$ converges to b as $t \to 0$, strongly in $L^p(\Omega)$. We can also see that v^+ is locally Hölder continuous for t in D_+ (in $L^p(\Omega)$), and so is $(v^+)^2$, since

$$\|v^+(t)^2 - v^+(s)^2 \|_{L^p} \le M (\| v^+(t) \|_{L^\infty} + \|v^+(s)\|_{L^\infty}) \times$$

$$\times \| v^+(t) - v^+(s) \|_{L^p}$$

(we simply write $v^+(t)$ for $v^+(t,b)$); note $v^+ \varepsilon Y_1$. Hence $v^+(t)$ is continuously differentiable in $t \varepsilon D_+$ (in $L^p(\Omega)$), and its derivative $dv^+(t)/dt$ is locally Hölder continuous (see, e.g. T. Kato [2]). We also see $v^+(t)$ is in $D(A)$, $Av^+(t)$ is locally Hölder continuous, and satisfies

$$dv^+(t)/dt + Av^+(t) = c(t)(v^+(t))^2 , \quad t \varepsilon D_+$$

with

$$v^+(0) = b.$$

By the regularity theorem on solutions of parabolic equations, v^+ is actually a unique classical solution of (7),(8),(9) that is analytic for $t \varepsilon D_+$ as a $W^{1,p}(\Omega)$-valued function. Hence $u^+(x,t) = y(t) + c(t)\, v^+(x,t)$ is a unique classical solution of (1), (2),(3) that is analytic for $t \varepsilon D_+$ as a $W^{1,p}(\Omega)$-valued function. Similarly we can show there is a unique classical solution u^- of (1),(2),(3) that is analytic for $t \varepsilon D_-$ as a $W^{1,p}(\Omega)$-valued function; $u^-(x,t)$ = the complex conjugate of $u^+(x,\bar{t})$, $t \varepsilon D_-$. This completes the proof of Theorem 1.

3. Proof of Theorem 2 If u_0 is a real constant, then $u^+(t)$ and $u^-(t)$ take the same real values on (α , ∞). We shall show $u^+(t)$ does not coincide with $u^-(t)$ for any real t $(> \alpha + \delta)$ if u_0 is not (real) constant, and if $\|Qu_0\|_{L^p} / |Pu_0|^2$ is sufficiently small. Let r_0 be a positive number such that $v^+(t,b)$ and $v^-(t,b)$ are both analytic in b , $\| b \|_{L^p} < r_0$;

Here and in what follows we shall use the same notations as in the proof of Theorem 1. Furthermore we can take r_0 so small that $v^+(t,b)$ is locally bounded in $\|b\|_{L^p} < r_o$ (in the norm of Y_1), as can be easily seen from the proof of Theorem 1. Hence $v^+(\cdot,b)$ has infinitely many Fréchet derivatives $\partial_b^N v^+(\cdot,b)[h,h,\cdots,h]$ ($h \in L^p(\Omega)$; $N = 1,2,\cdots$). In particular we have the expansion of the form:

$$v^+(\cdot,b) = v^+(\cdot,0) + \partial_b v^+(\cdot,0)[b] + \partial_b^2 v^+(\cdot,0)[b,b] + R^+(\cdot,b) \tag{18}$$

where

$$|||\, R^+(\cdot,b)\, |||_1 = O(\|b\|^3_{L^p}) . \tag{19}$$

Clearly

$$v_0^+(t) \equiv v^+(t,0) = 0 . \tag{20}$$

By (17) $v_1^+(t) \equiv \partial_b v^+(t,0)[b]$ satisfies

$$v_1^+(t) = e^{-tA}b + 2\int_0^t e^{-(t-s)A} c(s)v_0^+(s)v_1^+(s)ds$$

and hence, by (20),

$$v_1^+(t) = e^{-tA}b \tag{21}$$

We can also see $v_2^+(t) \equiv \partial_b^2 v^+(t,0)[b,b]$ satisfies

$$v_2^+(t) = 2\int_0^t e^{-(t-s)A}c(s)[v_0^+(s)v_2^+(s) + (v_1^+(s))^2]\, ds$$

$$= 2\int_{\Gamma+} e^{-(t-s)A}c(s)\,(v_1^+(s))^2 ds$$

by (20) and (21); the path of integration Γ_+ joing 0 to t is taken in D_+: The procedures above can be easily justified. If $v_0^-(t)$, $v_1^-(t)$, $v_2^-(t)$, $R_-(t,b)$ are similarly defined, we have

$$v_0^-(t) = 0 \; ; \qquad v_1^-(t) = e^{-tA}\, b$$

$$v_2^-(t) = 2 \int_{\Gamma_-} e^{-(t-s)A} c(s)(v_1^-(s))^2 ds \; .$$

Set $h(s) = (e^{-sA}b)^2$ ($s \in D_+$), and let Γ be as shown in Fig. 2 below.

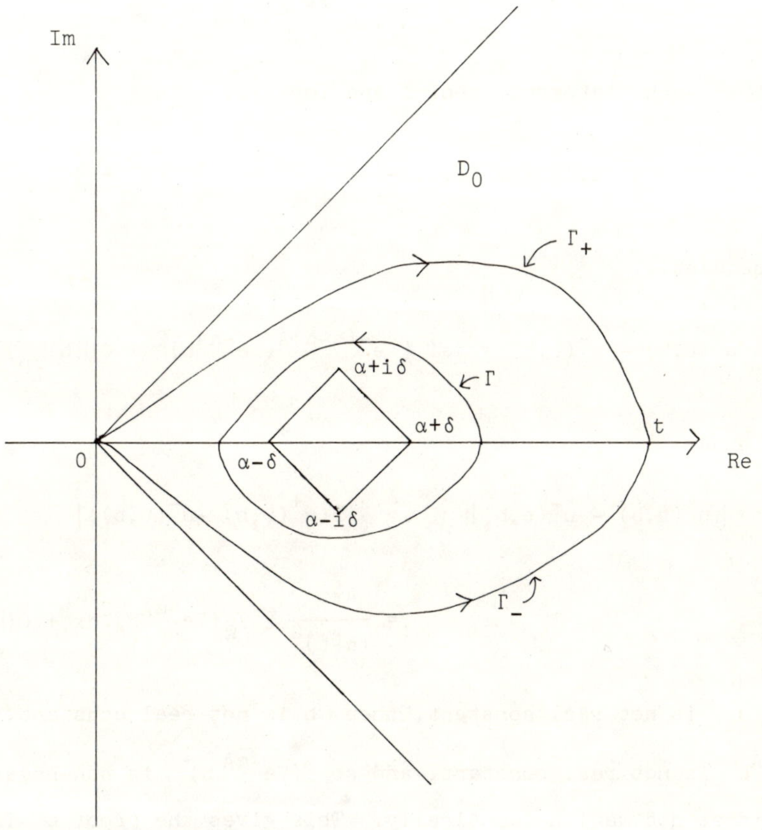

Fig. 2

Since $c(s)h(s)$ is holomorphic in a complex domain bounded by $-\Gamma_+ + \Gamma_-$, we have, for any fixed t ($> \alpha + \delta$),

$$v_2^+(t) - v_2^-(t) = - \int_\Gamma e^{-(t-s)A} c(s)h(s)ds$$

$$= - \int_\Gamma \frac{1}{(\alpha - s)^2} e^{-(t-s)A} h(s)\, ds$$

$$= - 2\pi i \frac{\partial}{\partial s} (e^{-(t-s)A} h(s))|_{s=\alpha}$$

$$= 4\pi i\, e^{-(t-\alpha)A} (\nabla e^{-\alpha A} b)^2$$

by the Cauchy integral theorem and formula. Hence

$$v^+(t,b) - v^-(t,b) = e^{-(t-\alpha)A} (\nabla e^{-\alpha A} b)^2 + O(\|b\|^3_{L^p}) .$$

Consequently,

$$u^+(t,b) - u^-(t,b) = 4\pi i \,[\, e^{-(t-\alpha)A} (e^{-\alpha A} b)^2 + O(\|b\|^3_{L^p})]/(\alpha - t)^2$$

and hence

$$\|u^+(t,b) - u^-(t,b)\|_{L^p} \geq |P(u^+(t,b) - u^-(t,b))|$$

(22)
$$= \frac{4\pi}{(\alpha - t)^2} \,[\, \int_\Omega (\nabla e^{-\alpha A} b)^2 dx + O(\|b\|^3_{L^p})].$$

If u_0 is not real constant, then b is not real constant. Hence, $e^{-\alpha A} b$ is not real constant, and so $(\nabla e^{-\alpha A} b)^2$ is non-negative, and does not vanish identically. This gives the proof of Theorem 2 by (22).

4. Remark We can also consider diffusion equations of more general type:

$$u_t = \Delta u + f(u) \tag{23}$$

where $f = f(z)$ is an entire function of z with some conditions. The results concerning the equations above shall be published elsewhere.

References

[1] Fujita, H., On some nonexistence and nonuniqueness theorems for nonlinear parabolic equations, in Proceedings of the Symposium on Pure Mathematics, American Math. Soc., Providence, R.I., 1970, vol. 18, Part 1, pp. 105-113.

[2] Kato, T., Perturbation theory for linear operators. Berlin-Göttingen-Heidelberg: Springer 1966.

[3] Tanabe, H., Equations of evolution. London-San Frncisco-Melbourne: Pitman 1979.

Lecture Notes in Num. Appl. Anal., **5**, 133–151(1982)
Nonlinear PDE in Applied Science. U.S.-Japan Seminar, Tokyo, 1982

Asymptotic Behavior of the Free Boundaries Arising in One Phase Stefan Problems in Multi-Dimensional Spaces

Hiroshi MATANO

Department of Mathematics
Faculty of Science
Hiroshima University
Hiroshima 730
JAPAN

We discuss the asymptotic behavior of weak solutions to one phase Stefan problems in exterior regions. It is shown that any weak solution eventually becomes classical after a finite period of time and that the shape of the free boundary approaches to that of a sphere as $t \to +\infty$.

1. Introduction.

A Stefan problem is a mathematical model for describing the melting of a body of ice in contact with a region of water. In one phase Stefan problems, the temperature of ice is supposed to be maintained at $0°C$. Hence the unknowns are the temperature distribution of water and the shape of the interface (free boundary) between ice and water.

In the case where the space dimension is one, the problem has been completely solved; and it is well known that, for any bounded smooth initial and boundary data, the solution is unique and exists globally in time in the classical sense (see Friedman [4]). However, in the case of multi-dimensional spaces, a classical solution cannot always be extended globally over the time interval $0 \leqq t < +\infty$, since cusp-like singularities sometimes appear on the free boundary. Such singularities occur, for instance, when a portion of ice is in the process of being separated from the rest of the ice by the growing water region.

Duvaut [3] has introduced a weak formulation of the above problem in terms of variational inequality, for which the solution exists globally in time. Following his formulation, Friedman, Kinderlehrer, Nirenberg, Caffarelli and others have studied the properties of weak solutions and obtained a number of regularity theorems. These studies have revealed that any singular point appearing on the free boundary should be a cusp-like singularity [1] and that the solution is sufficiently smooth, say C^∞, inside the water region and also up to the free boundary provided that this free boundary forms (locally) a C^1 hypersurface [8]. It is also known that the free boundary is C^∞ if it is C^1 or even simply Lipschitz ([8], [1]). Friedman and Kinderlehrer [5] have given an example in which the free boundary possesses no singular point for $0 \leqq t < +\infty$, in other words, the solution is classical for all time. This example, however, requires a strong geometric assumption on the initial data.

The aim of this paper is to prove that *any weak solution is "eventually" classical*. That is, the interface between the ice and water regions is sufficiently smooth for all large t ; hence so is the temperature distribution of water. It will also be shown that *the shape of the free boundary (i.e., the interface) approaches to that of a sphere in a certain manner as* $t \to +\infty$. Of course the radius of this sphere goes to infinity as $t \to +\infty$. Note that we shall impose no specific hypothesis on the geometric features of the initial and the boundary data. The only assumption required in this paper is that the temperature of water averaged on the prescribed boundary (that is, the surface of a heat supplying obstacle located in the midst of the water region) be bounded from below by a time-independent positive constant.

To state the above results more precisely, let us introduce some notation. First, Let G_1 be a bounded domain in $\mathbb{R}^n$ such that ∂G_1 is sufficiently smooth and that $\mathbb{R}^n \setminus G_1$ is connected. Let G be a bounded domain containing $\bar{G}_1$. The boundary of G is also assumed to be smooth. $\bar{G}_1$ represents the fixed obstacle, $G \setminus \bar{G}_1$ is the initial water region at $t = 0$ and $\mathbb{R}^n \setminus G$ denotes the initial ice region. As is mentioned above, the temperature of ice is supposed to be maintained at $0°C$. The water region at each $t \geqq 0$ and the interface between ice and water are denoted by $\Omega(t)$ and $\Gamma(t)$ respectively. From the definition it follows that

$$\Omega(0) = G \setminus \bar{G}_1 ,$$

$$\partial\Omega(t) = \Gamma(t) \cup \partial G_1 \quad (t \geqq 0).$$

The temperature of water will be denoted by $\theta = \theta(x,t)$. This is a nonnegative function defined on the closure of the set $\{ (x,t) \mid x \in \Omega(t),\ t \geqq 0 \}$ and supposed to satisfy the following initial and boundary conditions:

(1) $$\theta(x,0) = h(x), \qquad x \in \Omega(0) ,$$

(2) $$\theta(x,t) = g(x,t), \qquad x \in \partial G_1 ,\ t \geqq 0 ;$$

here h and g are given positive smooth functions satisfying

$$h(x) = \begin{cases} g(x,0) & (x \in \partial G_1) \\ 0 & (x \in \partial G) . \end{cases}$$

The equations governing the phenomenon are as follows:

(3) $$\theta_t = \Delta\theta , \qquad x \in \Omega(t) ,\ t > 0 ,$$

(4) $$\theta = 0 , \qquad x \in \Gamma(t) ,\ t > 0 ,$$

(5) $$\theta_t = \frac{1}{k}|\nabla\theta|^2 , \qquad x \in \Gamma(t) ,\ t > 0 ,$$

where k is a given positive constant. Thus the problem is to find $\theta(x,t)$ and $\Gamma(t)$ satisfying (1) ~ (5) and

(6) $$\Gamma(0) = \partial G .$$

Although the problem (1) ~ (6) does not necessarily have a global classical solution, it always has a weak solution (in the sense of Duvaut) that exists globally in time. In his weak formulation, the temperature θ is locally given by the form $\theta = u_t$, where $u = u(x,t)$ is a solution of a certain variational problem (see Section 3 for details). Caffarelli and Friedman [2] have proved that θ is continuous in $(\mathbb{R}^n \setminus G_1) \times [0, +\infty)$, where we understand that $\theta(x,t) = 0$ for $x \notin \Omega(t) \cup \bar{G}_1$, $t \geqq 0$. It is not difficult to see that θ is sufficiently smooth in the interior of the water region, since it satisfies there the heat equation $\theta_t = \Delta\theta$ in the sense of distributions. Further regularity of θ near the free boundary $\Gamma(t)$ depends on the regularity of $\Gamma(t)$. In other words, a weak solution is classical so long as $\Gamma(t)$ has no singular point ([1], [8]).

2. Main Theorems.

In what follows the pair $\theta(x,t)$, $\Gamma(t)$ will denote the weak solution to the problem (1) ~ (6). The notation in the preceding section will be used freely. The main theorem in this paper can now be stated as follows:

Theorem 1. *Suppose there exist positive constants* δ *and* M *such that*

$$(7) \qquad \delta \leqq \int_{\partial G_1} g(x,t)\, dx \leqq M$$

holds for all $0 \leqq t < +\infty$. *Then there is a positive number* T_0 *such that* $\Gamma(t)$ *is sufficiently smooth for all* $t \geqq T_0$.

Theorem 2. *Let the same assumption as in Theorem 1 hold. Assume further that* $n \geqq 3$. *Then there exist a positive number* T_0 *and a bounded convex set* W *in* $\mathbb{R}^n$ *such that for any* $t \geqq T_0$ *and any point* $x_0 \in \Gamma(t)$ *the inward normal line to* $\Gamma(t)$ *at* x_0 *intersects the set* W .

Combining Theorem 2 and the fact that the free boundary $\Gamma(t)$ is moving away from every finite region (that is, any point outside G_1 will eventually be swallowed by the water region), we get

Corollary 3. *Let $n \geqq 3$ and let* (7) *hold. Then, for any bounded set A in $\mathbb{R}^n$, there exists a positive number T_A such that $\Omega(t) \cup \bar{G}_1$ is star-shaped with respect to any point of A for each $t \geqq T_A$.*

Corollary 4. *Let $n \geqq 3$ and let* (7) *hold. Let x_1 be any point in $\mathbb{R}^n$ and put*

$$m(t) = \min_{x \in \Gamma(t)} |x - x_1| , \qquad M(t) = \max_{x \in \Gamma(t)} |x - x_1| .$$

Then $M(t) - m(t)$ remains bounded as $t \to +\infty$.

Remarks. (i) The condition (7) can be relaxed somewhat. For example, even if the integral of g tends to zero as $t \to +\infty$, the assertion of Theorem 1 still remains true so long as the rate of decay is moderate. However, if this value decays very rapidly as $t \to +\infty$, then the total amount of the heat energy to be supplied to the water region through the surface ∂G_1 becomes finite, which implies that only a finite portion of the ice will be melted. In such a case, as it seems, the conclusion of Theorem 1 is not likely to hold.

(ii) In the case $n = 2$, Corollary 3 still remains true, but the assertions in Theorem 2 and Corollary 4 should be weakened slightly. For example, the boundedness of $M(t) - m(t)$ must be replaced by

$$\lim_{t \to +\infty} \frac{m(t)}{M(t)} = 1 .$$

However, if $g(x,t) = \text{constant}$ on $\partial G_1 \times [0, +\infty)$, then Theorem 2 and its corollaries all hold true even for $n = 2$.

(iii) In the case where g is independent of t, the growth order of the radius of $\Gamma(t)$ can easily be calculated; namely we have

Corollary 5. *Let* $n \geqq 3$ *and let* g *in* (2) *be independent of* t. *Then*

$$M(t) - (Ct)^{1/n}$$

remains bounded as $t \to +\infty$, *where* $M(t)$ *is as in Corollary 4 and* C *is a constant defined by*

$$C = \frac{-1}{kb_n} \int_{\partial G_1} \frac{\partial \hat{\theta}}{\partial \gamma} dx ,$$

in which k *is as in* (5), b_n *is the volume of the* n*-dimensional unit ball, that is,* $b_n = \pi^{n/2}/\Gamma(n/2+1)$ *with* $\Gamma(\cdot)$ *being the usual gamma function,* $\partial/\partial\gamma$ *is the outward normal derivative on* ∂G_1 *and, finally,* $\hat{\theta} = \hat{\theta}(x)$ *is the solution of the boundary value problem*

$$(8) \qquad \begin{cases} \Delta\hat{\theta} = 0 & \text{in } \mathbb{R}^n \setminus \bar{G}_1 , \\ \hat{\theta} = g & \text{on } \partial G_1 , \\ \lim_{|x|\to\infty} \hat{\theta}(x) = 0 . \end{cases}$$

3. Preliminaries — Existence and Regularity Theorems.

In the weak formulation due to Duvaut, $\theta(x,t)$ is locally given by the form $\theta = u_t$, with u being a solution to a certain variational inequality. More precisely, let T be any positive number and R be a sufficiently large positive number. Set $B_R = \{ x \in \mathbb{R}^n \mid |x| < R \}$, $D = B_R \setminus \bar{G}_1$,

$$K = \{ v \in H^1(D\times(0,T)) \mid v \geqq 0 \} ,$$

$$\psi(x,t) = \int_0^t g(x,\tau)\, d\tau , \qquad x \in \partial G_1 ,\ 0 \leqq t \leqq T ,$$

$$(9) \qquad f(x) = \begin{cases} h(x) & (x \in G) \\ -k & (x \in D \setminus G) , \end{cases}$$

where k is the constant given in (5) and H^m is the Sobolev space of all L^2 functions whose derivatives up to order m belong to L^2. We are to find a solution $u \in K \cap L^2(0,T;H^2(D))$ to the variational inequality in the pointwise form

$$(10\text{ a}) \qquad (-\Delta u + u_t)(v-u) \geqq f(v-u) \quad \text{a.e. for } v \in K \,,$$

$$(10\text{ b}) \qquad \begin{cases} u = 0 & \text{on } D \times \{0\} \,, \\ u = \psi & \text{on } \partial G_1 \times (0,T) \,, \\ u = 0 & \text{on } \partial B_R \times (0,T) \,. \end{cases}$$

Given a solution u to (10), we put

$$(11) \qquad \Omega(t) = \{ x \in D \mid u(x,t) > 0 \} \,,$$

$$(12) \qquad \Gamma(t) = \partial\Omega(t) \setminus \partial G_1 \,,$$

$$(13) \qquad \theta(x,t) = \begin{cases} u_t(x,t) & (x \in D) \\ 0 & (x \in \mathbb{R}^n \setminus B_R) \,, \end{cases}$$

for $0 \leqq t \leqq T$.

It is known that the problem (10) admits a unique solution (see Proposition 3.1 below). Moreover, one can check that the above-defined $\theta(x,t)$, $\Gamma(t)$ are actually independent of the choice of T, R provided that R is chosen sufficiently large so that

$$(14) \qquad \Gamma(t) \cap \partial B_R = \emptyset$$

for $0 \leqq t \leqq T$. We therefore can define θ, Γ globally on the time interval $0 \leqq t < +\infty$. Hereafter by a *weak solution* we mean the solution pair θ, Γ defined in the above manner.

Let us now consider the problem (10) in more detail. It is easily seen that (10) possesses at most one solution. The existence of a solution can be shown by approximation with a suitable penalty function. The following construction of approximate solutions is due to Friedman and Kinderlehrer [5].

For $\varepsilon > 0$, define a function $\beta_\varepsilon(t) \in C^\infty(\mathbb{R}^n)$ with the properties

$$\beta_\varepsilon(t) = 0 \quad \text{if} \quad t \geqq \varepsilon ,$$

$$\beta_\varepsilon(0) = -1 ,$$

$$\beta_\varepsilon'(t) > 0 \quad \text{and} \quad \beta_\varepsilon''(t) \leqq 0 , \quad -\infty < t < \varepsilon .$$

Let $f(x)$ be as in (9) and let $f_\varepsilon(x)$ be a sequence of functions smooth in $\bar{D}$, uniformly bounded, decreasing to $f(x)$ as $\varepsilon \to 0$, and having uniformly bounded variation in D. Set

$$\eta_\varepsilon(x) = \varepsilon\eta(x) , \quad x \in \bar{D} ,$$

where $\eta \in C_0^\infty(\mathbb{R}^n)$ is a function satisfying

$$0 \leqq \eta \leqq 1 ,$$

$$\eta(x) = 1 \quad \text{in} \quad \{ x \in \mathbb{R}^n \mid \operatorname{dist}(x, G_1) < 2a \} ,$$

$$\eta(x) = 0 \quad \text{in} \quad \{ x \in \mathbb{R}^n \mid \operatorname{dist}(x, G_1) > 3a \}$$

for some positive constant a with $3a < \operatorname{dist}(\partial G, \partial G_1)$.

The approximate equation for (10) is an initial-boundary value problem of the form

$$(15)_\varepsilon \quad \begin{cases} -\Delta u + u_t + k\beta_\varepsilon(u) = f_\varepsilon & \text{in} \quad D \times (0,T) , \\ u = \eta_\varepsilon & \text{in} \quad D \times \{0\} , \\ u = \psi + \varepsilon & \text{on} \quad \partial G_1 \times (0,T) , \\ u = 0 & \text{on} \quad \partial B_R \times (0,T) , \end{cases}$$

where k is as in (5).

Proposition 3.1 (Friedman and Kinderlehrer). *For* $T > 0$, *there exists a unique solution* u *to the problem* (10) *with the properties*

$$u \in L^\infty(0,T;\, W^{2,p}(D)\,), \qquad 1 < p < \infty,$$

$$u_t \in L^\infty(0,T;\, L^\infty(D)\,),$$

$$u_t \geq 0 \qquad a.e. \;\; in \;\; D \times (0,T)\;,$$

where $W^{2,p}(D)$ *is the Sobolev space of* L^p *functions whose derivatives up to order 2 belong to* $L^p(D)$. *Furthermore, if* u_ε , $\varepsilon > 0$, *denotes the solution to the problem* $(15)_\varepsilon$, *then*

$$u_\varepsilon \to u \;\;(\text{ as } \;\; \varepsilon \to 0\;) \;\; \text{weakly in} \;\; W^{1,\infty}(D \times (0,T))$$

and weakly in $W^{2,p}(D)$ *for each* $t \in (0,T)$ *and* $1 < p < \infty$; *hence* $u_\varepsilon \to u$ *uniformly in* $D \times (0,T)$.

Remark 3.2. (i) Actually it can further be shown that $u \in L^\infty(0,T;W^{2,\infty}(D))$ and that u_t is continuous ([2]).

(ii) The region $\Omega(t)$ defined by (11) is easily seen to coincide with the set

$$\{\; x \in \mathbb{R}^n \setminus \bar{G}_1 \mid \; \theta(x,t) > 0 \;\}\;.$$

It therefore makes sense that we have called $\Omega(t)$ the water region.

Definition 3.3. Let $A \subset \mathbb{R}^n$ be a measurable set and let x_0 be any point in $\mathbb{R}^n$. We say x_0 has *positive Lebesgue density* with respect to A if

$$\liminf_{\rho \downarrow 0} \frac{\mu(A \cap B_\rho(x_0))}{\mu(B_\rho(x_0))} > 0\;,$$

where μ is the Lebesgue measure and $B_\rho(x_0) = \{\; x \in \mathbb{R}^n \mid \; |x - x_0| < \rho\;\}$.

Combining the C^∞ smoothness criterion of Kinderlehrer and Nirenberg [8] and the C^1 smoothness criterion of Caffarelli [1], we get

Proposition 3.4. *Let* t_0 *be any positive number and let* x_0 *be any point on* $\Gamma(t_0)$. *Suppose* x_0 *has positive Lebesgue density with respect to the set* $\mathbb{R}^n \setminus (\Omega(t_0) \cup \bar{G}_1)$. *Then there exists an* xt-*neighborhood* V *of the point* (x_0, t_0) *such that* $\Gamma(t) \cap V$ *is an* n-*dimensional* C^∞ *hypersurface transversal to the hyperplane* $t = t_0$. *Moreover,* θ *is* C^∞ *in* $V \cap \{(x,t) \mid x \in \Omega(t) \cup \Gamma(t),\ t > 0\}$.

4. Proof of the Main Theorems.

In order to prove that the solution is "eventually" classical, we must show that each point on $\Gamma(t)$ has positive Lebesgue density with respect to the ice region $\mathbb{R}^n \setminus (\Omega(t) \cup \bar{G}_1)$ if t is sufficiently large. Such a property of $\Gamma(t)$ does not follow simply from local regularity analysis; studies of the geometric features of $\Gamma(t)$ in some global aspect are needed.

The main tool employed is the "plane-reflection" method first introduced by Serrin [9] and later developed by Gidas, Ni and Nirenberg [6] and Jones [7]. In particular, we owe much of the discussion below to Jone's work [7], in which he has investigated the asymptotic behavior of radially expanding wave front solutions to the equation $u_t = \Delta u + f(u)$, $x \in \mathbb{R}^n$, $t > 0$. Of course his argument does not apply to our problem automatically, partly because of the existence of the obstacle G_1 and partly because of the presence of the interface $\Gamma(t)$ through which θ gains derivative gaps.

We begin with some notation and lemmas:

Notation. Let P be an $(n-1)$-dimensional hyperplane with $P \cap G_1 = \emptyset$; set

(16 a) S^+: the half space with boundary P such that $S^+ \supset G_1$,

(16 b) x^λ: reflection of a point $x \in \mathbb{R}^n$ in P.

The following is the key lemma for the proof of the main theorems:

Lemma A. *Assume* $n \geq 3$ *and that* (7) *holds. Then there exists a positive number* R_0 *such that for any hyperplane* P *with* $dist(P,G_1) \geq R_0$ *it holds that*

$$\theta(x,t) \geq \theta(x^\lambda,t) \tag{17}$$

for all $x \in S^+ \setminus G$, $t \geq 0$, *where* S^+ , x^λ *are as in* (16). *Moreover, we have*

$$\theta(x,t) > \theta(x^\lambda,t) \tag{18}$$

for all $x \in S^+ \cap \Omega(t) \setminus G$, $t > 0$.

Corollary A. *Assume* $n \geq 3$ *and that* (7) *holds; and let* P *be as in Lemma A. Then* $|\nabla_x \theta(x,t)| \neq 0$ *for any* $x \in P \cap \Omega(t)$, $t > 0$, *where* $\nabla_x = (\partial/\partial x_1, \cdots, \partial/\partial x_n)$.

Lemma B. *Assume* (7). *Then*

$$\lim_{t \to +\infty} dist(\Gamma(t),G_1) = \infty .$$

The proof of Lemma A will be carried out in the next section. Corollary A follows immediately from Lemma A and the strong maximum principle, since θ satisfies the heat equation $\theta_t = \Delta\theta$ in the water region $\Omega(t)$, $t > 0$. For the proof of Lemma B, see [5].

Proof of Theorem 1. For simplicity we assume $n \geq 3$. The case $n = 2$ follows from a similar (but slightly modified) argument.

Let $R_0 > 0$ be as in Lemma A and set

$$W = \overline{co}\,(\{ x \in \mathbb{R}^n \mid dist(x,G_1) \leq R_0\} \cup G) \ , \tag{19}$$

where $\overline{co}$ denotes the closed convex hull of a set. By virtue of Lemma B, there exists a positive number T_0 such that

$$\Gamma(t) \cap W = \emptyset$$

for all $t \geq T_0$.

Take any $t_0 \in [T_0, +\infty)$ and fix it. And let x_0 be any point on $\Gamma(t_0)$. In order to apply Proposition 3.4, we must check that x_0 has positive Lebesgue density with respect to the region $\mathbb{R}^n \setminus \Omega(t_0)$. This will be shown by constructing an open cone K_0 with vertex x_0.

Put

$$K_0 = \{ z \in \mathbb{R}^n \mid (z - x_0)\cdot(y - x_0) < 0 \text{ for all } y \in W \},$$

where $\cdot$ denotes the Euclidean inner product in $\mathbb{R}^n$. Since W is a compact convex set and since $x_0 \notin W$, K_0 is a non-empty open cone with vertex x_0; that is, K_0 is an open set such that $r(K_0 - x_0) = K_0 - x_0$ for all $r > 0$. Let z be any point in K_0 and set

$$P_\alpha = \{ y \in \mathbb{R}^n \mid (y - x_0 - \alpha\xi)\cdot\xi = 0 \},$$

where $\xi = z - x_0$. Applying Lemma A to $P = P_\alpha$ for each $\alpha \geqq 0$, we easily find that $\theta(x_0 + \alpha\xi, t_0)$ is monotone non-increasing in $\alpha \geqq 0$. Consequently

$$\theta(x_0 + \alpha\xi, t_0) \leqq \theta(x_0, t_0) = 0$$

for all $\alpha \geqq 0$. This shows that θ vanishes everywhere in K_0, hence x_0 has positive Lebesgue density with respect to $\mathbb{R}^n \setminus \Omega(t)$. The conclusion of Theorem 1 now follows by applying Proposition 3.4.

Proof of Theorem 2. Let W be as in (19) and set

$$\Gamma_\varepsilon = \{ x \in \mathbb{R}^n \setminus W \mid \theta(x, t_0) = \varepsilon \}$$

for each $\varepsilon > 0$. By virtue of Corollary A, $\nabla_x \theta(x, t_0)$ does not vanish on Γ_ε, hence Γ_ε is an $(n-1)$-dimensional smooth hypersurface. Moreover, as is easily seen, $\Gamma_\varepsilon \cap \partial W = \emptyset$ if ε is sufficiently small. We shall show that any ray inward normal to Γ_ε (i.e. the direction parallel to $\nabla_x \theta(x, t_0)$) intersects the set W. The conclusion of Theorem 2 then follows by letting $\varepsilon \to 0$ and using the fact that $\theta(x, t_0)$ is smooth in $\Omega(t_0)$ up to the boundary $\Gamma(t_0)$.

Let x_1 be any point on Γ_ε and let ℓ be a ray inward normal to Γ_ε at x_1. Assuming that

$$\ell \cap W = \emptyset , \tag{20}$$

we shall derive a contradiction.

Denote by ξ_1 a non-zero n-vector parallel to ℓ. By the definition of ℓ, we have

$$\xi_1 = c\nabla_x \theta(x_1, t_0) \tag{21}$$

for some $c > 0$. Since W is a compact convex set, (20) implies that there exists an n-vector ζ such that

$$\zeta \cdot (y - x_1) < 0 \quad \text{for all } y \in W , \tag{22 a}$$

$$\zeta \cdot \xi_1 > 0 . \tag{22 b}$$

Arguing as in the proof of Theorem 1, we see from (22 a) that θ is monotone non-increasing in the direction of ζ; more precisely, $\theta(x_1 + \alpha\zeta, t_0)$ is monotone non-increasing in $\alpha \geqq 0$. In particular, we have

$$\zeta \cdot \nabla_x \theta(x_1, t_0) \leqq 0 ,$$

hence $\zeta \cdot \xi_1 \leqq 0$ by virtue of (21). But this clearly contradicts (22), thus showing that the supposition (20) is false. This completes the proof of Theorem 2.

We omit the proof of Corollaries 3, 4 and 5.

5. Proof of Lemma A.

In this section we prove Lemma A, which, although simple, played a key role in the proof of the main theorems. We begin with some auxiliary lemmas:

Lemma 5.1. *Let* $\overline{\theta}(x,t)$, $\underline{\theta}(x,t)$ *be solutions to the following initial-boundary value problems:*

$$(23)\quad \begin{cases} \overline{\theta}_t = \Delta\overline{\theta} , & x\in \mathbb{R}^n\setminus\overline{G}_1 ,\ t>0 , \\ \overline{\theta}(x,0) = \begin{cases} h(x) & (x\in\overline{G}\setminus\overline{G}_1) \\ 0 & (x\in\mathbb{R}^n\setminus\overline{G}) , \end{cases} \\ \overline{\theta} = g , & x\in\partial G_1 ,\ t>0 , \end{cases}$$

$$(24)\quad \begin{cases} \underline{\theta}_t = \Delta\underline{\theta} , & x\in G\setminus\overline{G}_1 ,\ t>0 , \\ \underline{\theta}(x,0) = h(x) , & x\in\overline{G}\setminus G_1 , \\ \underline{\theta} = g , & x\in\partial G_1 ,\ t>0 . \end{cases}$$

Then

$$\overline{\theta}(x,t) \geqq \theta(x,t) \quad \textit{for } x\in\mathbb{R}^n\setminus G_1 ,\ t\geqq 0 ,$$

$$\underline{\theta}(x,t) \leqq \theta(x,t) \quad \textit{for } x\in\overline{G}\setminus G_1 ,\ t\geqq 0 .$$

Recalling that $\mathbb{R}^n\setminus\overline{G}_1\supset\Omega(t)\supset G\ \ \overline{G}_1$ for $t\geqq 0$ and that $\theta_t = \Delta\theta$ in $\Omega(t)$, $\theta = 0$ on $\Gamma(t)$, one can easily verify the assertion of Lemma 5.1. This lemma is a special case of a more general comparison principle in one phase Stefan problems; see [5;Lemma 2.5] and its subsequent remark. Suffice it to say that (23) (resp. (24)) derives from the Stefan problem (1) ~ (6) if one sets k = 0 (resp. k = ∞) in (5).

Corollary 5.2. *Assume* (7), *and let* G_2 *be a domain with a smooth boundary satisfying* $\overline{G}_1\subset G_2\subset\overline{G}_2\subset G$. *Then there exist constants* $\delta_1>0$, $M_1>0$ *such that*

$$\delta_1 \leqq \theta(x,t) \leqq M_1$$

for all $x\in\partial G_2$, $t\geqq 0$.

Corollary 5.3. *Assume* $n \geqq 3$ *and that* (7) *holds. Let* G_2 *be as in Corollary 5.2. Then there exists a positive constant* M_2 *such that*

$$\theta(x,t) \leqq M_2\phi(x) \quad \text{for all} \quad x \in \mathbb{R}^n \setminus G_2 ,$$

where ϕ *is the solution to the problem*

$$(25) \qquad \begin{cases} \Delta\phi = 0 & \text{in } \mathbb{R}^n \setminus G_2 , \\ \phi = 1 & \text{on } \partial G_2 , \\ \lim\limits_{|x|\to\infty} \phi(x) = 0 . \end{cases}$$

In particular, we have $\lim\limits_{|x|\to\infty} \theta(x,t) = 0$ *uniformly in* $t \geqq 0$.

The proof of these corollaries are straightforward, so we omit it. Note that the assumption $n \geqq 3$ in Corollary 5.3 cannot be dropped, since the problem (25) has no solution in the case $n = 1$ or $n = 2$.

Lemma 5.4. *Let* u_ε , $\varepsilon > 0$, *be the solution to the problem* $(15)_\varepsilon$ *and set* $\theta_\varepsilon = \partial u_\varepsilon/\partial t$. *Then* $u_\varepsilon \to u$, $\theta_\varepsilon \to \theta$ *decreasingly as* $\varepsilon \downarrow 0$, *provided that* f_ε *and* β_ε *are chosen appropriately.*

Proof. Define β_ε by

$$\beta_\varepsilon(s) = \beta(\frac{s}{\varepsilon}) ,$$

where $\beta(s)$ is a function with the properties

$$\beta(s) = 0 \quad \text{if} \quad s \geqq 1 ,$$

$$\beta(0) = -1 ,$$

$$\beta'(s) > 0 \quad \text{and} \quad \beta''(s) \leqq 0 , \qquad -\infty < s < 1 .$$

It is clear that β_ε has the properties required in Section 3. Next we choose f_ε appropriately so that $\varepsilon\Delta\eta + f_\varepsilon$ decreases as $\varepsilon \downarrow 0$ and that f_ε satisfies the conditions listed above $(15)_\varepsilon$. The existence of such f_ε is obvious.

The decreasing property of u_ε (as $\varepsilon \downarrow 0$) follows immediately from that of $\psi + \varepsilon, \eta_\varepsilon$, f_ε and $-\beta_\varepsilon(s)$ ($s \geqq 0$) . In order to show that θ_ε is decreasing as $\varepsilon \downarrow 0$, we differentiate $(15)_\varepsilon$ by t , to get

$$\begin{aligned} -\Delta\theta_\varepsilon + \frac{\partial\theta_\varepsilon}{\partial t} + k\beta'_\varepsilon(u_\varepsilon)\theta_\varepsilon &= 0 && \text{in } D\times(0,T) , \\ \theta_\varepsilon(x,0) &= \varepsilon\Delta\eta(x) - k\beta(\eta(x)) + f_\varepsilon(x) && \text{in } D , \\ \theta_\varepsilon &= g && \text{on } \partial G_1 \times (0,T) , \\ \theta_\varepsilon &= 0 && \text{on } \partial B_R \times (0,T) . \end{aligned}$$

By virtue of the assumption above, the initial data $\theta_\varepsilon(x,0)$ decreases as $\varepsilon \downarrow 0$. Moreover, since $\beta''_\varepsilon \leqq 0$ and since u_ε is decreasing as $\varepsilon \downarrow 0$, $\beta'_\varepsilon(u_\varepsilon)$ is increasing as $\varepsilon \downarrow 0$. Applying the maximum principle, we see that θ_ε is decreasing as $\varepsilon \downarrow 0$; this completes the proof of Lemma 5.4. (Note that the convergence $\theta_\varepsilon \to \theta$ does not take place everywhere on $D\times[0,T]$; in fact, $\theta_\varepsilon(x,0)$ does not converges to $\theta(x,0)$ for some $x \in G$.)

<u>Proof of Lemma A.</u> Let u_ε , θ_ε be as in Lemma 5.4. By virtue of Corollary 5.3, there exists a positive constant R_1 such that

$$\sup_{t\geqq 0,\ |x|\geqq R_1} \theta(x,t) < \delta_1 ,$$

where δ_1 is as in Corollary 5.2. By a similar argument, we see that there is a positive constant, again denoted by R_1 , such that

$$(26) \qquad \sup_{\substack{(x,t)\in D\times[0,T] \\ |x| \geqq R_1}} \theta_\varepsilon(x,t) < \delta_1$$

for any sufficiently small ε .

Now take a positive constant R_0 such that the set

$$W = \overline{co}\,(\{x \in \mathbb{R}^n \mid \operatorname{dist}(x,G_1) \leqq R_0\})$$

contains both G and $\{x \in \mathbb{R}^n \mid |x| \leqq R_1\}$. And let P be a hyperplane with $\operatorname{dist}(P,G_1) \geqq R_0$; and set $D' = \{x \in S^+ \mid x^\lambda \in D\}$, where x^λ, S^+ are as in (16). We shall show that

$$\theta_\varepsilon(x^\lambda,t) \leqq \theta_\varepsilon(x,t) \tag{27}$$

for all $x \in D' \setminus G_2$, where G_2 is as in Corollary 5.2.

Put

$$v(x,t) = u_\varepsilon(x,t) - u_\varepsilon(x^\lambda,t) ,$$

$$w(x,t) = \frac{\partial v}{\partial t}(x,t) = \theta_\varepsilon(x,t) - \theta_\varepsilon(x^\lambda,t) .$$

The functions v, w satisfy the degenerated parabolic system

$$\text{(28 a)} \qquad \begin{cases} v_t = w & \text{in } (D'\setminus G_2) \times (0,T) \\ w_t = \Delta w - k\beta'(u_\varepsilon)w - k\theta_\varepsilon(x^\lambda,t)\xi(x,t)v & \text{in } (D'\setminus G_2) \times (0,T) \end{cases}$$

together with the initial and the boundary conditions

$$\text{(28 b)} \qquad \begin{cases} v \geqq 0 & \text{in } D'\setminus G_2 , \\ w \geqq 0 & \text{in } D'\setminus G_2 , \\ w \geqq 0 & \text{on } \partial(D'\setminus G_2) \times (0,T) , \end{cases}$$

where

$$\xi(x,t) = \frac{\beta_\varepsilon'(u_\varepsilon(x,t)) - \beta_\varepsilon'(u_\varepsilon(x^\lambda,t))}{u_\varepsilon(x,t) - u_\varepsilon(x^\lambda,t)}$$

Note that the last inequality in (28 b) follows from (26), Corollary 5.2 and the decreasing property of θ_ε (Lemma 5.4). Since $\beta_\varepsilon'' \leqq 0$, we have

(29) $\xi \leqq 0$.

It is now clear from (28), (29) and the maximum principle that $w \geqq 0$ in $(D'\setminus G_2)\times[0,T]$, which implies that (27) holds. Letting $\varepsilon \downarrow 0$, we get

(30) $$\theta(x^\lambda,t) \leqq \theta(x,t)$$

a.e. in $(D'\setminus G_2)\times[0,T]$; hence (30) holds everywhere in $(D'\setminus G_2)\times[0,T]$ by virtue of the continuity of θ ([2]). Considering that T and R (where $D = B_R\setminus G_1$) can be chosen arbitrarily large, we see that (30) holds for all $x\in S^+\setminus G_2$, $t \geqq 0$. Thus the former part of Lemma A is proved. The latter part is now obvious from the strong maximum principle. This completes the proof of Lemma A.

Acknowledgement

The author would like to express his gratitude to Professor Ei-Ichi Hanzawa for stimulating discussions.

References

[1] Caffarelli, L.A., The regularity of free boundaries in higher dimensions, Acta Math. 139 (1977) 155 - 184.

[2] Caffarelli, L.A. and Friedman, A., Continuity of the temperature in the Stefan problem, Indiana Univ. Math. J. 28 (1979) 53 - 70.

[3] Duvaut , G., Résolution d'un problème de Stefan (Fusion d'un bloc de glace à zero degré), C. R. Acad. Sci. Paris 276 (1973) 1461 - 1463.

[4] Friedman, A., Partial differential equations of parabolic type (Prentice-

Hall, Englewood Cliffs, 1964).

[5] Friedman, A. and Kinderlehrer, D., A one phase Stefan problem, Indiana Univ. Math. J. 24 (1975) 1005 - 1035.

[6] Gidas, G., Ni, Wei-Ming and Nirenberg, L., Symmetry and related properties via the maximum principle, Commun. Math. Phys. 68 (1979) 209 - 243.

[7] Jones, C. K. R. T., Asymptotic behavior of a reaction-diffusion equation in higher space dimensions, Technical Report, Inst. Appl. Math. Stat. Univ. British Columbia (Oct. 1980).

[8] Kinderlehrer, D. and Nirenberg, L., The smoothness of the free boundary in the one phase Stefan problem, Comm. Pure. Appl. Math. 31 (1978) 257 - 282.

[9] Serrin, J., A symmetry problem in potential theory, Arch. Rat. Mech. Anal. 43 (1971) 304 - 318.

Lecture Notes in Num. Appl. Anal., **5**, 153–170 (1982)
Nonlinear PDE in Applied Science. U.S.-Japan Seminar, Tokyo, 1982

INITIAL BOUNDARY VALUE PROBLEMS FOR THE EQUATIONS OF COMPRESSIBLE VISCOUS AND HEAT-CONDUCTIVE FLUID

Akitaka Matsumura and Takaaki Nishida

Department of Applied Mathematics and Physics, Kyoto University
Kyoto 606, Japan

Department of Mathematics
Kyoto University
Kyoto 606, Japan

The equations of motion of compressible viscous and heat-conductive fluid are investigated for initial boundary value problems on the interior and exterior domain of any bounded region and also on the half space. The global solution in time is proved to exist uniquely and approach the stationary state as $t \to +\infty$, provided the prescribed initial data and the external force are sufficiently small.

§ 1. Introduction

The motion of a compressible, viscous and heat-conductive Newtonian fluid is described by five conservation laws:

$$(1.1)\quad \begin{cases} \rho_t + (\rho u^j)_{x_j} = 0 \\ u^i_t + u^j u^i_{x_j} + \frac{1}{\rho} p_{x_i} = \frac{1}{\rho}(\mu(u^i_{x_j} + u^j_{x_i}) + \mu'(u^k_{x_k})\delta^{ij})_{x_j} + f^i, \quad i = 1,2,3, \\ \theta_t + u^j\theta_{x_j} + \frac{\theta p_\theta}{\rho c} u^j_{x_j} = \frac{1}{\rho c}((\kappa\theta_{x_j})_{x_j} + \Psi), \end{cases}$$

where $x = (x_1,x_2,x_3) \in R^3$, $t > 0$, ρ is the density, $u = (u^1,u^2,u^3)$ is the velocity, θ is the absolute temperature, $p = p(\rho,\theta)$ is the pressure, $f = (f^1,f^2,f^3)$ is the external force, $\mu = \mu(\rho,\theta)$ and $\mu'(\rho,\theta)$ are viscosity coefficients, $\kappa = \kappa(\rho,\theta)$ is the coefficient of heat conductivity, $c = c(\rho,\theta)$ is the heat capacity at constant volume and $\Psi = \frac{\mu}{2}(u^i_{x_j} + u^j_{x_i})^2 + \mu'(u^k_{x_k})^2$

is the dissipation function. First we assume the following basic assumptions on (1.1):

(A.1) The external force f^i is generated by a potential function $\Phi(x)$, i.e., $f^i = -\Phi_{x_i}$.

(A.2) μ, μ', κ, p and c are smooth functions of $\rho,\theta > 0$, and

$$\mu, \kappa, c, p, p_\rho, p_\theta > 0, \quad \mu' + \frac{2}{3}\mu \geqq 0.$$

We consider the system (1.1) in the following domain Ω.

Case 1. $\Omega = \Omega_1$: a bounded open set in R^3 with C^∞ boundary $\partial\Omega$.

Case 2. $\Omega = \Omega_2$: compliment of $\overline{\Omega}_1$ or the half space $R^3_+ = \{x \in R^3, x_3 > 0\}$.

For both cases we discuss the existence of a global solution in time under the suitable initial and boundary conditions since the local existence theorem is established by Tani [14] under full generality. In this note we shall summarize the results in [9] and [10].

Case 1. (Interior Problem) Let us consider the system (1.1) in Ω_1 with the initial condition

$$(1.2) \qquad (\rho,u,\theta)(0,x) = (\rho_0,u_0,\theta_0)(x) \qquad x \in \Omega,$$

and the boundary conditions

$$(1.3)_1 \qquad u(t,x) = 0, \quad \theta(t,x) = \overline{\theta}, \quad x \in \partial\Omega, \quad t \geqq 0,$$

where $\overline{\theta}$ is any fixed positive constant. In order to state the theorem precisely, let us list up further assumptions:

$(A.4)_1$ $(\rho_0,u_0,\theta_0) \in H^3(\Omega)$, $\Phi \in H^4(\Omega)$ and $(\rho_0,\theta_0)(x) > 0$ for $x \in \overline{\Omega}$,

(A.5) (compatibility condition)

$$u_0 \in H^1_0(\Omega), \quad \theta_0 - \overline{\theta} \in H^1_0(\Omega),$$

$$-(p_0)_{x_i} + (\mu_0(u^i_{0,x_j} + u^j_{0,x_i}) + \mu'_0 u^k_{0,x_k}\delta^{ij})_{x_j} - \rho_0\Phi_{x_i} \in H^1_0(\Omega),$$

$$-\theta_0(p_0)_0 u^j_{0,x_j} + (\kappa_0\theta_{0,x_j})_{x_j} + \Psi_0 \in H^1_0(\Omega),$$

where $p_0 = p(\rho_0,\theta_0)$, $\mu_0 = \mu(\rho_0,\theta_0),\cdots$, and so on. Here $H^\ell(\Omega)$ denotes the Sobolev's space on Ω

$$H^\ell(\Omega) = \{f \in L^2(\Omega),\ D^k f = \{\partial^{|\alpha|} f/\partial x_1^{\alpha_1}\,\partial x_2^{\alpha_2}\,\partial x_3^{\alpha_3},\ |\alpha|=k\} \in L^2(\Omega),\ 1 \leq k \leq \ell\}$$

with the norm $\|f\|_\ell = (\sum_{k=0}^{\ell} \int_\Omega |D^k f|^2 dx)^{\frac{1}{2}}$, and $H^1_0(\Omega)$ denotes the completion of $C^\infty_0(\Omega)$ in $H^1(\Omega)$. Define a positive constant $\bar\rho$ by

$$\bar\rho = \frac{1}{V(\Omega)} \int_\Omega \rho_0(x)dx$$

where $V(\Omega)$ represents the volume of Ω. We call $(\hat\rho(x), \hat u(x), \hat\theta(x)) \in C^1(\bar\Omega) \times (C^2(\bar\Omega))^2$ an equilibrium state of the problem (1.1)(1.2)(1.3)$_1$ when $(\tilde\rho,\tilde u,\tilde\theta)$ satisfies (1.1) and the additional conditions

$$\tilde u\big|_{\partial\Omega} = 0, \quad \tilde\theta\big|_{\partial\Omega} = \bar\theta, \quad \frac{1}{V(\Omega)}\int_\Omega \tilde\rho\, dx = \bar\rho .$$

Then we have

<u>Theorem 1.</u> Under the assumptions (A.1)(A.2)(A.3)(A.4)$_1$(A.5), there exist positive constant ε, α and C such that if $\|\rho_0-\bar\rho,\ u_0,\ \theta_0-\bar\theta\|_3 + \|\Phi\|_4 \leq \varepsilon$, then the problem (1.1)(1.2)(1.3)$_1$ has a unique global solution in time (ρ,u,θ) and a unique equilibrium state $(\tilde\rho,\tilde u,\tilde\theta) = (\tilde\rho,0,\bar\theta)$ satisfying

$$\rho \in C^0(0,+\infty;\ H^3(\Omega)) \cap C^1(0,+\infty;\ H^2(\Omega)),$$

$$(u,\theta-\bar\theta) \in C^0(0,+\infty;\ H^3(\Omega) \cap H^1_0(\Omega)) \cap C^1(0,+\infty;\ H^1_0(\Omega)),$$

and

$$\sup_{x\in\Omega} |(\rho,u,\theta)(t) - (\tilde\rho,0,\bar\theta)| \leq Ce^{-\alpha t} .$$

<u>Case 2.</u> (Exterior or Half-space Problem) In this case let us consider the system (1.1) in Ω_2 with the initial condition (1.2) and the boundary conditions

$(1.3)_2$ $\quad u(t,x) = 0, \quad \theta(t,x) = \overline{\theta} \quad$ for $\quad x \in \partial\Omega, \quad t \geqq 0,$

$$u(t,x) \to 0, \quad \theta(t,x) \to \overline{\theta}, \quad \rho(t,x) \to \overline{\rho} \quad \text{as} \quad |x| \to +\infty \quad \text{for} \quad t \geqq 0,$$

where $\overline{\rho}$ and $\overline{\theta}$ are any fixed positive constants. Let us assume that

$(A.4)_2$ $\quad (\rho_0-\overline{\rho}, u_0, \theta_0-\overline{\theta}) \in H^3(\Omega), \quad \Phi \in H^5(\Omega),$

and the compatibility condition (A.5). In this case we call $(\tilde{\rho}(x), \tilde{u}(x), \tilde{\theta}(x))$ an equilibrium state of the problem $(1.1)(1.2)(1.3)_2$ when $(\tilde{\rho},\tilde{u},\tilde{\theta})$ satisfies (1.1) $(1.3)_2$ and $(\tilde{\rho}-\overline{\rho}, \tilde{u}, \tilde{\theta}-\overline{\theta}) \in H^2(\Omega)$. Then we have

Theorem 2. Under the assumptions $(A.1)(A.2)(A.3)(A.4)_2(A.5)$, there exist a positive constant ε such that if $\|\rho_0-\overline{\rho}, u_0, \theta_0-\overline{\theta}\|_3 + \|\Phi\|_5 \leqq \varepsilon$, then the problem $(1.1)(1.2)(1.3)_2$ has a unique global solution in time (ρ,u,θ) and a unique equilibrium state $(\tilde{\rho},\tilde{u},\tilde{\theta}) = (\tilde{\rho},0,\overline{\theta})$ satisfying

$$\rho - \overline{\rho} \in C^0(0,+\infty; H^3(\Omega)) \cap C^1(0,+\infty; H^2(\Omega)),$$

$$(u,\theta-\overline{\theta}) \in C^0(0,+\infty; H^3(\Omega) \cap H^1_0(\Omega)) \cap C^1(0,+\infty; H^1_0(\Omega)),$$

and

$$\sup_x |(\rho,u,\theta)(t) - (\tilde{\rho},0,\overline{\theta})| \to 0 \quad \text{as} \quad t \to +\infty .$$

The proof of both Theorems are given by an energy method similar to those of our previous papers [7][8] on the initial value problem. However the initial boundary value problems require a new a-priori estimates of the solution near the boundary. In the following sections, we shall show rough sketch of the proof only for the half space case because the other cases are proved along similar strategy. Details are to be appeared in [10]. Finally we should mention that for one-dimensional model system of (1.1) we can see more precise results in Kawashima-Nishida [5] and Okada-Kawashima [11][12].

§ 2. Stationary solution

Let us write the equations and conditions for the stationary solution $(\tilde{\rho},\tilde{u},\tilde{\theta})$:

$$(2.1) \qquad (\tilde{\rho}\tilde{u}^j)_{x_j} = 0,$$

$$(2.2) \qquad \tilde{\rho}\tilde{u}^j\tilde{u}^i_{x_j} + \tilde{p}_{x_i} + \tilde{\rho}\Phi_{x_i} - (\tilde{\mu}(\tilde{u}^i_{x_j} + \tilde{u}^j_{x_i}) + \tilde{\mu}'\tilde{u}^k_{x_k}\delta^{ij})_{x_j} = 0, \quad i = 1,2,3,$$

$$(2.3) \qquad \tilde{\rho}\tilde{c}\,\tilde{u}^j\tilde{\theta}_{x_j} + \tilde{\theta}\tilde{p}_\theta u^j_{x_j} - (\tilde{\kappa}\tilde{\theta}_{x_j})_{x_j} - \tilde{\Psi} = 0,$$

$$(2.4) \qquad \tilde{u}\big|_{\partial\Omega} = \tilde{u}\big|_{\infty} = 0, \qquad \tilde{\theta}\big|_{\partial\Omega} = \tilde{\theta}\big|_{\infty} = \bar{\theta}, \qquad \tilde{\rho}\big|_{\infty} = \bar{\rho},$$

where $\tilde{p} = p(\tilde{\rho},\tilde{\theta})$ etc. The stationary problem (2.1)-(2.4) has a unique solution as

<u>Lemma 2.1.</u> Under the assumptions A.1 ~ A.3 there exist positive constant ε and C such that if $\|\Phi\|_\ell \leq \varepsilon$, $\ell = 3,4$, or 5, the problem (2.1)-(2.4) has a unique solution $(\tilde{\rho}(x),0,\bar{\theta})$ in a small neighborhood of $(\bar{\rho},0,\bar{\theta})$ in $(H^2)^3$ satisfying

$$(2.5) \qquad \|\tilde{\rho} - \bar{\rho}\|_\ell \leq C\|\Phi\|_\ell, \quad \ell = 3,4, \text{ or } 5 \text{ respectively},$$

where $\tilde{\rho}(x)$ is determined by (1.8), i.e.,

$$(2.6) \qquad \int_{\bar{\rho}}^{\tilde{\rho}(x)} \frac{p_\rho(\eta,\bar{\theta})}{\eta}\, d\eta + \Phi(x) = 0.$$

<u>Proof.</u> Since we consider a small neighborhood of $(\bar{\rho},0,\bar{\theta})$ in $(H^2)^3$, by Sobolev's lemma we may suppose $|\tilde{\rho}-\bar{\rho}|$, $|\tilde{u}|$, $|\tilde{\theta}-\bar{\theta}| < \frac{1}{2}\min\{\bar{\rho},\bar{\theta}\}$. Then we can estimate the equalities:

$$\int [2.1] \times \int_{\bar{\rho}}^{\tilde{\rho}} \frac{p_\rho(\eta,\bar{\theta})}{\eta}\, d\eta\, dx = 0,$$

$$(2.7) \qquad \int [2.2]^i \times \tilde{u}^i dx = 0,$$

$$\int [2.3] \times (\tilde{\theta}-\bar{\theta})dx = 0,$$

where [2.1], $[2.2]^i$ and [2.3] denote the terms on the left hand side of (2.1), $(2.2)^i$ and (2.3) respectively. Take the sum of (2.7) and integrate it by parts using the mean value theorem and Lemma 4.1. We obtain the inequality:

$$\|D\tilde{u}\|^2 + \|D\tilde{\theta}\|^2 \leq C\{\|D\tilde{\rho}\| + \|\tilde{u}\|_1 + |\tilde{\theta}-\bar{\theta}|_{C^0} + \|\tilde{\theta}-\bar{\theta}\|\}(\|D\tilde{u}\|^2 + \|D\tilde{\theta}\|^2).$$

Therefore if $\|D\tilde{\rho}\|$, $\|\tilde{u}\|_1$, $\|\tilde{\theta}-\bar{\theta}\|_2$ is small, we can conclude

$$\tilde{u} = 0, \quad \tilde{\theta} = \bar{\theta}. \tag{2.8}$$

If we substitute (2.8) in (2.2), we have

$$\left\{\int_{\bar{\rho}}^{\tilde{\rho}} \frac{p_\rho(\eta,\bar{\theta})}{\eta} d\eta + \Phi\right\}_{x_i} = 0$$

which implies (2.6).

§ 3. Local and global existence

Let us rewrite the problem (1.1), (1.2) by the change of variables $(\rho,u,\theta) \to (\rho + \tilde{\rho},u,\theta + \bar{\theta})$ using (2.6) as follows:

$$L_u^0(\rho,u) \equiv \rho_t + u^j\rho_{x_j} + \bar{\rho}u^j_{x_j} = f^0, \tag{3.1^0}$$

$$L^i(\rho,u,\theta) \equiv u^i_t - \hat{\mu}u^i_{x_jx_j} - (\hat{\mu}+\hat{\mu}')u^j_{x_ix_j} + p_1\rho_{x_i} + p_2\theta_{x_i} = f^i, \quad i = 1,2,3, \tag{3.1^i}$$

$$L^4(u,\theta) \equiv \theta_t - \hat{\kappa}\theta_{x_jx_j} + p_3u^j_{x_j} = f^4. \tag{3.1^4}$$

$$(u,\theta)\big|_{\partial\Omega} = (u,\theta)\big|_{\infty} = 0, \tag{3.2}$$

$$(\rho,u,\theta)(0) = (\rho_0,u_0,\theta_0), \tag{3.3}$$

where we denote the constant for the function g of ρ and θ by $\bar{g} = g(\bar{\rho},\bar{\theta})$, and also $\hat{\mu} = \bar{\mu}/\bar{\rho}$, $\hat{\mu}' = \bar{\mu}'/\bar{\rho}$, $\hat{\kappa} = \bar{\kappa}/\bar{\rho}\bar{c}$, $p_1 = \bar{p}_\rho/\bar{\rho}$, $p_2 = \bar{p}_\theta/\bar{\rho}$ and $p_3 = \bar{\theta}\bar{p}_\theta/\bar{\rho}\bar{c}$. The terms on the right hand side of equations (3.4) are nonlinear and have the form:

$$
\begin{aligned}
f^0(\rho,u,\theta) &\equiv (\bar{\rho}-\rho-\tilde{\rho})u^j_{x_j} - \tilde{\rho}_{x_j}u^j ,\\
f^i(\rho,u,\theta) &\equiv -u^j u^i_{x_j} + \Big(\frac{\mu}{\rho+\tilde{\rho}} - \hat{\mu}\Big)u^i_{x_j x_j} + \Big(\frac{\mu+\mu'}{\rho+\tilde{\rho}} - \hat{\mu} - \hat{\mu}'\Big)u^j_{x_i x_j} + \\
&\quad + \frac{1}{\rho+\tilde{\rho}}\{\mu_{x_j}(u^i_{x_j} + u^j_{x_i}) + \mu'_{x_j}u^k_{x_k}\delta^{ij}\} + \Big(p_1 - \frac{p_\rho}{\rho+\tilde{\rho}}\Big)\rho_{x_i} + \\
&\quad + \Big(p_2 - \frac{p_\theta}{\rho+\tilde{\rho}}\Big)\theta_{x_i} + \Big(\frac{\tilde{\rho}p_\rho(\rho+\tilde{\rho},\theta+\bar{\theta})}{(\rho+\tilde{\rho})p_\rho(\tilde{\rho},\bar{\theta})} - 1\Big)\Phi_{x_i}, \quad i = 1,2,3,\\
f^4(\rho,u,\theta) &\equiv -u^j\theta_{x_j} + \Big(\frac{\kappa}{(\rho+\tilde{\rho})c} - \hat{\kappa}\Big)\theta_{x_j x_j} + \\
&\quad + \Big(p_3 - \frac{(\theta+\bar{\theta})p_\theta}{(\rho+\tilde{\rho})c}\Big)u^j_{x_j} + \frac{1}{(\rho+\tilde{\rho})c}(\kappa_{x_j}\theta_{x_j} + \Psi).
\end{aligned}
\tag{3.4}
$$

Next we choose a constant E_0 by use of Sobolev's lemma such that

$$\|g\|_{C^0} \leq \frac{1}{2}\min(\bar{\rho},\bar{\theta}) \quad \text{for any} \quad g \in H^2, \quad \|g\|_2 \leq E_0 .$$

Then the solution of (3.1)-(3.3) is sought in the set of functions $X(0,\infty; E)$ for some $E \leq E_0$, where for $0 \leq t_1 \leq t_2 < \infty$, we define

$$
\begin{aligned}
X(t_1,t_2; E) = \{(\rho,u,\theta) :\ & \rho \in C(t_1,t_2; H^3), \quad D\rho \in L_2(t_1,t_2; H^2),\\
& \rho_t \in C(t_1,t_2; H^2) \cap L_2(t_1,t_2; H^2),\\
& u,\theta \in C(t_1,t_2; H^3 \cap H^1), \quad D(u,\theta) \in L_2(t_1,t_2; H^3),\\
& u_t,\theta_t \in C(t_1,t_2; H^1) \cap L_2(t_1,t_2; H^2) \ \text{and}\ N(t_1,t_2) \leq E\},
\end{aligned}
$$

where

$$N^2(t_1,t_2) \equiv \sup_{t_1 \leqq t \leqq t_2} \| \rho,u,\theta(t) \|_3^2 + \| \rho_t(t) \|_2^2 + \| u_t,\theta_t(t) \|_1^2 + \tag{3.5}$$
$$+ \int_{t_1}^{t_2} \| D\rho(s) \|_2^2 + \| \rho_t(s) \|_2^2 + \| D(u,\theta)(s) \|_3^2 + \| u_t,\theta_t(s) \|_2^2 ds .$$

Here and in what follows we do not write Ω in $H^{\ell}(\Omega)$.

We will obtain the global solution by a combination of a local existence theorem and some a priori estimates for the solution in X, namely that for the norm N.

Proposition 3.1. (local existence)

Suppose the problem (3.1)-(3.3) has a unique solution $(\rho,u,\theta) \in X(0,h;\ E_0)$ for some $h \geqq 0$ and consider the problem (3.1)-(3.3) for $t \geqq h$. Then there exist positive constants τ, ε_0 and $C_0(\varepsilon_0\sqrt{1+C_0^2} \leqq E_0)$ independent of h such that if $N(h,h)$, $\| \Phi \|_5 \leqq \varepsilon_0$, the problem has a unique solution

$$(\rho,u,\theta) \in X(h,h+\tau;\ C_0N(h,h)).$$

The proof is the same as that for the interior problem in [9] and is omitted. Although the local existence theorem by Tani [14] is more general, we need it in the form of Proposition 3.1 to extend the solution globally in time by use of L^2 energy method.

Proposition 3.2. (a priori estimates)

Suppose the problem (3.1)-(3.3) has a solution $(\rho,u,\theta) \in X(0,h;\ E_0)$ for given $h > 0$. Then there exist positive constants ε_1 and C_1 $(\varepsilon_1 \leqq \varepsilon_0,\ \varepsilon_1 C_1 \leqq E_0)$ which are independent of h such that if $N(0,h)$, $\| \Phi \|_5 \leqq \varepsilon_1$, it holds

$$N(0,h) \leqq C_1 N(0,0).$$

If Proposition 3.1 and 3.2 are known, the global existence of unique solution can be proved as follows: Choose the initial data (ρ_0,u_0,θ_0) and the potential function Φ so small that it holds

$$N(0,0) \leqq \min\{\varepsilon_0,\ \varepsilon_1/C_0,\ \varepsilon_1/C_1\sqrt{1+C_0^2}\,\} \quad \text{and} \quad \|\Phi\|_5 \leqq \varepsilon_1 .$$

Then Proposition 3.1 with h=0 gives a local solution $(\rho,u,\theta) \in X(0,\tau;\ C_0N(0,0))$. Since $C_0N(0,0) \leqq \varepsilon_1 \leqq \varepsilon_0$, Proposition 3.2 with $h=\tau$ implies $N(0,\tau) \leqq C_1N(0,0)$. Then Proposition 3.1 with $h=\tau$ implies the existence of solution

$$(\rho,u,\theta) \in X(\tau,2\tau;\ C_0N(\tau,\tau)),$$

$$\in X(0,2\tau;\ \sqrt{1+C_0^2}\ N(0,\tau)).$$

Hence, since $\sqrt{1+C_0^2}\ N(0,\tau) \leqq C_1\sqrt{1+C_0^2}\ N(0,0) \leqq \varepsilon_1$, Proposition 3.2 with $h = 2\tau$ gives $N(0,2\tau) \leqq C_1N(0,0)$, and Proposition 3.1 with $h = 2\tau$ gives

$$(\rho,u,\theta) \in X(2\tau,3\tau;\ C_0N(2\tau,2\tau)),$$

$$\in X(0,3\tau;\ \sqrt{1+C_0^2}\ N(0,2\tau)).$$

Repetition of this process yields

Proposition 3.3. (global existence)

There exist positive constants ε and $C(\varepsilon C \leqq E_0)$ such that if $N(0,0)$, $\|\Phi\|_5 \leqq \varepsilon$, then the initial boundary value problem (3.1)-(3.3) has a unique solution

$$(\rho,u,\theta) \in X(0,\infty;\ CN(0,0)).$$

§ 4. A priori estimates for the half space case $\Omega = R_+^3$.

First we recall some inequalities of Sobolev type.

Lemma 4.1. It holds

$$\|f\|_{C^\sigma(\Omega)} \leqq C\|f\|_{H^2(\Omega)}, \qquad 0 \leqq \sigma < 1/2 ,$$

$$\|f\|_{L_p(\Omega)} \leqq C\|f\|_{H^1(\Omega)}, \qquad 2 \leqq p < 6 , \tag{4.1}$$

$$\| f \|_{L_6(\Omega)} \leqq C \| Df \|_{L_2(\Omega)} .$$

Proof. See for example [3], [4].

Next we note some estimates of elliptic system of equations for our domain, when we regard equation $(3.1)^i$, $i = 1,\cdots,4$, as elliptic with respect to x variables, i.e.,

$$\hat{\mu} u^i_{x_j x_j} + (\hat{\mu}+\hat{\mu}')u^j_{x_i x_j} = u^i_t + p_1 \rho_{x_i} + p_2 \theta_{x_i} - f^i , \quad i = 1,2,3,$$

$$(4.2) \qquad \hat{\kappa}\theta_{x_j x_j} = \theta_t + p_3 u^j_{x_j} - f^4 ,$$

$$u\big|_{\partial\Omega} = u\big|_{\infty} = \theta\big|_{\partial\Omega} = \theta\big|_{\infty} = 0.$$

Lemma 4.2. We have for $k = 2,3$

$$(4.3) \qquad \| D^k u \| \leqq C\{ \| u_t \|_{k-2} + \| D(\rho,\theta) \|_{k-2} + \| f \|_{k-2} + \| u \| \},$$

$$(4.4) \qquad \| D^k \theta \| \leqq C\{ \| \theta_t \|_{k-2} + \| Du \|_{k-2} + \| f \|_{k-2} + \| D\theta \| \}.$$

The first estimate is well known, e.g. [1]. The second one is given in [5].

The last estimate for an elliptic system concerns Stokes equation in Ω which comes from $(3.1)^i$, $i = 0,\cdots,3$.

$$\bar{\rho} u^j_{x_j} = h ,$$

$$(4.5) \qquad -\hat{\mu} u^i_{x_j x_j} + p_1 \rho_{x_i} = g^i , \quad i = 1,2,3,$$

$$u\big|_{\partial\Omega} = a, \quad u\big|_{\infty} = 0.$$

Lemma 4.3. For $k = 2,3,4$ it holds

$$(4.6) \qquad \| D^k u \|^2 + \| D^{k-1}\rho \|^2 \leqq C\{ \| h \|^2_{k-1} + \| g \|^2_{k-2} + \| a \|^2_{H^{k-1/2}(\partial\Omega)} + \| Du \|^2 \},$$

where the last term on the right hand side is necessary in the case of exterior domain.

Proof. See for example Solonnikov [13] and Cattabriga [2].

Now we begin to obtain the energy estimate for solution of equation $(3.4)^i$, $i = 0,\cdots,4$, with (3.5).

Lemma 4.4. We have for $\ell = 0$ and 1

$$\| \partial_t^\ell(\rho,u,\theta)(t) \|^2 + \int_0^t \| D\partial_t^\ell(u,\theta)(s) \|^2 + \| \partial_t^\ell \frac{d\rho}{dt}(s) \|^2 ds$$

$$\leq C\{\| \partial_t^\ell(\rho,u,\theta)(0) \|^2 + \int_0^t |A_\ell| + \| \partial_t^\ell f^0 \|^2 ds\} \tag{4.7}$$

where

$$A_0 = \int \frac{p_1}{\bar\rho}\rho(f^0 - u^j\rho_{x_j}) + u^i f^i + \frac{p_2}{p_3}\theta f^4\, dx,$$

$$A_1 = \int \frac{p_1}{\bar\rho}\rho_t(f^0 - u^j\rho_{x_j})_t + u^i_t f^i_t + \frac{p_2}{p_3}\theta_t f^4_t\, dx, \tag{4.8}$$

$$\frac{d\rho}{dt} \equiv \rho_t + u^j\rho_{x_j} = f^0 - \bar\rho u^j_{x_j}.$$

We have also for $k = 0$ and 1

$$\| D\partial_t^k(u,\theta)(t) \|^2 + \int_0^t \| \partial_t^{k+1}(\rho,u,\theta)(s) \|^2 ds$$

$$\leq C\{\| D\partial_t^k(u,\theta)(0) \|^2 + \| \partial_t^k\rho(0) \|^2 + \| \partial_t^k\rho(t) \|^2 \tag{4.9}$$

$$+ \int_0^t \| D\partial_t^k(u,\theta)(s) \|^2 + \| \partial_t^k(f^0 - u^j\rho_{x_j}) \|^2 + \| \partial_t^k f(s) \|^2 ds\}.$$

Proof. Compute the integral

$$\int_0^t \int_\Omega \frac{p_1}{\bar\rho}\rho(L^0 - f^0) + u^i(L^i - f^i) + \frac{p_2}{p_3}\theta(L^4 - f^4)dxdt = 0.$$

Integration by parts using the boundary condition gives

$$\frac{1}{2}\int_\Omega \frac{p_1}{\bar\rho}\rho^2 + |u|^2 + \frac{p_2}{p_3}\theta^2 dx + \int_0^t\int_\Omega \hat\mu|Du|^2 + (\hat\mu+\hat\mu')(u^j_{x_j})^2 + \frac{p_2}{p_3}\hat\kappa|D\theta|^2 dxdt$$

$$= \frac{1}{2}\int \frac{p_1}{\bar\rho}\rho_0^2 + |u_0|^2 + \frac{p_2}{p_3}\theta_0^2 dx + \int_0^t A_0 dt ,$$

where A_0 is defined by (4.8). If we use the notation $d\rho/dt$ in (4.8), we can obtain (4.7), $\ell = 0$ from this equality. The time derivative can be treated similarly, because it has the same boundary conditions. Next compute the integral

$$\int_0^t\int_\Omega \rho_t(L^0 - f^0) + u^i_t(L^i - f^i) + \theta_t(L^4 - f^4)dxdt = 0$$

Integration by parts gives by use of Schwarz inequality

$$\int_\Omega \hat\mu|Du|^2 + (\hat\mu+\hat\mu')(u^j_{x_j})^2 + \hat\kappa|D\theta|^2 + p_1\rho u^j_{x_j} dx + \frac{1}{2}\int_0^t\int_\Omega \rho_t^2 + |u_t|^2 + \theta_t^2 dxdt$$

$$\leq \int_\Omega \hat\mu|Du_0|^2 + (\hat\mu+\hat\mu')(u^j_{0,x_j})^2 + \hat\kappa|D\theta_0|^2 + p_1\rho_0 u^j_{0,x_j} dx$$

$$+ C\int_0^t\int_\Omega |D(u,\theta)(s)|^2 + |\tilde f|^2 dxdt ,$$

where $\tilde f = (f^0 - u^j\rho_{x_j}, f^1, f^2, f^3, f^4)$. If we use Schwarz inequality for the term $\rho u^j_{x_j}$, we obtain (4.10), $\ell = 0$. The estimate (4.9), $\ell = 1$ is obtained similarly.

Since the tangential derivatives of the solution of (3.1) satisfy the same boundary conditions (3.2)(3.3), we can obtain the estimates for these similarly to the above lemma 4.4. Let us denote the tangential derivatives by $\partial = (\partial_{x_1}, \partial_{x_2})$ and integrate the equality for each $k = 1,2,3$ by use of integration by parts

$$\frac{p_1}{\bar\rho}\partial^k(L^0 - f^0)\partial^k\rho + \partial^k(L^i - f^i)\partial^k u^i = 0.$$

Thus we have

<u>Lemma 4.5.</u> For $k = 1,2,3$

$$\|\partial^k(\rho,u)(t)\|^2 + \int_0^t \|D\partial^k u(s)\|^2 + \|\partial^k \frac{d\rho}{dt}(s)\|^2 ds$$

(4.10)

$$\leq C\{\|\partial^k(\rho,u)(0)\|^2 + \int_0^t \|\partial^k f^0(s)\|^2 + \|\partial^{k-1}f(s)\|^2 + |A_{k+1}|ds\},$$

where

$$(4.11) \qquad A_{k+1} = \int_\Omega \frac{p_1}{\bar\rho}\partial^k\rho\partial^k(f^0 - u^j\rho_{x_j})dx \quad \text{for each} \quad k = 1,2,3.$$

Then we have to obtain the estimates for the normal derivatives of solution. To do that we use the following equations from (3.1).

$$(4.12) \qquad \begin{aligned} &(\tfrac{d\rho}{dt})_{x_3} + \bar\rho u^j_{x_j x_3} = f^0_{x_3}, \\ &u^3_t - \hat\mu\Delta u^3 - (\hat\mu+\hat\mu')u^j_{x_j x_3} + p_1\rho_{x_3} + p_2\theta_{x_3} = f^3 . \end{aligned}$$

If we eliminate the term $u^3_{x_3x_3}$ from these, we have

$$(4.13) \qquad \begin{aligned} \frac{2\hat\mu+\hat\mu'}{\bar\rho}(\tfrac{d\rho}{dt})_{x_3} + p_1\rho_{x_3} &= -u^3_t - p_2\theta_{x_3} + \frac{2\hat\mu+\hat\mu'}{\bar\rho}f^0_{x_3} + f^3 \\ &+ \hat\mu(u^3_{x_1x_1} + u^3_{x_2x_2}) + \hat\mu'(u^1_{x_1} + u^2_{x_2})_{x_3}, \end{aligned}$$

where we note the second derivatives of u at the last two terms on the right hand side contain one tangential derivative. Multiply (4.13) by ρ_{x_3} and $(d\rho/dt)_{x_3}$ respectively and integrate them respectively. We obtain after integration by parts

$$\begin{aligned} &\int \frac{2\hat\mu+\hat\mu'}{2\bar\rho}\rho^2_{x_3}dx + \int_0^t\int p_1\rho^2_{x_3}dxdt \\ &= \int \frac{2\hat\mu+\hat\mu'}{2\bar\rho}\rho^2_{0,x_3}dx + \int_0^t\int \frac{2\hat\mu+\hat\mu'}{2\bar\rho}(-u^j_{x_j}\rho_{x_3} + 2u^j_{x_3}\rho_{x_j})\rho_{x_3}dxdt \\ &+ \int_0^t\int \{-u^3_t - p_2\theta_{x_3} + \hat\mu(u^3_{x_1x_1} + u^3_{x_2x_2} + u^1_{x_1x_3} + u^2_{x_2x_3})\}\rho_{x_3} + \\ &\qquad + (\frac{2\hat\mu+\hat\mu'}{\bar\rho}f^0_{x_3} + f^3)\rho_{x_3}dxdt \end{aligned}$$

$$\leqq \frac{p_1}{2}\int_0^t\int \rho_{x_3}^2\,dxdt + C\{\|D\rho(0)\|^2 + \int_0^t |B_{0,0}|dt +$$

$$+ \int_0^t \|u_t\|^2 + \|D\partial u\|^2 + \|D\theta\|^2 + \|Df^0\|^2 + \|f^3\|^2 ds\ ,$$

and

$$\int_0^t\int \frac{2\hat{\mu}+\hat{\mu}'}{\bar{\rho}}\{(\frac{d\rho}{dt})_{x_3}\}^2 dxdt + \frac{p_1}{2}\int \rho_{x_3}^2\,dx$$

$$= \frac{p_1}{2}\int \rho_{0,x_3}^2\,dx + \int_0^t\int \frac{p_1}{2}(-u^j_{x_j}\rho_{x_3} + u^j_{x_3}\rho_{x_j})\rho_{x_3}\,dxdt$$

$$+ \int_0^t\int \frac{2\hat{\mu}+\hat{\mu}'}{\bar{\rho}}(\frac{d\rho}{dt})_{x_3}\{-u^3_t - p_2\theta_{x_3} + \hat{\mu}(u^3_{x_1x_1} + u^3_{x_2x_2} + u^1_{x_1x_3} + u^2_{x_2x_3})$$

$$+ \frac{2\hat{\mu}+\hat{\mu}'}{\bar{\rho}} f^0_{x_3} + f^3)dxdt$$

$$\leqq \frac{2\hat{\mu}+\hat{\mu}'}{2\bar{\rho}}\int_0^t\int (\frac{d\rho}{dt})^2_{x_3}\,dxdt + C\{\|D\rho(0)\|^2 + \int_0^t |B_{0,0}|dt$$

$$+ \int_0^t \|u_t\|^2 + \|D\theta\|^2 + \|D\partial u\|^2 + \|Df^0\|^2 + \|f^3\|^2 dxdt\}$$

respectively. Thus we have obtained the following

Lemma 4.6. For $k + \ell = 0,1,2$ it holds

$$\|\partial^k\partial_3^{\ell+1}\rho(t)\|^2 + \int_0^t \|\partial^k\partial_3^{\ell+1}\rho(s)\|^2 + \|\partial^k\partial_3^{\ell+1}(\frac{d\rho}{dt})(s)\|^2 ds$$

$$\leqq C\{\|D\rho(0)\|^2_{k+\ell} + \int_0^t \|\partial^{k+1}\partial_3^{\ell}Du\|^2 + \|\partial^k\partial_3^{\ell}u_t\|^2$$

$$+ \|D(u,\theta)(s)\|^2_{k+\ell} + \|f^0\|^2_{k+\ell+1} + \|f\|^2_{k+\ell} + |B_{k,\ell}|ds\},$$

where

$$(4.14)\qquad B_{k,\ell} \equiv \int \{\partial^k\partial_3^{\ell}(\frac{d\rho}{dt})_{x_3} - \partial^k\partial_3^{\ell}\rho_{tx_3}\}\partial^k\partial_3^{\ell}\rho_{x_3}\,dx\ ,$$

and here the summation is not taken for k and ℓ.

Last we use lemma 4.3 for Stokes equation (4.6) with $u|_{\partial\Omega} = 0$, where h and g^i have the following explicit forms.

$$h = f^0 - \frac{d\rho}{dt}$$

(4.15)

$$g^i = -u_t^i + \frac{\hat{\mu}+\hat{\mu}'}{\bar{\rho}} h_{x_i} - p_2\theta_{x_i} + f^i , \quad i = 1,2,3.$$

Lemma 4.7. For $k + \ell = 0,1,2,$ we have

(4.16)

$$\| D^{2+\ell}\partial^k u \| + \| D^{1+\ell}\partial^k \rho \| \leq C\{\| u_t \|_{k+\ell} + \| \partial^k(\frac{d\rho}{dt}) \|_{1+\ell} + \| D\theta \| + \| f^0 \|_{1+k+\ell} + \| f \|_{k+\ell}\}.$$

Now we can combine the above lemmas 4.1 ~ 4.7 to obtain necessary a-priori estimates. Although we omit the details, by combining step by step lemma 4.4, $\ell = 0$, $k = 0$, $\ell = 1$, lemma 4.2 for θ, $k = 2$ lemma 4.5, $k = 1$, lemma 4.6, $k+\ell = 0$, lemma 4.2 for u, $k = 2$, lemma 4.7, $k+\ell = 0$, lemma 4.2 for θ, $k = 3$, lemma 4.5, $k = 2$, lemma 4.6, $k = 1$, $\ell = 0$, lemma 4.7, $k = 1$, $\ell = 0$, lemma 4.6, $k = 0$, $\ell = 1$ and lemma 4.7, $k = 0$, $\ell = 1$, we can obtain the H^2 version of norm $N(0,t)$, i.e,

(4.17)

$$\| (\rho,u,\theta)(t) \|_2^2 + \| \rho_t(t) \|_1^2 + \| (u_t,\theta_t)(t) \|^2 + \int_0^t \| \rho_t(s),D\rho(s) \|_1^2 + \| (u_t,\theta_t)(s) \|_1^2 + \| D(u,\theta)(s) \|_2^2 + \| \frac{d\rho}{dt} (s) \|_2^2 ds$$

$$\leq C\{\| \rho_0,u_0,\theta_0 \|_2^2 + \sup_{0\leq s\leq t} \{\| f^0-u^j\rho_{x_j} \|_1^2, \| f \|^2\} + \int_0^t \| f^0 \|_2^2 + \| f_t^0 \|^2 + \| f \|_1^2 + \| u^j\rho_{x_j} \|_1^2 + \sum_{k=0}^{3} |A_k| + \sum_{k+\ell=0}^{1} |B_{k,\ell}| ds\}.$$

To elevate the differentiability once to obtain the estimate of norm $N(0,t)$ we can repeat the above argument beginning from lemma 4.4, $k = 1$ and by use of lemma 4.2, $k = 3$, lemma 4.5, $k = 3$, lemma 4.6, $k+\ell = 2$ and lemma 4.7, $k+\ell = 2$. Therefore we arrive at the estimate for $N(0,t)$.

$$N(0,t)^2 \equiv \|\rho,u,\theta(t)\|_3^2 + \|\rho_t(t)\|_2^2 + \|u_t,\theta_t(t)\|_1^2$$

$$+ \int_0^t \|\rho_t, D\rho(s)\|_2^2 + \|u_t,\theta_t(s)\|_2^2 + \|D(u,\theta)(s)\|_3^2 ds$$

$$(4.18) \qquad \leq C\{\|\rho,u,\theta(0)\|_3^2 + \sup_{0\leq s\leq t}\{\|f^0-u^j\rho_{x_j}(s)\|_2^2, \|f(s)\|_1^2\} +$$

$$+ \int_0^t \|f^0(s)\|_3^2 + \|f(s)\|_2^2 + \|f_t(s)\|^2 + \|(u^j\rho_{x_j})_t\|^2 +$$

$$+ \|u^j\rho_{x_j}\|_2^2 + \sum_{k=0}^{4}|A_k| + \sum_{k+\ell=0}^{2}|B_{k,\ell}|ds\} .$$

Last we have to show

Lemma 4.8.

$$\sup_{0\leq s\leq t}\{\|f^0-u^j\rho_{x_j}(s)\|_2^2, \|f(s)\|_1^2\} + \int_0^t \|f^0(s)\|_3^2 + \|(f^0-u^j\rho_{x_j})_t(s)\|^2 +$$

$$(4.19) \qquad + \|f_t(s)\|^2 + \|\tilde{f}(s)\|_2^2 + \sum_{k=0}^{4}|A_k| + \sum_{k+\ell=0}^{2}|B_{k,\ell}|ds$$

$$\leq C_2(N(0,t) + \|\Phi\|_5)N^2(0,t).$$

It is proved by use of lemma 4.1 and integration by parts. We show only the term A_0 and omit the proof of the other terms which can be treated similarly. Let us recall (4.9) and compute the following

$$\left|\int \rho(f^0-u^j\rho_{x_j})dx\right| = \left|\int \rho\{(\bar{\rho}-\tilde{\rho})u^j - \rho u^j\}_{x_j}dx\right| = \left|\int \rho_{x_j}\{(\bar{\rho}-\tilde{\rho})u^j - \rho u^j\}dx\right|$$

$$\leq \|D\rho\|\{(\int (\bar{\rho}-\tilde{\rho})^2|u|^2dx)^{1/2} + (\int \rho^2|u|^2dx)^{1/2}\}$$

$$\leq \|D\rho\|\{\|\bar{\rho}-\tilde{\rho}\|_{L_3}\|u\|_{L_6} + \|\rho\|_{L_3}\|u\|_{L_6}\}$$

$$\leq C\|D\rho\|\|Du\|\{\|\Phi\|_3 + \|\rho\|_1\} \leq C\, N(0,t)^2\{\|\Phi\|_3 + N(0,t)\}.$$

$$\left|\int u^i\{(\frac{\mu}{\rho+\tilde{\rho}} - \hat{\mu})u^i_{x_j x_j} + \frac{1}{\rho+\tilde{\rho}}\mu_{x_j} u^i_{x_j}\}dx\right|$$

$$\leq \int |u^i_{x_j}(\frac{\mu}{\rho+\tilde{\rho}} - \hat{\mu})u^i_{x_j}| + |u^i(\frac{1}{\rho+\tilde{\rho}})_{x_j} u^i_{x_j}|dx$$

$$\leq |\frac{\mu}{\rho+\tilde{\rho}} - \hat{\mu}|_0 \|Du\|^2 + \|(\frac{1}{\rho+\tilde{\rho}})_{x_j}\|_{L_3}\|u\|_{L_6}\|Du\|$$

$$\leq C\, N(0,t)^2\{\|\Phi\|_3 + N(0,t)\}.$$

The remaining terms in A_0 can be treated in the same way as above.

Finally we note that the inequalities (4.18)(4.19) easily imply the desired a-priori estimates, in fact, we may choose ε_1 so small that

$$N^2(0,t) \leq C\|\rho_0,u_0,\theta_0\|_3^2 \quad \text{for} \quad N(0,t),\ \|\Phi\|_5 \leq \varepsilon_1 .$$

Thus the proof of Theorem 2 for the half space is completed.

References

[1] Agmon, S., Douglis, A. and Nirenberg, L., Estimates near the boundary for solutions of elliptic partial differential equations satisfying general boundary conditions II, Comm. Pure Appl. Math., 17 (1964) 35-92.

[2] Cattabriga, Su un problema al contorno relativo al sistema di equazioni di Stokes, Rend. Mat. Sem. Univ. Padova, 31 (1961) 308-340.

[3] Finn, R., On the exterior stationary problem for the Navier-Stokes equations, and associated perturbation problems, Arch. Rat. Mech. Anal., 19 (1965) 363-406.

[4] Heywood, J.G., A uniqueness theorem for non-stationary Navier-Stokes flow past an obstacle, Ann. Scuola Norm. Sup. Pisa IV 6 (1979) 427-445.

[5] Kawashima, S. and Nishida, T., Global solutions to the initial value problem for the equations of one-dimensional motion of viscous polytropic gases, J. Math. Kyoto Univ., 21 (1981) 825-837.

[6] Ladyzhenskaya, O.A., The Mathematical Theory of Viscous Incompressible Flow, Gordon and Breach, New York, 1969.

[7] Matsumura, A., An energy method for the equations of motion of compressible viscous and heat-conductive fluids, Univ. of Wisconsin-Madison, MRC Technical Summary Report #2194, 1981.

[8] Matsumura, A. and Nishida, T., The initial value problem for the equations of motion of compressible viscous and heat-conductive fluids, Proc. Japan Acad. Ser. A, 55 (1979) 337-342.

[9] ....., Initial boundary value problems for the equations of motion of general fluids, Proc. of 5th Internat. Symp. on Computing Methods in Appl. Sci. and Engin., Dec. 1982, INRIA, Versailles, France, (1982) 389-406.

[10] ....., Initial boundary value problems for the equations of motion of compressible viscous and heat-conductive fluids, to appear in Communications in Mathematical Physics.

[11] Okada, M. and Kawashima, S., On the equations of one-dimensional motion of compressible viscous fluids, to appear in J. Math. Kyoto Univ.

[12] ....., Smooth Global Solutions to the one-dimensional equations in Magnetohydrodynamics, to appear in Proc. Japan Acad.

[13] Solonnikov, V.A., A priori estimates for certain boundary value problems, Sov. Math. Dokl., 2 (1961) 723.

[14] Tani, A., On the first initial-boundary value problem of compressible viscous fluid motion, Publ. RIMS, Kyoto Univ., 13 (1977) 193-253.

Lecture Notes in Num. Appl. Anal., **5**, 171–188(1982)
Nonlinear PDE in Applied Science. U.S.-Japan Seminar, Tokyo, 1982

Integral Representation of Solutions for Equations of Mixed Type in a Half Space

Sadao Miyatake

Department of Mathematics
Faculty of Science
Kyoto University
Kitashirakawa Sakyo-ku
Kyoto 606
JAPAN

We consider boundary value problems for equations of mixed type $u_{xx} + q(x)u_{tt} = 0$ in a half space $(0,\infty) \times R$. We discuss the existence of solutions written in the integral form $u(x, t) = \int e^{i\tau t} v(x, \tau^2)\hat{g}(\tau)d\tau$, the estimate for $v(x, \alpha)$ and function spaces for $g(t)$ and $u(x, t)$. We put $w(x, \alpha) = v'(x, \alpha)/v(x, \alpha_2)$ and consider non-linear equation of type $w' = q(x)\alpha - w^2$ in Riemann sphere, where a topological method is useful. As for estimates we need a special device concerning the energy method.

§1. Introduction and statements of results.

This note is concerned with the following problem

$$\text{(P)} \qquad \begin{cases} u_{xx} + q(x)u_{tt} = 0 \ , & \text{for } (x, t) \in (0,\infty) \times R \\ u(0,t) = g(t) \text{ and } \lim\limits_{x\to\infty} u(x, t) = 0, & \text{for } t \in R, \end{cases}$$

where the coefficient $q(x)$ depends only on x and satisfies

(C) $q(x)$ is a real valued piecewise continuous bounded function satisfying

$\lim\limits_{x\to\infty} q(x) > 0.$

Remark that $q(x)$ may change its sign.

The boundary data $g(t)$ is assumed to be written in a form

$$(1) \qquad g(t) = \frac{1}{2\pi} \int_\Gamma e^{i\tau t}\hat{g}(\tau)d\tau \ .$$

Namely $g(t)$ is a summation of exponential function $e^{i\tau t}$ with density function $\hat{g}(\tau)$ defined on complex path Γ. Γ will be defined later. Naturally the formula (1) is not anything but Laplace inversion formula if $g(t)$ and Γ satisfy suitable conditions. From the linear property of (P) we seek the solution given by

(2) $$u(x, t) = \frac{1}{2\pi}\int_{\Gamma} e^{i\tau t} v(x, \tau^2)\hat{g}(\tau)d\tau .$$

Here $v(x, \alpha)$ is a solution of

$(\tilde{P})$ $$\begin{cases} v_{xx} - q(x)\alpha v = 0 \quad \text{for } x \in (0, \infty) \\ v(o, \alpha) = 1 \quad \text{and} \quad v(\infty, \alpha) = 0. \end{cases}$$

We have denoted $\alpha = \tau^2$. Thus Γ is a curve in complex plane such that $\Gamma^2 = \{\alpha: \alpha = \tau^2, \ \tau \in \Gamma\}$ is involved in the following domain $\mathcal{D}(q)$ which we describe in Theorem 1

Theorem 1. Suppose (C). The problem $(\tilde{P})$ has unique solution $u(x, \alpha)$ if and only if α belongs to the following set

$$\mathcal{D}(q) = \mathbb{C} - \left\{\bigcup_{j=1}^{\infty}\{d_j\} \cup (-\infty, 0]\right\},$$

where d_j, $(0 < d_1 < d_2 < \ldots)$ are positive constants tending to ∞ if $q(x)$ changes its sign, while $\bigcup_{j=1}\{d_j\}$ is replaced by empty set ϕ if $q(x) \geq 0$ in $(0, \infty)$. Moreover $v(x, \alpha)$ is analytic in $\alpha \in \mathcal{D}(q)$ for each $x \in [0, \infty)$, each d_j being a simple pole. Finally $v(x, \alpha)$ is bounded in a neighbourhood of $\alpha = 0$ for $\alpha > 0$.

Remark 1. If $q(x) \geq 0$ for $x \in (0, \infty)$, we can take Γ as real axis. In case where $q(x) \equiv 1$, (2) becomes Poisson formula by Fubini theorem with $v(x, \tau^2) = e^{-|\tau|x}$ and $\hat{g}(\tau) = \int_{-\infty}^{\infty} e^{-i\tau s} g(s)ds$.
If $q(x)$ changes its sign, we take Γ as a curve satisfying $\Gamma^2 \subset \mathcal{D}(q) \cup \{0\}$. For example Γ coincides real axis in a neighbourhood of origin and another parallel line in a neighbourhood of infinity.

Next we estimate $v(x, \alpha)$ in order to give exact meanings to the formula (2) in function spaces.

Theorem 2. Suppose (C). Then for any $\gamma > 0$, there exists positive constant C_γ such that we have

(E) $|v(x, \tau^2)| \leq C_\gamma (|\tau| + 1)^{5/2}$, if $|\mathrm{Im}\,\tau| = \gamma > 0$.

In general we will see $C_\gamma = 0(1/\gamma^2)$ through the process of proofs. By Theorem 2 we can discuss function spaces for $g(t)$ and $u(x, t)$.

At first we introduce function spaces $L^2_{\gamma,k}$, $L^1_{\gamma,k}$, $\tilde{L}^2_{\gamma,k}$, $\tilde{L}^1_{\gamma,k}$ and $\tilde{B}_{\gamma,k}$ with following norms respectively.

$$\| g \|_{L^p_{\gamma,k}} = \sum_{j=1}^{k} \left\{ \int_{-\infty}^{0} |e^{-\gamma s} \frac{d^j}{dt^j} g(s)|^p ds + \int_0^\infty |\frac{d^j}{dt^j} g(s)|^p ds \right\}^{1/p}$$

$$\| h \|_{\tilde{L}^p_{\gamma,k}} = \sum_{j=1}^{k} \left\{ \int_{-\infty}^{0} |\frac{d^j}{dt^j} h(s)|^p ds + \int_0^\infty |e^{-\gamma s} \frac{d^j}{dt^j} h(s)|^p ds \right\}^{1/p},$$

$(p = 1, 2)$.

$$\| h \|_{\tilde{B}_{\gamma,k}} = \sum_{j=1}^{k} \left\{ \sup_{s \in (-\infty,0)} |\frac{d^j}{dt^j} h(s)| + \sup_{s \in (0,\infty)} |e^{-\gamma s} \frac{d^j}{dt^j} h(s)| \right\},$$

$(k = 0, 1, 2, \ldots)$.

If $0 < \gamma_1 < \gamma_2$, then it follows

$$L^p_{\gamma_2,k} \subset L^p_{\gamma_1,k} \subset H^p_k \subset \tilde{L}^p_{\gamma_1,k} \subset \tilde{L}^p_{\gamma_2,k}.$$

Then we fix a smooth path $\Gamma = \Gamma_1 + \Gamma_2 + \Gamma_3 + \Gamma_4$, where

$$\Gamma_1 = \{ \tau : \mathrm{Im}\,\tau = 0, \ -a \leq \mathrm{Re}\,\tau \leq a \}, \quad a < d_1^{1/2}$$

$$\Gamma_2 = \{ \tau : -\mathrm{Im}\,\tau = \gamma, \ |\mathrm{Re}\,\tau| \geq 2a \},$$

$$\Gamma_3 \subset \{ \tau : 0 < -\mathrm{Im}\,\tau < \gamma, \ a < \mathrm{Re}\,\tau < 2a \},$$

$$\Gamma_4 \subset \{ \tau : 0 < -\mathrm{Im}\,\tau < \gamma, \ -2a < \mathrm{Re}\,\tau < -a \}.$$

Denote by $\bar{\Gamma}$ the curve $\{\tau : \bar{\tau} \in \Gamma\}$. Remark that

$$g(t) = \frac{1}{2\pi} \int_\Gamma e^{i\tau t} \hat{g}(\tau) d\tau, \quad \text{if } g(t) \in L^1_{\gamma,2},$$

where $\hat{g}(\tau) = \int_{-\infty}^{\infty} e^{-i\tau t} g(t) dt$. If $g(-t) \in L^1_{\gamma,2}$,

$g(t) = \frac{1}{2\pi} \int_{\bar{\Gamma}} e^{i\tau t} \hat{g}(\tau) d\tau$. Now let us put

$$u_+(x, t) = \frac{1}{2\pi} \int_{\Gamma} e^{i\tau t} v(x, \tau^2) \hat{g}(\tau) d\tau$$

$$u_-(x, t) = \frac{1}{2\pi} \int_{\bar{\Gamma}} e^{i\tau t} v(x, \tau^2) \hat{g}(\tau) d\tau$$

Using these notations we can state

Theorem 3. Suppose (C). If $g(t) \in L^1_{\gamma, k+4}$, (respectively $g(-t) \in L^1_{\gamma, k+4}$) for certain $\gamma > 0$, then $u_+(x, t)$ belongs to $\tilde{B}_{\gamma,k}$, (resp. $\check{u}_-(x, t) = u_-(x, -t) \in \tilde{B}_{\gamma,k}$), $(k = 0, 1, 2, \ldots)$. In case of $k \geq 2$, $u_+(x, t)$, (resp. $u_-(x, t)$) is a C^2 solution of (P). $u_+(x, t)$ satisfies following estimates if right hand sides are bounded.

$$(E)_1 \qquad \sup_{x \in (0,\infty)} \| u_+(x, .) \|_{\tilde{B}_{\gamma,k}} \leq C_1(\gamma, k) \| g \|_{L^1_{\gamma,k+4}},$$

$$(E)_2 \qquad \sup_{x \in (0,\infty)} \| u_+(x, .) \|_{\tilde{L}^2_{\gamma,k}} \leq C_2(\gamma, k) \| g \|_{L^2_{\gamma,k+3}},$$

where $K = 0, 1, 2, \ldots$ and $C_j(\gamma, k)$ are constants. $\check{u}_-(x, t) = u_-(x, -t)$ satisfies the same inequalities replaced $g(t)$ by $\check{g}(t) = g(-t)$.

Remark 2. The analyticity of $v(x, \tau^2)$ says that $u(x, t)$ in (2) is invariant even if we change Γ in a bounded domain which excludes $d_j^{1/2}$ and $iR - \{0\}$. Incidentally we give another remark. Let Γ be smooth closed curve such that Γ is the boundary of a bounded domain $\mathcal{D}$. Then $u(x, t)$ of (2) is in general non trivial solution of (P) with $g \equiv 0$, if $\hat{g}(\tau)$ is a given analytic function. In fact, since $\tau = \pm d_j^{1/2}$ is a simple pole of $v(x, \tau^2)$, $u(x, t)$ is described by residue calculus. This solution is of type $\sum_j e^{i(\pm d_j^{1/2})t} a_{\pm j}(x)$ and does not belong to function spaces $\tilde{L}^2_{\gamma,k}$ in Theorem 3.

As for equations of mixed type we know the Tricomi equation, for which were investigated usually local singular boundary value problems such as Tricomi problem and Frankl problem. Intensive references are found in [1], [2], and [3]. Global treatments as our problem (P) could not be found by the author. The method in this paper is continued from [4] and [5].

§ 2. Plan of proofs.

From the condition (C) there exists a positive number x_0 such that

(2, 1) $\quad q(x) > m_0 > 0$ in (x_0, ∞) and $|q(x)| < M$ for all $x \in (0, \infty)$.

If the solution $v(x,\alpha)$ of $(\tilde{P})$ satisfies $v(x,\alpha) \neq 0$, then putting

(2, 2) $\quad w(x,\alpha) = v'(x,\alpha) / v(x,\alpha)$

we see that $w(x,\alpha)$ is a solution of

$$(P)_1 \quad \begin{cases} w' = q(x)\alpha - w^2 & \text{in } (0,\infty), \\ \mathrm{Re} \int_0^{\infty} w(s,\alpha)ds = -\infty . \end{cases}$$

Conversely if $w(x,\alpha)$ is a solution of $(P)_1$ then

(2, 3) $\quad v(x,\alpha) = \exp \int_0^x w(s,\alpha)ds$

is a solution of $(\tilde{P})$ for any fixed α. Thus we consider $(P)_1$ from topological viewpoints as will be shown in lemmas in next section.

2.1. Here we expain our steps in the proofs of Theorem 1.

(I) Replacing $(0,\infty)$ by (x_0,∞), we show that there exists unique solution $w(x,\alpha)$ of $(P)_1$ for all $\alpha \in C - (-\infty, 0]$.

Put $\tilde{v}(x,\alpha) = \exp \int_{x_0}^{x} w(s,\alpha)ds$, then $\tilde{v}(x,\alpha)$ is unique solution of

$\tilde{v}''(x,\alpha) = q(x)\alpha\,\tilde{v}(x,\alpha)$ in (x_0,∞) satisfying $\tilde{v}(x_0,\alpha) = 1$ for any $\alpha \in C - (-\infty, 0]$.

(II) For any $x \in (x_0,\infty)$, $w(x,\alpha)$ is analytic in $C - (-\infty, 0]$, thus so is $\tilde{v}(x,\alpha)$.

(III) Then we extend $w(x,\alpha)$ to $[0,\infty)$ satisfying $w' = q(x)\alpha - w^2$ for any $\alpha \in C - (-\infty,\infty)$, and define $\tilde{v}(x,\alpha) = \exp \int_{x_0}^{x} w(s,\alpha)ds$.

For $\alpha \in (0,\infty)$, we extend $\tilde{v}(x,\alpha)$ satisfying $\tilde{v}'' = q(x)\alpha\tilde{v}$ in $(0,\infty)$. In this case we prove $\tilde{v}(0, d_j) = 0$, $\frac{d}{d\alpha}\tilde{v}(0, d_j) \neq 0$, $(j = 1, 2, 3, \ldots)$, $\underline{\lim}_{\alpha\downarrow 0} \tilde{v}(0,\alpha) > 0$ and $\overline{\lim}_{\alpha\downarrow 0} \tilde{v}(x,\alpha) < \infty$.

(VI) Put $v(x,\alpha) = \dfrac{\tilde{v}(x,\alpha)}{\tilde{v}(0,\alpha)}$ for $\alpha \in \mathcal{D}(q)$. Then $v(x,\alpha)$ satisfies the desired conditions.

2.2. In order to obtain the estimate in Theorem 2, we use the identity

(2, 4)
$$w(x,\alpha)|v(x,\alpha)|^2 - w(x_1,\alpha)|v(x_1,\alpha)|^2 = \int_{x_1}^{x} |v'(y,\alpha)|^2 dy + \alpha\int_{x_1}^{x} q(y)|v(y,\alpha)|^2 dy, \quad 0 \leq x_1 < x,$$

which follows from $\int v''(y,\alpha)\overline{v(y,\alpha)}dy = \alpha\int q(y)|v(y,\alpha)|^2 dy$, and (2, 2). Now we introduce following coordinates depending on α.

(2, 5)
$$w = w_\alpha \alpha + w_\beta \beta, \qquad w_\alpha = \frac{\mathrm{Im}(w\bar{\beta})}{\mathrm{Im}(\alpha\bar{\beta})}, \quad w_\beta = \frac{\mathrm{Im}(w\bar{\alpha})}{\mathrm{Im}(\beta\bar{\alpha})}$$

where β satisfies $\beta^2 = \alpha$ and $\arg\beta = \frac{1}{2}\arg\alpha$, being $0 < |\arg\alpha| < \pi$.

w_β is said simply to be β - component of w. Denote $s(x,\alpha) = -(w(x,\alpha))_\beta$. Take β - component of both sides of (2, 4), then it follows

(2, 6)
$$s(x_1,\alpha)|v(x_1,\alpha)|^2 = s(x,\alpha)|v(x,\alpha)|^2 + l_\beta\int_{x_1}^{x} |v'(y,\alpha)|^2 dy,$$

where $l_\beta = \dfrac{\mathrm{Im}\,\bar{\alpha}}{\mathrm{Im}(\beta\bar{\alpha})} = \dfrac{\mathrm{Im}\,\alpha}{|\alpha|\,\mathrm{Im}\,\beta} > 0$. Therefore

(2, 7)
$$|v(x,\alpha)| \leq \left|\frac{s(0,\alpha)}{s(x,\alpha)}\right|^{1/2} v(0,\alpha) = \left|\frac{s(0,\alpha)}{s(x,\alpha)}\right|^{1/2}.$$

Hence it suffices to estimate $s(0,\alpha)$ and $1/s(s,\alpha)$. For this purpose we can use the equation

$$s'(x,\alpha) = \mathrm{Im}(w'\bar{\alpha})/\mathrm{Im}(\alpha\bar{\beta}) = -\mathrm{Im}(w^2\bar{\alpha})/\mathrm{Im}(\alpha\bar{\beta}).$$

The details will be shown in next section.

2.3. Here we prove Theorem 3 in view of Theorem 1 and 2. The estimate $(E)_1$ follows directly from the estimate (E). $\lim_{x\to\infty} u(x, t) = 0$ follows from Lebesque theorem. To obtain $(E)_2$ we use the following localization of $u(x, t)$: $u(x, t) = \sum_{j=1}^{4} u_j(x, t)$, where

$$u_j(x, t) = \frac{1}{2\pi}\int_\Gamma e^{i\tau t}\chi_j(\tau)v(x, \tau^2)\hat{g}(\tau)d\tau$$

$$= \frac{1}{2\pi}\int_{-\infty}^{\infty}\left(\int_\Gamma e^{i\tau(t-s)}\chi_j(\tau)v(x, \tau^2)d\tau\right)g(s)ds$$

Here $\chi_j(\tau)$ are smooth functions defined on Γ as follows.

$\chi_j(\tau) \geq 0$. $\sum_{j=1}^{4}\chi_j(\tau) \equiv 1$ on Γ. supp $\chi_1 \subset (-a, a)$. $\chi_1 \equiv 1$ in a neighbourhood of origin. supp $\chi_2 \subset \{\tau\in\Gamma, \ |\mathrm{Re}\,\tau| > 2a\}$

supp $\chi_3 \subset \{\tau\in\Gamma, \ -3a < \mathrm{Re}\,\tau < -\frac{1}{2}a\}$. supp $\chi_4 \subset \{\tau\in\Gamma, \ \frac{1}{2}a < \mathrm{Re}\,\tau < 3a\}$.

The estimates for $u_1(x, t)$ and $u_2(x, t)$ follow from Plancherel's theorem. In fact we have $\| u_1(x, t)\|_{L^2} \leq C \| g(t) \|_{L^2}$ and

$\| e^{-\gamma t} u_2(s, t) \|_{L^2} \leq \sum_{j=0}^{3} \| e^{-\gamma t} \frac{\partial^j}{\partial t^j} g(t) \|_{L^2}$ for all $x \in [0,\infty)$

Thus $(E)_2$ holds for k=o and similarly for $k \geq 1$.

As for $u_3(x, t)$ and $u_4(x, t)$ we use integration by parts in view of smoothness of $v(x, \tau^2)$ on the support of $\chi_j(\tau)$.

In fact for j = 3 and 4 , $u_j(x, t) = \frac{1}{2\pi}\int_{-\infty}^{\infty} K_j(t, s)\, g(s)ds$, where

$$K_j(t, s) = \frac{1}{(t-s)^k}\int_\Gamma e^{i\tau(t-s)}\left(i\frac{\partial}{\partial\tau}\right)^k(\chi_j(\tau)\, v(x, \tau^2))d\tau \ , \quad (k = 0, 1, 2, \ldots). \tag{2, 8}$$

Here we put, for convenience,

$$e(t, \gamma) = \begin{cases} e^{-t}, & t < 0 \\ 1, & t > 0, \end{cases} \qquad \tilde{e}(t, \gamma) = \begin{cases} 1, & t < 0 \\ e^{-t}, & t > 0. \end{cases} \tag{2, 9}$$

Then we have $\|g\|^2_{L^2_\gamma,k} = \sum_{j=0}^{k} \|e(.,\gamma)\frac{d^j}{dt^j}g\|_{L^2}$, etc, and

$$(2,10)\quad \begin{cases} (1)\ \tilde{e}(t,\gamma) \leq 1 \leq e(t,\gamma),\ \tilde{e}(t,\gamma) \leq e^{-\gamma t} \leq e(t,\gamma) \\ (2)\ e(t,\gamma)^{-1} = \tilde{e}(-t,\gamma) \\ (3)\ |e^{i\tau t}e(t,\gamma)| \leq 1 \quad \text{for } \tau \in \Gamma. \end{cases}$$

Denote $\tilde{K}_j(t, s) = \tilde{e}(t,\gamma)\, K_j(t, s)\, e(s,\gamma)^{-1}$. Then we have from (2, 8) and (2, 10)

$$(2,\ 11)\qquad |\tilde{K}_j(t, s; x)| \leq \frac{C}{|t-s|^2+1}$$

From (2, 11) we have $\|\tilde{K}_j(t, .; x)\|_{L^1} \leq C\pi$ and $\|\tilde{K}_j(., s; x)\|_{L^1} \leq C\pi$ for any fixed t, s and x. Since

$$(2,\ 12)\quad \tilde{e}(t,\gamma)\, u_j(x, t) = \int_{-\infty}^{\infty} \tilde{K}_j(t, s; x)\left\{e(s,\gamma)\, g(s)\right\} ds$$

we have $\|\tilde{e}(.,\gamma)\, u_j(x, .)\|_{L^2}$

$$\leq \int \|\tilde{K}_j(t,.; x)\|_{L^1} \left(\int |\tilde{K}_j(t, s; x)|\, |e(s,\gamma)g(s)|^2 ds\right)dt$$

Therefore $\|\tilde{e}(.,\gamma)\, u_j(x, .)\|_{L^2} \leq C^2\pi^2 \|e(.,\gamma)\, g(.)\|_{L^2}$, which means (E_2) for $k = 0$. In the same way we have $(E)_2$ for $k \geq 1$. (q.e.d.)

§3. Detailed proofs and viewpoints.

At first we prove

Lemma 1. Let $\tilde{q}(x)$ be complex valued piecewise continuous function satisfying $\mathrm{Re}\,\tilde{q}(x) > 0$ for all $x \in [0,\infty)$. Then there exists a solution $w(x)$ of $w' = \tilde{q}(x) - w^2$ satisfying $\mathrm{Re}\, w(x) < 0$ for all $x \in [0,\infty)$.

Corollary. Suppose the same conditions as in Lemma 1. Then $u'' = \tilde{q}(x)u$ has a solution $u(x)$ satisfying $|u(x)| > |u(x_1)|$ if $0 \leq x < x_1$.

Proof of Lemma 1. From $w' = \tilde{q}(x) - w^2$ follws

$$(3.1)\qquad \tilde{w}' = 1 - \tilde{q}(x)\tilde{w}^2$$

if we put $\tilde{w} = 1/w$ in the case where $w \neq 0$. Thus we regard $w' = q(x) - w^2$ as a differential equation on Riemann sphere with two local coordinates :

$$\tilde{C} = \{w : w \in C\} \cup \{\tilde{w} : \tilde{w} = 1/w \in C\}.$$

Put $\Omega = \{w : \mathrm{Re}\, w < 0\} \cup \{\tilde{w} : \mathrm{Re}\, \tilde{w} < 0\}$, then Ω is compact. Remark that the vector $\tilde{q}(x) - w^2$ faces the exterior on $\partial\Omega$, i.e.

$\mathrm{Re}\,(\tilde{q}(x) - w^2) > 0$ and $\mathrm{Re}\,(1 - \tilde{q}(x)\tilde{w}^2) > 0$ for all $w \in \partial\Omega$ and $x \in [0,\infty)$.

Now assume that for every $w_0 \in \Omega$ the solution $w(t)$ of $w' = \tilde{q}(x) - w^2$ with $w(0) = w_0$ goes out of Ω. Namely there exists a positive number x_0 uniquely such that $w(x) \in \Omega$ in $(0, x_0)$, $w(x_0) \in \partial\Omega$ and $w(x) \in C\bar{\Omega}$ in (x_0, ∞). Denote by f the mapping from $\bar{\Omega}$ to $\partial\Omega$: $w_0 \to w(x_0)$. Then f is continuous and $f|_{\partial\Omega}$ is identity mapping. This contradicts to well-known property of the continuous function on the contractible domain. (q.e.d.)

Identifying C and R^2, we can regard $w' = \tilde{q}(x) - w^2$ and $\tilde{w}' = 1-\tilde{q}(x)\tilde{w}^2$ as a system of differential equations with values on a real compact two dimensional manifold. The above proof gives directly following ststements

Lemma A. Let M be a real manifold of n-dimensions. $U' = Q(x, U)$ is a differential equation with values on M, where $Q(x, U)$ is piecewise continuous on x and smooth on U. Let Ω be an open contractible set in M such that $\bar{\Omega}$ is compact and $\partial\Omega$ is a piecewise smooth hypersurface i.e. a union of a finite number of parts of hypersurfaces. Assume that on $Q(x, U)$ faces the exterior strictly i.e. $\nu.Q(x, U) > 0$ for all $x \in [0,\infty)$ and $U \in \partial\Omega$, where ν stands for unit outer normals of $\partial\Omega$. Then there exists at least one solution $U(x)$ of $U' = Q(x, U)$ such that $U(x) \in \Omega$ for all $x \in [0,\infty)$.

Evidently we have

Lemma B. Suppose same conditions as in Lemma A, replacing $\nu.Q(x, U) > 0$ by $\nu.Q(x, U) < 0$. Then from $U(0) \in \Omega$ follows

$U(x) \in \Omega$ for all $x \in (0, \infty)$.

As for Lemma A we can find general statements in n dimensional space of T. Wazewski [6] by virtue of retract method, (cf. for example [7]). Here we state another direct extension, which is useful in our reasoning.

Lemma $\tilde{A}$. Suppose the same conditions on M, $U' = Q(x, U)$ and replacing $\nu.Q(x, U) > 0$ by $\nu.Q(x, U) \geq 0$ Then there exists at least one solution $U(x)$ of $U' = Q(x, U)$ such that $U(x) \in \bar{\Omega}$ for all $x \in [0, \infty)$. At that time we have the alternative as follows : $U(x)$ stays in Ω for all $x \in [0, \infty)$ or $U(x)$ belongs to $\partial\Omega$ for all $x \in [x_1, \infty)$ for some $x_1 > 0$.

Proof of Lemma $\tilde{A}$. Take a smooth vector field $Q_1(U)$ satisfying $\nu.Q_1(U) > 0$ on $\partial\Omega$. Let $U_\varepsilon(x)$ be a solution of $U' = Q(x, U) + \varepsilon Q_1(U)$ such that $U_\varepsilon(x) \in \Omega$ for all $x \in [0, \infty)$. Since $\bar{\Omega}$ is compact $U_{\varepsilon_j}(0)$ has a limit U_0 as a suitabe sequence ε_j tends to zero. The solution $U(x)$ of $U' = Q(x, U)$ with $U(0) = U_0$ satisfies $U(x) \in \bar{\Omega}$ for $x \in [0, \infty)$ since $U(x) = \lim U_{\varepsilon_j}(x)$.
In the same way we have the above alternative.

Remark In our problems the latter case of the above alternative does not occur.

Now we make Lemma 1 more precise.

Lemma 2. Let $\tilde{q}(x)$ satisfy $0 < m < \mathrm{Re}\, \tilde{q}(x)$ and $|\tilde{q}(x)| < M$. Then there exists a solution $w(x)$ of $w' = \tilde{q}(x) - w^2$ satisfying $\mathrm{Re}\, w < -m^{1/2}$ and $|w| < (2M)^{1/2}$ for all $x \in [0, \infty)$. This solution $w(x)$ is unique one satisfying $\mathrm{Re}\, w(x) < 0$ for $x \in [0, \infty)$.

Proof. Let Ω be an open bounded set in $\{ w \in C : \mathrm{Re}\, w < 0 \}$ surrounded by the following S_1, S_2 and S_3 : $S_1 = \{ w : \mathrm{Re}\, w = -m^{1/2} \}$, $S_2 = \{ \tilde{w} = 1/w \ : \ \tilde{w} = t(-\varepsilon_1) + (1 - t)i\varepsilon_1 \ , \ 0 \leq t \leq 1 \}$ and $S_3 = \{ \tilde{w} = 1/w \ : \ \tilde{w} = t(-\varepsilon_1) - (1 - t)i\varepsilon_1 \ , \ 0 \leq t \leq 1 \}$, where $\varepsilon_1 = (2M)^{-1/2}$.

Now we identify Riemann sphere and real compact manifold of two dimension. Then Ω and $w' = \tilde{q}(x) - w^2$ satisfy all the conditions in Lemma A, since we can show $\operatorname{Re}(\tilde{q}(x) - w^2) > 0$ on S_1 and

$$(3.2)\quad |\tan\arg(1 - \tilde{q}(x)\tilde{w}^2)| < \frac{M\varepsilon_1^2}{1-M\varepsilon_1^2} = 1 \quad \text{on } S_2 \cup S_3 .$$

The uniqueness follows if we apply Limma B to $w' = \tilde{q}(x) - w^2$ replacing Ω by $-\Omega$, and take account of the linear property of solutions of $v'' = q(x)v$ given by (2.3). (q.e.d.)

Now we proceed to our case where $\tilde{q}(x) = q(x)\alpha$, $q(x)$ being a real valued function. Then we have more precise results.

Lemma 3. Suppose that a real valued function $q(x)$ satisfies $0 < m_0 < q(x) < M_0$ and that α is a number satisfying $\operatorname{Re}\alpha > 0$ and $\operatorname{Im}\alpha > 0$. Then there exists a solution $w(x)$ of $w' = q(x)\alpha - w^2$ satisfying $\operatorname{Re} w < -(m_0 \operatorname{Re}\alpha)^{1/2}$, $|w| < (2M_0|\alpha|)^{1/2}$, $\operatorname{Im} w < 0$ and

$$(3.3)\qquad (m_0\operatorname{Re}\alpha)^{1/2}(\operatorname{Re}\beta)^{-1} \le s(x,\alpha) \equiv \frac{\operatorname{Im}(w(x,\alpha)\bar{\alpha})}{\operatorname{Im}(\alpha\bar{\beta})}$$

for all $x \in [0,\infty)$, where $\beta = \alpha^{1/2}$, $\arg\beta$ being $\frac{1}{2}\arg\alpha$.

At first we must give an explanation of $s(x,\alpha)$. Remark that any complex number w is described uniquely

$$(3.4)\qquad w = r\alpha - s\beta,$$

where r and s are real numbers. We have

$$(3.4)'\qquad r = \frac{\operatorname{Im}(w\bar{\beta})}{\operatorname{Im}(\alpha\bar{\beta})} \quad \text{and} \quad s = \frac{\operatorname{Im}(w\bar{\alpha})}{\operatorname{Im}(\alpha\bar{\beta})} .$$

In order to show $w(x,\alpha) \ne 0$, we will verify $s(x,\alpha) > 0$ in our arguments

Proof of Lemma 3. Let Ω be an open bounded domain in $\{w : \operatorname{Re} w < 0\}$ surrounded by S_1, S_2, $S_3 = \{w : \operatorname{Im} w = 0\}$ and $S_4 = \left\{w : (m_0\operatorname{Re}\alpha)^{1/2}(\operatorname{Re}\beta)^{-1} = \frac{\operatorname{Im}(w\bar{\alpha})}{\operatorname{Im}(\alpha\bar{\beta})}\right\}$, where S_1 and S_2 are the same ones as in the proof of Lemma 2 with $m = m_0 \mathrm{Re}\,\alpha$ and $M = M_0 |\alpha|$. Then we can apply Lemma A as in the case of Lemma 2. Thus we have Lemma 3. (q.e.d.)

Now return to our assumption (2.1) and apply Lemma 2 and Lemm 3 replaced $(0,\infty)$ by (x_0,∞), then $w' = q(x)\alpha - w^2$ has a solution satisfying, for $x \in [x_0,\infty)$ and $\mathrm{Re}\,\alpha > 0$,

$$(3.5)\quad \begin{cases} \text{i)} \quad \mathrm{Re}\, w < -(m_0 \mathrm{Re}\,\alpha)^{1/2}, \quad |w| < (2M|\alpha|)^{1/2} \\ \text{ii)} \quad (m_0 \mathrm{Re}\,\alpha)^{1/2}(\mathrm{Re}\,\beta)^{-1} < s(x,\alpha) < (2M|\alpha|)^{1/2} \dfrac{\mathrm{Im}\,\alpha}{\mathrm{Im}(\alpha\bar{\beta})} \quad \text{if } \mathrm{Im}\,\alpha \neq 0. \end{cases}$$

In fact (3.5) ii) follows from (3.4)' and $-1 = \dfrac{\mathrm{Im}\,\beta}{\mathrm{Im}(\alpha\bar{\beta})}\alpha - \dfrac{\mathrm{Im}\,\alpha}{\mathrm{Im}(\alpha\bar{\beta})}\beta$.

Here we consider the case where $\mathrm{Re}\,\alpha \leq 0$ and $\mathrm{Im}\,\alpha \neq 0$. At first we state

<u>Lemma 4.</u> Let $\tilde{q}(x)$ be a bounded piecewise continuous function satisfying $\mathrm{Im}\,\tilde{q}(x) \leq -\delta < 0$, (respectively $\mathrm{Im}\,\tilde{q}(x) \geq \delta > 0$) for all $x \in [0,\infty)$. Then there exists a solution $w(x)$ of $w' = q(x) - w^2$ satisfying $\mathrm{Im}\, w(x) > \delta_1 > 0$, (resp. $\mathrm{Im}\, w(x) < -\delta_1 < 0$) and $|w| < M$ for some constants δ_1 and M. The solution is unique one satisfying $\mathrm{Im}\, w(x) > 0$, (resp. $\mathrm{Im}\, w(x) < 0$) for all $x \in [0,\infty)$.

<u>Proof.</u> Let Ω be an open bounded domain in $\{w : \mathrm{Im}\, w > 0\}$ surrounded by S_1, S_2, S_3 and S_4, where $\varepsilon_i > 0$, $(i = 1, 2, 3)$ and

$$S_1 = \left\{\tilde{w} = \frac{1}{w} : \tilde{w} = v(t;\varepsilon_1,\varepsilon_2), \quad 0 \leq t \leq 1\right\}$$

$$S_2 = \left\{\tilde{w} = \frac{1}{w} : \tilde{w} = (s, -\varepsilon_1), \quad s \leq -\varepsilon_1 \cot\theta\right\}$$

$$S_3 = \left\{\tilde{w} = \frac{1}{w} : \tilde{w} = (r, -\varepsilon_2), \quad \varepsilon_2 \cot\theta \leq r\right\}$$

$$S_4 = \left\{w : w = (t, \varepsilon_3), \quad t_1 \leq t \leq t_2\right\}.$$

Here θ, t_1, t_2 and $v(t;\varepsilon_1,\varepsilon_2)$ satisfy the followings:

(i) $\tan 2\theta = -\sup(\operatorname{Im} \tilde{q}(x))/\sup |\tilde{q}(x)|$

(ii) $v(t;\ \varepsilon_1, \varepsilon_2) = t\, \varepsilon_1(-\cot\theta - i) + (1-t)\, \varepsilon_2(\cot\theta - i)$

(iii) $\operatorname{Im}(t_1 + i\varepsilon_3)^{-1} = -\varepsilon_1 \quad , \quad t_1 < 0$

(iv) $\operatorname{Im}(t_2 + i\varepsilon_3)^{-1} = -\varepsilon_2 \quad , \quad t_2 > 0.$

First we see that $S_2 \cap \partial\Omega$ and $S_3 \cap \partial\Omega$ satisfy the condition in Lemma $\tilde{A}$. Take ε_1 sufficiently small and put $\varepsilon_2 = 2\varepsilon_1$, so that $S_1 \cap \partial\Omega$ satisfies the condition. Finally we can choose positive constant ε_3 such that $\tilde{q}(x) - w^2$ faces the exterior on $\partial\Omega$ for all $x \in [0,\infty)$. The uniqueness holds from the linear property of solutions of $v'' = \tilde{q}(x)\, v$ as follows. The integration by parts of $(v'' - \tilde{q}(x)v)\bar{v}$ yields

(3.6)
$$w(x_2)\,|v(x_2)|^2 = w(x_1)\,|v(x_1)|^2 + \int_{x_1}^{x_2} |v'(s)|^2 ds + \int_{x_1}^{x_2} \tilde{q}(s)|v(s)|^2 ds.$$

Take the imaginary part of (3.6) with $v(x) = \exp\int_0^x w(s)ds$ and $v_1(x) = \exp\int_0^x w_1(s)ds$, where $w_1(x)$ is a solution staying in $-\Omega$. Then we see

(3.7) $$\lim_{x\to\infty} |v(x)| = 0 \quad \text{and} \quad \int_{x_1}^{\infty} |v(s)|^2\, ds < \infty \ .$$

and $\lim_{x\to\infty} |v_1(x)| = \infty$. If there exists another solution $w_2(x)$ of $w' = \tilde{q}(x) - w^2$ such that $\operatorname{Im} w_2 > 0$ for $x \in [0,\infty)$, then from (3.6) $v_2(x) = \exp\int_0^x w_2(s)ds$ must satisfy $\int_{x_1}^{x} |v(s)|^2 ds < \infty$, which is a contradiction since the solution space is two dimensional. (q.e.d.)

Therefore we have unique solution $w(x,\alpha)$ of $w' = q(x)\alpha - w^2$ in (x_0, ∞) for all $\alpha \in C - (-\infty, 0]$ satisfying

(3.8) $\lim_{x\to\infty} \tilde{v}(x,\alpha) = 0$, where $\tilde{v}(x,\alpha) = \exp\int_{x_0}^{x} w(s,\alpha)ds,\quad x > x_0$.

Here we have

Proposition 1. Suppose (2.1) Then above $w(x,\alpha)$ is analytic in $C - (-\infty, 0]$ for any fixed $x \in [x_0, \infty)$.

Proof. At first let us see that $w(x_0, \alpha)$ is continuous. In fact it suffices to apply Heine-Borel theorem on a compact set $\{w : \mathrm{Re}\, w < 0\} \cup \{\tilde{w} : \tilde{w} = 1/w ,\ \mathrm{Re}\, \tilde{w} < 0\} - \{w : |w(x_0, \alpha_0) - w| < \varepsilon\}$, in case where $\mathrm{Re}\,\alpha > 0$. If $\mathrm{Re}\,\alpha \leq 0$ and $\mathrm{Im}\,\alpha \neq 0$ we define the above compact set similarly using $\mathrm{Im}\, w$ instead of $\mathrm{Re}\, w$. We use only the uniqueness of $w(x,\alpha)$ and the continuity of solutions for data and parameter α. To prove analyticity we put

$$w_h(x,\alpha) = (w(x,\alpha + h) - w(x,\alpha))/h \quad ,$$

which satisfies $\frac{d}{dx} w_h(x,\alpha) = q(x) - \{w(x, \alpha + h) - w(x,\alpha)\}\, w_h(x,\alpha)$.

Thus it follows

$$(3.9)\qquad w_h(x,\alpha) = -\{\tilde{v}(x, \alpha + h)\tilde{v}(x,\alpha)\}^{-1}\int_x^{\infty} \tilde{v}(s, \alpha + h)\tilde{v}(s,\alpha)q(s)ds.$$

Therefore we can prove from (3.6) and (3.7)

$$(3.10)\qquad \frac{\partial}{\partial\alpha} w(x,\alpha) = -\tilde{v}(x,\alpha)^{-2}\int_x^{\infty} \tilde{v}(s,\alpha)^2 q(s)ds. \qquad \text{(q.e.d.)}$$

Evidently $\tilde{v}(x,\alpha)$ is analytic in α for all $x \in [x_0,\infty)$. Now we extend $w(x,\alpha)$ to $x \in [0,\infty)$ as a solution of $w' = q(x)\alpha - w^2$. In order to show this possibility we state

Lemma 5. Suppose (C). Then for all $\alpha \in \{\alpha : \mathrm{Im}\,\alpha \neq 0\}$ $w' = q(x)\alpha - w^2$ has unique bounded solution $w(x,\alpha)$ satisfying

$$(3.11)\qquad \mathrm{Im}\, w(x,\alpha)\bar{\alpha} \gtrless 0 \quad \text{for all } x \in [0,\infty) \quad \text{if } \mathrm{Im}\,\alpha \gtrless 0 .$$

Proof. It suffices to prove in the case $\mathrm{Im}\,\alpha > 0$, since $\bar{w}' = q(x)\bar{\alpha} - \bar{w}^2$ follows from $w' = q(x)\alpha - w^2$. Let Ω be

$\{w : \operatorname{Im} w\bar{\alpha} > 0\} \cup \{\tilde{w} = 1/w ; \operatorname{Im} \tilde{w}\alpha < 0\}$. Then we can apply Lemma $\tilde{A}$ as before. The uniqueness follows from Lemmas 4, and Lemma B. (q.e.d.)

Taking account of Lemma 3 and the uniqueness in Lemma 2 and Lemma 4 we see that $w(x,\alpha)$ in Lemma 5 is the desired extension. Thus

$\tilde{v}(x,\alpha) = \exp \int_{x_0}^{x} w(s,\alpha)ds$ is bounded, non zero and analytic in α for

$x \in [0,\infty)$ if $\operatorname{Im} \alpha \neq 0$. For real positive α we extend $\tilde{v}(x,\alpha)$ directly as a solution of $\tilde{v}'' = q(x)\alpha\tilde{v}$ with $\tilde{v}(x_0,\alpha) = 1$ and $\tilde{v}'(x_0,\alpha) = w(x_0,\alpha)$.

Then $\tilde{v}(x,\alpha)$ is analytic in $\alpha \in C - (-\infty, 0]$ for any fixed $x \in [0,\infty)$. Since $v(x,\alpha) = \tilde{v}(x,\alpha)/\tilde{v}(0,\alpha)$, we need a lemma concerning zeros of $\tilde{v}(0,\alpha)$.

Lemma 6. Suppose that $q(x)$ satisfies (C) and $q(x_1) < 0$ for certain $x_1 \in (0, x_0)$. Then $\{\alpha : 0 < \alpha , \tilde{v}(0,\alpha) = 0\}$ equals $\{\alpha = d_j ; 0 < d_1 < d_2 < d_3 < \ldots, \lim d_j = \infty\}$, where d_j ,($j = 1, 2, \ldots$) is determined by $q(x)$. $\tilde{v}(0,\alpha)$ has a simple zero at each $\alpha = d_j$ and satisfies

(3.12) $\quad 0 < \underline{\lim}_{\alpha\downarrow 0} v(0,\alpha) < \overline{\lim}_{\alpha\downarrow 0} v(0,\alpha) < \infty$.

Proof. Put $\tilde{\tilde{v}}(x,\alpha) = v(x_0 + \frac{x}{\alpha}, \alpha)$ Then v satisfies

(3.13) $\quad \frac{d^2}{dx^2}\tilde{\tilde{v}}(x,\alpha) = q(x,\alpha)\tilde{\tilde{v}}(x,\alpha)$,

$\tilde{\tilde{v}}(0,\alpha) = 1$ and $\lim_{x\to\infty} v(x,\alpha) = 0$ for any $\alpha > 0$, where $q(x,\alpha) = q(x_0 + \frac{x}{\alpha})$,

(3.13)' $\quad m_0 < q(x,\alpha) < M$ for $x \in [0,\infty)$ and all $\alpha > 0$.

Applying Lemma A to $\tilde{\tilde{w}}' = q(x,\alpha) - \tilde{\tilde{w}}^2$ with $\Omega = (-M^{1/2}, -m_0^{1/2})$ in R, we have a solution $\tilde{\tilde{w}}(x,\alpha)$ staying in Ω for $x \in [0,\infty)$. Then we have $\tilde{\tilde{v}}(x,\alpha) = \exp \int_0^x w(s,\alpha)ds$ from the uniqueness in Lemma 2.

Remark that for all $\alpha > 0$, $(\frac{d}{dx}\tilde{\tilde{v}})(0,\alpha) = \tilde{\tilde{w}}(0,\alpha) \in (-M^{1/2}, -m_0^{1/2})$.

Since $|q(x,\alpha)| < M$ for $x \quad [-\alpha x_0, 0]$, $\tilde{v}(0,\alpha) = \tilde{\tilde{v}}(-\alpha x_0, \alpha)$ satisfies (3.12). If α becomes larger the solution $\tilde{\tilde{v}}(x,\alpha)$ oscillates in the interval where $q(x,\alpha)$ is negative valued. From this fact we can see later that $\{\alpha : 0 < \alpha\ ,\ \tilde{v}(0,\alpha) = 0\}$ is an infinite set. Now suppose $\tilde{v}(0, d) = 0$. Then $\tilde{v}'(0, d) = \frac{d}{dx}\tilde{v}(0, d) \neq 0$. Similarly to (3.6) we have for $\alpha \in C - (-\infty, 0]$

$$-\tilde{v}'(0,\alpha)\,\overline{\tilde{v}(0,\alpha)} = \int_0^\infty |\tilde{v}'(s,\alpha)|^2 ds + \alpha\int_0^\infty q(s)|\tilde{v}(s,\alpha)|^2 ds$$

Therefore $\int_0^\infty q(s)\tilde{v}(s,d)^2 ds = -d^{-1}\int_0^\infty \tilde{v}'(s, d)^2 ds \neq 0$. On the other hand, (3.10) and $w(0,\alpha) = \tilde{v}'(0,\alpha)/\tilde{v}(0,\alpha)$ yield

$$\int_0^\infty q(s)\tilde{v}(s,\alpha)^2 ds = -\tilde{v}(0,\alpha)^2 \frac{\partial}{\partial\alpha} w(0,\alpha)$$

$$= \tilde{v}'(0,\alpha)\frac{d}{d\alpha}\tilde{v}(0,\alpha) - \frac{d}{d\alpha}\tilde{v}'(0,\alpha)\tilde{v}(0,\alpha)$$

for $\alpha \in C - (-\infty, \infty)$. Making α tend to d we have

$$\frac{d}{d\alpha}\tilde{v}(0, d) = -\tilde{v}'(0, d)^{-1} d^{-1} \int_0^\infty \tilde{v}'(s,d)^2 ds \neq 0.$$

Thus we have Lemma 6. (q.e.d.)

From Lemma 6 $v(x,\alpha)$ has simple poles at $\alpha = d_j$ for $x \notin X_j = \{x : x > 0\ ,\ \tilde{v}(x, d_j) = 0\}$, which constitutes of $j-1$ points. Combining these arguements we have Theorem 1.

<u>Proof of Theorem 2.</u> We show the following two estimates.

$$\text{(3.14)} \qquad \inf_{x\in[0,\infty)} s(x,\alpha) \geq \left(\frac{|\mathrm{Re}\,\beta|}{(m_0|\mathrm{Re}\,\alpha|)^{1/2}} + \frac{|\alpha||\mathrm{Im}\,\beta|}{|\mathrm{Im}\,\alpha|} x_0\right)^{-1} \equiv s_m.$$

$$\text{(3.15)} \qquad \sup_{x\in[0,\infty)} s(x,\alpha) \leq \frac{2M|\alpha|^2}{|\mathrm{Im}\,\alpha||\mathrm{Im}\,\beta|^2} s_m^{-1} \equiv s_M ,$$

for $\alpha \in \{\alpha \in C : 0 < |\arg \alpha| < \frac{\pi}{4}\}$. From (3.14), (3.15) and (2.7), we obtain the estimate (E) in Theorem 2, if we use $|\alpha| = |\sigma - i\gamma|^2 = O(|\sigma|^2)$, $\operatorname{Re}\alpha = O(|\sigma|^2)$, $|\operatorname{Im}\alpha| = O(|\sigma|)$, $|\operatorname{Re}\beta| = O(|\sigma|)$ and $|\operatorname{Im}\beta| = O(1)$ for fixed γ. In order to prove (3.14) we consider $\frac{d}{dx} s(x,\alpha)$. From (2.8) and $w^2 = (r\alpha - s\beta)^2$, follows

$$\frac{d}{dx} s = -|\alpha|(\operatorname{Im}\alpha)(\operatorname{Im}\beta)^{-1} r^2 + 2|\alpha| rs$$

$$\leq |\alpha|(\operatorname{Im}\beta)(\operatorname{Im}\alpha)^{-1} s^2$$

Thus from $(1/s)' \geq -|\alpha|(\operatorname{Im}\beta)(\operatorname{Im}\alpha)^{-1}$ we have

$$\frac{1}{s(x)} \leq \frac{1}{s(x_0)} + \frac{|\alpha| \operatorname{Im}\beta}{\operatorname{Im}\alpha}(x_0 - x) \quad , \quad 0 \leq x < x_0 .$$

From (3.5), $1/s(x) \leq (m_0 \operatorname{Re}\alpha)^{-1/2}(\operatorname{Re}\beta)$, for $x \in [x_0, \infty)$.

Thus we have (3.14). To show (3.15) we prove that $w(x,\alpha)$ stays in the following $\tilde{\Omega}$ for all $x \in [0,\infty)$. Here $\tilde{\Omega}$ $(= \tilde{\Omega}_\alpha)$ is an open bounded domain surrounded by $S_1 = \{w \,;\, s = -\operatorname{Im}(w\bar{\alpha})/\operatorname{Im}(\beta\bar{\alpha}) = s_m\}$ and $S_2 = \{\tilde{w} = 1/w \,;\, w = \varepsilon_0(-i\bar{\alpha}) + r\bar{\alpha} \,,\, r \in R\}$, where we take ε_0 sufficiently small so that following two conditions are fulfilled:

(I) The vector field $1 - q(x)\alpha\tilde{w}^2$ faces the exterior on $S_2 \cap \partial\tilde{\Omega}$.

(II) $\tilde{\Omega}$ involved the domain defined by (3.5) i).

The conditions $|\arg(1 - q(x)\alpha\tilde{w}^2)| < |\arg\bar{\alpha}|$ and $|\tilde{w}| < (2M|\alpha|)^{-1/2}$ for all $\tilde{w} \in S_2 \cap \partial\tilde{\Omega}$ are sufficient. By elemental calculus we can take

$$\varepsilon_0 = \frac{|\operatorname{Im}\alpha \ \operatorname{Im}\beta|}{2M|\alpha|^3} s_m .$$ Then we have $w(x,\alpha) \in \tilde{\Omega}$ for all $x \in [0,\infty)$,

and $$s_M = \sup_{w \in \tilde{\Omega}} \frac{\operatorname{Im}(w\bar{\alpha})}{\operatorname{Im}(\alpha\bar{\beta})} = (|\alpha| \operatorname{Im}\beta)^{-1} \operatorname{Im}(\varepsilon_0^{-1}(i\alpha|\alpha|^{-2})\bar{\alpha})$$

$$= \varepsilon_0^{-1} \, |\alpha|^{-1} |\operatorname{Im}\beta|^{-1} = \frac{2M\,|\alpha|^2}{|(\operatorname{Im}\alpha)(\operatorname{Im}\beta)^2|} \, s_m^{-1}$$. This completes the proof of Theorem 2.

Final comments. It is interesting that unstable solutions of $v'' = q(x)\,v$ are useful to construct the solution of (P). As for Theorem 3, the uniqueness of (P) should be discussed more precisely. The formula (2) is useful to see the singularities of solutions of (P). The argument in this paper will be valid for another type of problems, for example, (P) with $q(x)$ satisfying $\overline{\lim}_{x\to\infty} q(x) < 0$ and (P) replaced $\frac{\partial^2}{\partial x^2}$ by Δ in R^n . They will be discussed elsewhere.

Bibliography

[1] L. Bers, Mathematical aspects of subsonic and transsonic gas dynamics, Wiley, New York, (1958)

[2] A.R. Manwell, The tricomi equation with applications to the theory of plane transsonic flow, Pitman Adv. Publ. (1979)

[3] M.M. Smirnov, Equations of mixed type, American Math. Soc. Monographs. 51(1978)

[4] S. Miyatake, Construction of eigenfunctions for Sturm-Liouville operator in $(-\infty, \infty)$. to appear Kyoto J. Math.

[5] S. Miyatake, Théorème d'existence des solutions des systèmes d'équations différentielles et applications, Séminaire J.vaillant Univ. Paris VI (1980-81)

[6] T.Wazewski, Une methode topologique de l'examen du phénomène asymptotique relativement aux équation différentielles ordinaires, Lincei-Rend. Sc. vol III, 1947

[7] S.R.Bernfeld-V. Lakshmikantham, An introduction to nonlinear boundary value problems, Academic Press, 1974.

Lecture Notes in Num. Appl. Anal., **5**, 189–204 (1982)
Nonlinear PDE in Applied Science. U.S.-Japan Seminar, Tokyo, 1982

Yielding and Unloading in Semidiscrete Problem of Plasticity

Tetsuhiko Miyoshi

Department of Mathematics
Kumamoto University
Kumamoto 860, JAPAN

Introduction

In formulating the elastic plastic problem it is usually assumed that each element of the material is either in the *elastic state* or in the *plastic state* and that these states continue for a while after they have been chosen. The transition from the elastic state to the plastic state is called the *yielding*. The reverse change is a case of the *unloading*. Therefore the material undergoes the elastic and plastic deformations repeating yielding and unloading alternately. Mathematically, this is nothing but to assume the existence of the classical solution, which is proved only for some special cases.

For the semidiscrete system, that is, for the case that only the spatial region is discrete,it is possible to show the existence of the classical solution. In this paper we discuss the way to get this solution. This kind of systems is essential for deriving the approximate methods in engineering. Especially, the *explicit* integration schemes start from this system. Also, if we want the solution of the fully continuous problem, it is only necessary to pass to the limit with respect to the spatial approximation.

The key to prove the existence of the classical solution is to guess the behaviour of the solution after the time at which the state of the element may change. Assume that there are N elements which may yield or unload at time t=

t_0. The number of the combinations of the possible states after t_0 is 2^N. Hence if there exists a unique classical solution, then only one in 2^N possibilities must take place, and what we want to show is that this is actually realized. We will show that the next state of each element is uniquely determined by using the data of the solution before t_0 and the given data at t_0 which is independent of the solution.

The mechanism of this determination in dynamic problem is different from that in the quasi-static problem. In the former the existence of the inertia term is the key (see [1] for the details). In the latter, a certain potential relating to each order of derivatives of the solution plays the essential role. In this paper we discuss the quasi-static case. We will explain our basic idea taking a finite element approximation of a 2-dimensional problem as an example. The result of the present paper is announced in [2] with a brief proof.

1. Semidiscrete finite element approximation

Let Ω be a region in $x=(x_1,x_2)$ plane which is composed of the finite number of triangles. Each triangle is called a finite element or simply an element. Let $\{\phi_p\}$ be the usual piecewise linear finite element basis. We seek the function u_i (i=1,2) of the form

$$u_i(t) = \sum_{p\varepsilon P} u_i^p(t)\, \phi_p(x).$$

Here P denotes the set of all the vertexes of the triangles of Ω excepting P_u on which the boundary values of u_i are given. We assume that, for the sake of simplicity, P_u includes at least two adjecient vertexes and $u_i=0$ on P_u.

$\{u_i^p(t)\}$ are determined by the following system of equations.

$$(1.1)_a \qquad \sum_j (\sigma_{ij}, \phi_{p,j}) = (b_i, \phi_p) \qquad p \in P,$$

$$(1.1)_b \qquad \dot\sigma = D\dot\varepsilon, \quad \dot\alpha = 0 \qquad \text{if } f(\sigma-\alpha) < \bar\sigma, \text{ or } f(\sigma-\alpha) = \bar\sigma \text{ and } \partial f^*\dot\sigma < 0,$$

$$(1.1)_c \qquad \dot\sigma = (D - D')\dot\varepsilon, \quad \dot\alpha = (\sigma-\alpha)\frac{\partial f^*\dot\sigma}{f} \qquad \text{if } f(\sigma-\alpha) = \bar\sigma \text{ and } \partial f^*\dot\sigma \geqslant 0,$$

where (,) denotes the $L^2(\Omega)$ inner product of functions. We use this notation for both the single and the vector functions. The above system is derived from the Prandtl-Reuss flow rule and the Ziegler's hardening assumption.

NOTATIONS :

$$\sigma = \sigma(t,x) = (\ \sigma_{11},\ \sigma_{22},\ \sigma_{12}\ ) \quad (\ \sigma_{21} = \sigma_{12})$$

$$\varepsilon = \varepsilon(t,x) = (\ \varepsilon_{11},\ \varepsilon_{22},\ \varepsilon_{12})$$

$$\alpha = \alpha(t,x) = (\ \alpha_{11},\ \alpha_{22},\ \alpha_{12})$$

$$\varepsilon_{11} = u_{1,1} \quad \varepsilon_{22} = u_{2,2} \quad \varepsilon_{12} = u_{1,2} + u_{2,1}$$

$$f = f(\sigma - \alpha), \quad f(\tau) = \tau_{11}^2 + \tau_{22}^2 - \tau_{11}\ \tau_{22} + 3\ \tau_{12}^2$$

η ; positive function (assumed to be constant)

D ; elastic stress-strain matrix

$D' = D\partial f \partial f^* D/(\ \eta + \partial f^* D \partial f\ )$ $\quad \partial f^*$: transposed ∂f

$\partial f = (\ \partial f/\partial\sigma_{11},\ \partial f/\partial\sigma_{22},\ \partial f/\partial\sigma_{12}\ )(\sigma - \alpha)$

$u_{,i} = \partial u/\partial x_i$, $\quad \dot{u} = du/dt$, $\quad b_i = b_i(t,x)$; given function

$\bar{\sigma}$: given positive constant.

Now, since u_i is piecewise linear with respect to x, ε is constant on each element. Hence σ and α which are determined by $(1.1)_b$ or $(1.1)_c$ are also constant on each element. We assume that b_i are continuous and piecewise analytic with respect to t. Under these assumptions we seek a continuous (u,σ,α) satisfying the equations (1.1) in $I = [0, T]$ and the initial condition $(u,\sigma,\alpha)=0$.

2. Determination of the first derivative

We say that an element is elastic (resp. plastic) if $(1.1)_b$ (resp. $(1.1)_c$) is satisfied on this element. Let E be the set of all elements of Ω. Since we started from the zero initial condition, all elements of E are elastic until some elements satisfy $f(\sigma) = \bar{\sigma}$ at, for example, $t = t_0$. It is clear that our problem has a unique piecewise analytic solution $(u,\sigma,0)$ in $[0, t_0)$. Let E_0 be the set of all elements which satisfy $f(\sigma)=\bar{\sigma}$ at $t=t_0$. $E- E_0$ is clearly still elastic after t_0, since the solution must be continuous. Hence the next problem is whether the elements of E_0 yield at $t=t_0$ or still remain elastic.

The key is to guess the sign of $\partial f^* \dot{\sigma}$ at $t=t_0+0$, since if this is positive (resp. negative) then the stress point σ moves to the outside (resp. inside) of the *yield surface* $f(\sigma) = \bar{\sigma}$. We note that the signs of $\partial f^* \dot{\sigma}$ and $\partial f^* D\dot{\varepsilon}$ for $t> t_0$ are same, since

$$\partial f^* \dot{\sigma} = \partial f^* D\dot{\varepsilon} \quad \left(= \partial f^* D\dot{\varepsilon} \left(1- \frac{\partial f^* D \partial f}{\eta + \partial f^* D\, \partial f} \right) \right) \text{ if elastic (if plastic).}$$

Hence we shall consider the following system which must be satisfied by the first derivatives of the solution (if it exists).

$$(2.1)_a \qquad \sum_j (\sigma^o_{ij}, \phi_{p,j}) = (\dot{b}_i(t_0+0), \phi_p) \qquad p \in P$$

$$(2.1)_b \qquad \sigma^o = D\varepsilon^o \qquad \text{for } E- E_0$$

$$(2.1)_c \quad \begin{cases} \sigma^o = D\varepsilon^o & \text{in } D_- = \{ u^o \ ; \ \partial f^*(t_0)D\varepsilon^o < 0 \} \\ \sigma^o = (D - D')\varepsilon^o & \text{in } D_+ = \{ u^o \ ; \ \partial f^*(t_0)D\varepsilon^o \geq 0 \} \end{cases} \quad \text{for } E - E_0$$

where

$$\varepsilon^o = \varepsilon(u^o) \qquad u^o_i = \sum_{p \in P} u^{p,o}_i \ \phi_p \quad (i=1,2)$$

$$D' = D'(t_0).$$

Theorem 2.1. Problem (2.1) has a unique solution (u^o, σ^o). This u^o minimizes the functional

$$(2.2) \qquad F_1(u^o) = \frac{1}{2}(\sigma^o,\varepsilon^o) - (\dot{b}(t_0+0), u^o),$$

under the subsidiary conditions $(2.1)_b$ and $(2.1)_c$.

Proof. (i) Let (,) be the $L^2(e)$ inner product of (vector) functions ($e \varepsilon E$) and define

$$F^e = \frac{1}{2} (\sigma^o, \varepsilon^o)_e - (\dot{b}(t_0+0), u^o)_e .$$

F^e is a C_1-class function of u^o and so is $F_1 = \Sigma F^e$ too. This is clear for $e \varepsilon E - E_0$. Let e be an element of E_0. Then in D_- we have

$$\frac{\partial F^e}{\partial u_i^{p,o}} = (D\varepsilon^o, \frac{\partial \varepsilon^o}{\partial u_i^{p,o}})_e - (\dot{b}(t_0+0), \frac{\partial u^o}{\partial u_i^{p,o}})_e$$

and similarly in D_+

$$\frac{\partial F^e}{\partial u_i^{p,o}} = ((D - D')\varepsilon^o, \frac{\partial \varepsilon^o}{\partial u_i^{p,o}})_e - (\dot{b}(t_0+0), \frac{\partial u^o}{\partial u_i^{p,o}})_e$$

These derivatives coincide on the plane between D_- and D_+ since

$$D'\varepsilon^o = \frac{D\partial f \partial f^* D}{\eta + \partial f^* D \partial f} \varepsilon^o = 0$$

on this hyperplane. This proves the C_1 continuity of F^e.

(ii) $D - D'$ is positive definite. This is easily proved by the facts that η is positive constant and $c_1 \leqslant \partial f^* D \partial f \leqslant c_2$, $(D\partial f \partial f^* D\varepsilon, \varepsilon) \leqslant \partial f^* D \partial f (D\varepsilon, \varepsilon)$, where c_1 and c_2 are positive constants. Therefore $F_1(u^o)$ is bounded below by the Korn's inequality (we do not need this inequality as far as we treat the semi-discrete system). Hence $F_1(u^o)$ has a minimum point which is also a stationary point. However, if u^o is a stationary point of F_1, then at this point $\sigma^o = \sigma^o(u^o)$ determined by $(2.1)_b$ or $(2.1)_c$ must satisfy the stationary condition which is equivalent to $(2.1)_a$. Hence the problem (2.1) has a solution which minimizes F_1.

(iii) To prove the uniqueness of the solution it suffices to show the uniqueness of the stationary point of $F_1(u^o)$. For each element $e \varepsilon E_0$ we consider the hyperplane in u^o-space :

$$\pi_e = \{ u^o ; \partial f^*(t_0+0)D\varepsilon^o = 0 \}.$$

Let $\{R_\lambda\}$ be the partition of the u^o-space by these planes. In each R_λ the σ^o-ε^o relation is definite for all elements of E. Also, in R_λ $F_1(u^o)$ is a positive definite quadratic form of u^o and the stationary point is at most one.

Now assume that there are two stationary points $u^1 \varepsilon R_\lambda$ and $u^2 \varepsilon R_\mu$ ($\lambda \neq \mu$). Consider the line

$$l : u^1 + t (u^2 - u^1) \qquad t \varepsilon [0,1].$$

This line goes through at least two regions of $\{R_\lambda\}$ when t moves from 0 to 1. Then the function

$$g(t) = F_1(u^1 + t (u^2 - u^1))$$

is smooth, and a non- degenerate quadratic on t in each region. Therefore if u^1 is a stationary point, that is, if $F_1(u^1)$ is the minimum, then g(t) must be strictly increasing in [0, 1], which contradicts that u^2 is another stationary point. This completes the proof.

We want to show that the solution (u^o,σ^o) of the problem (2.1) is the first derivative at $t=t_0+0$ of the solution of (1.1) provided (1.1) has a solution. By Theorem 2.1 we can determine the sign of $\partial f^*(t_0)D\dot{\varepsilon}$ for the element of E_0. We denote by E^e and E^p the set of all elements of E_0 for which this sign is negative and nonnegative, respectively, and solve the following initial value problem set up at $t=t_0$.

$$(2.3)_a \qquad \sum_j (\sigma_{ij}, \phi_{p,j}) = (b_i, \phi_p) \qquad p \varepsilon P$$

$$(2.3)_b \qquad \begin{cases} \dot{\sigma} = D\dot{\varepsilon}, \quad \dot{\alpha} = 0 & \text{for } E - E^p \\ \dot{\sigma} = (D - D')\dot{\varepsilon}, \quad \dot{\alpha} = (\sigma - \alpha) \dfrac{\partial f^* \dot{\sigma}}{f} & \text{for } E^p, \end{cases}$$

where $\dot{\varepsilon} = \varepsilon(\dot{u})$, $D' = D'(t)$, $f = f(\sigma - \alpha)$ and $(u,\sigma,\alpha)(t_0) = (u,\sigma,\alpha)(t_0-0)$.

Theorem 2.2. The initial value problem (2.3) has a unique analytic solution (u,σ,α) in a certain neighborhood of t_0, and $(\dot{u},\dot{\sigma})(t_0+0)=(u^o, \sigma^o)$.

Proof. Differentiate the both sides of $(2.3)_a$ with respect to t and denote the resulting equation by $(2.3)_a^{(1)}$. By $(2.3)^{(1)}$ we denote the system $(2.3)_a^{(1)}$ and $(2.3)_b$. Substituting the $\dot{\sigma}$-$\dot{\varepsilon}$ relation of $(2.3)_b$ into $(2.3)_a^{(1)}$ and solving the resulting equation with respect to $\dot{u}$, we have $\dot{u}$ as an analytic function of σ, α and t in a certain neighborhood of t_0. Therefore $(2.3)_b$ can be regarded as a system of ordinary differential equations of the form

$$\frac{d}{dt}\begin{vmatrix}\sigma\\ \alpha\end{vmatrix} = X(\sigma,\alpha,t)$$

where X is an analytic vector function. This system has a unique analytic solution under the given initial condition. Hence the problem for (2.3) has a unique analytic solution (u,σ,α) in a certain neighborhood of t_0. Furthermore, the solution (u^o, σ^o) of the problem (2.1) satisfies $[2.3)^{(1)}$ at t_0+0. Here $[2.3)^{(1)}$ is the system composed of $(2.3)_a^{(1)}$ and the $\dot{\sigma}$-$\dot{\varepsilon}$ relation of $(2.3)_b$. Since $[2.3)^{(1)}$ has a unique solution on $(\dot{u},\dot{\sigma})(t_0+0)$, we have the theorem.

Let E_1 be the set of all the elements of E^p for which $\partial f^*(t_0)D\dot{\varepsilon}(t_0+0)=0$ holds for the solution of (2.3). If E_1 is empty, then the next state is completely determined for all the elements. Because, the elements of $E-E_0$ are still elastic after t_0 and for those of E^e holds

$$f^2(\sigma) - f^2(\sigma(t_0)) = 2\int_{t_0}^{t_0+\delta} f(\sigma)\partial f^*\dot{\sigma}\, ds < 0 \qquad (\delta>0)$$

for small δ and for those of E^p we have

$$\partial f^*\big|_{(\sigma-\alpha)}\dot{\sigma} = \partial f^*\big|_{(\sigma-\alpha)}D\dot{\varepsilon}\ (1-\theta) > 0 \text{ and } f(\sigma-\alpha) = \overline{\sigma}$$

for a while after t_0, where $0<\theta<1$. In other words, the $\dot{\sigma}$-$\dot{\varepsilon}$ relation of the elements of E^e and E^p are already chosen correctly and we could determine which elements yielded at t_0. We here emphasize that this determination is dependent only on the data at $t=t_0-0$ and the given function b.

If, however, E_1 is not empty we have to guess the sign of $d/dt(\partial f^*\dot{\sigma})$ at $t = t_0+0$. In this case the following theorem is important. Replace some elements of E_1 from E^p to E^e and solve the initial value problem (2.3) for this *new* E^p. Let the new system be denoted by {2.3}. Then {2.3} has a unique solution (u,σ,α) under the same initial condition at $t=t_0$, and moreover we have

Theorem 2.3. For every element of E, the value $(\dot{u},\dot{\sigma},\dot{\alpha})(t_0+0)$ is determined independently of the choice of the next $\dot{\sigma}-\dot{\varepsilon}$ relation of E_1.

Proof. Samely as before let $[2.3]^{(1)}$ be the system which is composed of $\{2.3\}_a^{(1)}$ and the $\dot{\sigma}-\dot{\varepsilon}$ relation of $\{2.3\}_b$. Since the solution (u^o,σ^o) of(2.1) satisfies $D'(t_0)\varepsilon^o=0$ for the elements of E_1, (u^o,σ^o) is a solution of $[2.3]^{(1)}$ at $t = t_0+0$. But the solution of $[2.3]^{(1)}$ is unique with respect to $(\dot{u},\dot{\sigma})(t_0+0)$. Therefore $(\dot{u},\dot{\sigma})(t_0+0) = (u^o,\sigma^o)$ and follows the theorem since $\dot{\alpha}(t_0+0)=0$ for E_1.

Thanks to this theorem it is assured that the elements of $E-E_1$ behave so as to satisfy the subsidiary condition of (1.1) *for any choice* of the next state of E_1. In other words, the next state of the element of $E-E_1$ is already determined. Hence we can exclude them from our consideration.

3. Determination of the higher derivatives

The next state of the element of E_1 is determined by the sign of $d/dt(\partial f^*\dot{\sigma})$ at $t=t_0+0$. It is easy to see that this sign and that of $d/dt(\partial f^* D\dot{\varepsilon})$ is same at t_0+0 for the elements of E_1. Hence we consider the following problem which must be satisfied by the second derivatives of the solution.

$$(3.1)_a \qquad \sum_j (\sigma^o{}_{ij}, \phi_{p,j}) = (\frac{d^2}{dt^2} b_i(t_0+0), \phi_p) \qquad p \in P$$

$$(3.1)_b \quad \begin{cases} \sigma^o = D\varepsilon^o & \text{for the elastic element of } E - E_1 \\ \sigma^o = \{(D - D')\varepsilon^o - \frac{d}{dt}(D')\dot{\varepsilon}\}_{t_0+0} & \text{for the plastic element of } E - E_1 \end{cases}$$

and for the element of E_1

$$(3.1)_c \begin{cases} \sigma^o = D\varepsilon^o & \text{in } D_- = \{u^o;\ \partial f^*(t_0)D\varepsilon^o + r_1 < 0\} \\ \sigma^o = \{(D - D')\varepsilon^o - \frac{d}{dt}(D')\dot{\varepsilon}\}_{t_0+0} & \text{in } D_+ = \{u^o;\ \partial f^*(t_0)D\varepsilon^o + r_1 \geqslant 0\}, \end{cases}$$

where ε^o, u^o, ∂f and D' are defined as before, and

$$r_1 = \{\frac{d}{dt}(\partial f^*)D\dot{\varepsilon}\}_{t_0+0}.$$

Note that $(\dot{u},\dot{\sigma},\dot{\alpha})(t_0+0)$ is the derivative of the solution of (2.3).

Theorem 3.1. Problem (3.1) has a unique solution (u^o,σ^o). This u^o minimizes the following functional $F_2(u^o)$ under the subsidiary conditions $(3.1)_b$ and $(3.1)_c$.

$$(3.2) \qquad F_2(u^o) = \frac{1}{2}(\sigma^o,\varepsilon^o) - \frac{1}{2}(\lambda_2,\bar{\varepsilon}) - (\frac{d^2}{dt^2} b(t_0+0),\ u^o),$$

where $\lambda_2 = \{dD'/dt\ \dot{\varepsilon}\}_{t_0+0}$ and

$$\bar{\varepsilon} = \begin{cases} 0 \ (\text{resp. } \varepsilon^o) & \text{for the elastic (resp. plastic) element of } E\text{-}E_1 \\ \left.\begin{matrix} \varepsilon^o_* & \text{in } D_- \\ \varepsilon^o & \text{in } D_+ \end{matrix}\right\} & \text{for the element of } E_1. \end{cases}$$

Here ε^o_* is an arbitrary fixed vector included in the hyperplane of u^o-space

$$(3.3) \qquad \pi_e : \{u^o;\ \partial f^*(t_0)D\varepsilon^o + r_1 = 0\}.$$

Proof. (i) $F_2(u^o)$ is a continuous function. To prove this let F^e be the e - part of F_2 as before ($e \ \varepsilon E$). For $e \ \varepsilon\ E\text{-}E_1$ its continuity is clear. For $e\ \varepsilon E_1$ the discontinuity of F^e might appear across the plane π_e. However, at $t=t_0+0$ it holds that on π_e

$$D'\varepsilon^o + \frac{d}{dt}(D')\dot{\varepsilon} = \frac{D\ \partial f \partial f^* D}{\eta + \partial f^* D \partial f}\varepsilon^o + \frac{D\partial f\ \frac{d}{dt}(\partial f^*)D}{\eta + \partial f^* D\ \partial f}\dot{\varepsilon}$$

$$(3.4) \qquad = \frac{D\partial f}{\eta + \partial f^* D\partial f}(\partial f^* D\varepsilon^o + \frac{d}{dt}(\partial f^*)D\dot{\varepsilon}) = 0.$$

Hence σ^o is continuous with respect to u^o and the first term of F^e is continuous. The jump at π_e of the second term of F^e is $1/2(\lambda_2, \varepsilon_*^o - \varepsilon^o)_e$. But if ε_*^o and ε^o belong to π_e, then $D'(\varepsilon_*^o - \varepsilon^o) = 0$ at $t=t_0+0$ by (3.4). Therefore, since D' is symmetric, we have at $t=t_0+0$

$$(\lambda_2, \varepsilon_*^o - \varepsilon^o)_e = (\frac{d}{dt}(D')\dot{\varepsilon}, \varepsilon_*^o - \varepsilon^o)_e$$

$$= -(D'\varepsilon^o, \varepsilon_*^o - \varepsilon^o)_e = -(\varepsilon^o, D'(\varepsilon_*^o - \varepsilon^o))_e = 0.$$

This implies the continuity of the second term of F^e and hence of F^e itself.

(ii) $F_2(u^o)$ is a C_1-class function. To prove this we check the three cases. First let e be an elastic element of $E - E_1$. Then F^e is smooth, since

$$F^e = \frac{1}{2}(D\varepsilon^o, \varepsilon^o)_e - (\frac{d^2}{dt^2} b(t_0+0), u^o)_e.$$

Secondary let e be the plastic element of $E - E_1$. Then

$$F^e = \frac{1}{2}(\sigma^o, \varepsilon^o)_e - \frac{1}{2}(\{\frac{d}{dt}(D')\dot{\varepsilon}\}(t_0+0), \varepsilon^o)_e - (\frac{d^2}{dt^2} b(t_0+0), u^o)_e.$$

Therefore

$$\frac{\partial F^e}{\partial u_i^{p,o}} = ((D - D')\varepsilon^o, \frac{\partial \varepsilon^o}{\partial u_i^{p,o}})_e - (\{\frac{d}{dt}(D')\dot{\varepsilon}\}(t_0+0), \frac{\partial \varepsilon^o}{\partial u_i^{p,o}})_e$$

$$(3.5) \qquad - (\frac{d^2}{dt^2} b(t_0+0), \frac{\partial u^o}{\partial u_i^{p,o}})_e$$

$$= (\sigma^o, \frac{\partial \varepsilon^o}{\partial u_i^{p,o}})_e - (\frac{d^2}{dt^2} b(t_0+0), \frac{\partial u^o}{\partial u_i^{p,o}})_e.$$

Since σ^o is continuous, this equality implies the smoothness of F^e. Finally let e be an element of E_1. Then the following is clear in D_-.

$$\frac{\partial F^e}{\partial u_i^{p,o}} = (\sigma^o, \frac{\partial \varepsilon^o}{\partial u_i^{p,o}})_e - (\frac{d^2}{dt^2} b(t_0+0), \frac{\partial u^o}{\partial u_i^{p,o}})_e.$$

On the other hand, the relation (3.5) is valid in D_+. Hence the C_1 continuity of $F_2(u^o)$ is proved.

(iii) $F_2(u^o)$ is a positive definite quadratic form which is bounded below. Hence the minimizing point u^o of F_2 exists and (u^o,σ^o), where σ^o is determined by the unique σ^o-ε^o relation, is the solution of (3.1). These are proved by just the same way as in Theorem 2.1. This completes the proof of the theorem.

The solution (u^o, σ^o) of the problem (3.1) is the second derivatives of the true solution. Strictly speaking, this is in the sense of the following theorem. Devide $E_1 = E_1^e + E_1^p$, where

$$E_1^e = \{ e \; \varepsilon E_1 \; ; \; \partial f^*(t_0)D \; \varepsilon^o + r_1 < 0 \}$$

$$E_1^p = \{ e \; \varepsilon E_1 \; ; \; \partial f^*(t_0)D \; \varepsilon^o + r_1 \geqslant 0 \},$$

and solve (2.3) replacing E^p by the *new* $E^p = E^p - E_1^e$. Let (u,σ,α) be its solution. By Theorem 2.3, $(\dot{u},\dot{\sigma},\dot{\alpha})(t_0+0)$ is same to that of the solution of the problem (2.3) with *old* E^p.

Theorem 3.2. Let (u^o, σ^o) be the solution of (3.1). Then hold

(i) $(u^o, \sigma^o) = (\ddot{u},\ddot{\sigma})(t_0+0)$,

(ii) Let E_2 be the set of elements of E_1^p such that

$$\partial f^*(t_0)D \; \varepsilon^o + r_1 = 0. \tag{3.6}$$

Then, for every element, $(\ddot{u},\ddot{\sigma},\ddot{\alpha})(t_0+0)$ is determined independently of the choice of the next $\dot{\sigma}-\dot{\varepsilon}$ relation of E_2.

Proof. By $(2.3)_a^{(2)}$ and $(2.3)_b^{(1)}$ we denote the equations obtained by differenciating twice the both sides of $(2.3)_a$ and once the $\dot{\sigma}$-$\dot{\varepsilon}$ relation of $(2.3)_b$ with respect to time t, respectively, where E^p is replaced by the new E^p. By $(2.3)^{(2)}$ we denote the system of these equations. Then $(\ddot{u},\ddot{\sigma})(t_0+0)$ and (u^o,σ^o) satisfy $[2.3)^{(2)}$, where [ ) has the same meaning as before. Also the solution of [2.3] $^{(2)}$ is unique at t_0+0. Hence (i) holds. The proof of (ii) is exactly the same to that of Theorem 2.3. Note that for the element of E_2 the equality

$$\{ D' \varepsilon^o + \frac{d}{dt} (D')\dot{\varepsilon} \}_{t_0+0} = 0$$

holds for the solution (u^o, σ^o) of (3.1)

If, furthermore, E_2 is not empty, we repeat this discussion until E_K becomes empty for a certain $K < \infty$. It might happen that there are some elements for which the equality

$$\{ \frac{d^k}{dt^k} (\partial f^* \dot{\sigma})\}_{t_0+0} = 0$$

holds for all k. But this means that the stress point moves along the yield surface for a while after t_0: $f(\sigma) = \bar{\sigma}$. Hence we assign the plastic $\dot{\sigma}-\dot{\varepsilon}$ relation to these elements. For the completeness we descrive below the procedure to determine the derivatives of the solution at t_0+0 when E_k ($k \geqslant 2$) is not empty.

Assume that the derivatives of order k are already determined independently of the choice of the $\dot{\sigma}$-$\dot{\varepsilon}$ relation of E_k:

$$E_k = \{ e \,\varepsilon E_{k-1} \;;\{ \frac{d^{k-1}}{dt^{k-1}}(\partial f^* D\dot{\varepsilon}) \}_{t_0+0} = 0 \}$$

Let us define (formally)

$$\lambda_{k+1} = \{\frac{d^k}{dt^k}(D'\dot{\varepsilon}) - D' \frac{d^k}{dt^k}\dot{\varepsilon}\}_{t_0+0} ,$$

$$r_k = \{\frac{d^k}{dt^k}(\partial f^* D\dot{\varepsilon}) - \partial f^* D \frac{d^k}{dt^k}\dot{\varepsilon}\}_{t_0+0}.$$

Let (u^o,σ^o) be the solution of the following problem set up at $t=t_0+0$.

$$(3.7)_a \qquad \sum_j (\sigma^o_{ij}, \phi_{p,j}) = (\frac{d^{k+1}}{dt^{k+1}} b_i, \phi_p) \qquad p \in P,$$

$$(3.7)_b \quad \begin{cases} \sigma^o = D\varepsilon^o & \text{for the elastic element of } E - E_k \\ \sigma^o = (D - D')\varepsilon^o - \lambda_{k+1} & \text{for the plastic element of } E - E_k \end{cases}$$

and for the element of E_k

$$(3.7)_c \quad \begin{cases} \sigma^o = D\varepsilon^o & \text{in } D_- = \{ u^o \ ; \ \partial f^*(t_0)D\varepsilon^o + r_k < 0 \} \\ \sigma^o = (D - D')\varepsilon^o - \lambda_{k+1} & \text{in } D_+ = \{ u^o \ ; \ \partial f^*(t_0)D\varepsilon^o + r_k \geq 0 \} \end{cases}$$

Theorem 3.3. The problem (3.7) has a unique solution (u^o,σ^o). This u^o minimizes the following functional F_{k+1} under the subsidiary conditions $(3.7)_b$ and $(3.7)_c$.

$$F_{k+1}(u^o) = \frac{1}{2}(\sigma^o, \varepsilon^o) - \frac{1}{2}(\lambda_{k+1}, \bar{\varepsilon}) - (\frac{d^{k+1}}{dt^{k+1}} b(t_0+0), u^o)$$

where

$$\bar{\varepsilon} = \begin{cases} 0 \ (\text{resp. } \varepsilon^o) & \text{for the elastic (resp. plastic) elements of } E - E_k \\ \left.\begin{array}{ll} \varepsilon^o_* & \text{in } D_- \\ \varepsilon^o & \text{in } D_+ \end{array}\right\} & \text{for the element of } E_k \end{cases}$$

and ε^o_* is an arbitrary vector in u^o-space included in the hyperplane

$$\pi_e : \{u^o \ ; \ \partial f^*(t_0)D\varepsilon^o + r_k = 0 \}.$$

Classify $E_k = E^e_k + E^p_k$ in such a way that $\partial f^* D\varepsilon^o + r_k$ is negative for E^e_k and non-negative for E^p_k at $t=t_0+0$, and solve (2.3) of the preceding stage replacing E^p by the new $E^p = E^p - E^e_k$. Let (u,σ,α) be the solution. Then (u^o,σ^o) is the derivative of order k+1 of (u,σ) at t_0+0. The derivative of order k+1 of (u,σ,α) is determined independently of the choice of the next $\dot{\sigma}-\dot{\varepsilon}$ relation of E_{k+1}.

Summarizing the above results we have

Theorem 3.4. The $\dot{\sigma} - \dot{\varepsilon}$ relation of each element of E is determined uniquely after t_0 and the problem (1.1) has a unique analytic solution in a certain time interval $[t_0, t_0+\delta)$ $(\delta>0)$.

So far we have discussed only the initial yielding. However, the above procedure and the results are valid almost word for word to the subsequent yieldings and also to the case that the unloading may occur. Since the boundedness of the solution is assured by the energy inequality we can continuate the solution over the given time interval I. In fact we have

Theorem 3.5. Let (u,σ,α) be the solution of (1.1) in a time interval $I' \subset I$. Then there is a constant C which is independent of I' such that

$$(3.8) \qquad \|\dot{u}\|,\ \|\dot{\varepsilon}\|,\ \|\dot{\sigma}\|,\ \|\dot{\alpha}\| \leqslant C.$$

Proof. In I' we have

$$(\dot{\sigma}, \dot{\varepsilon}) = (\dot{b}, \dot{u}).$$

Therefore hold

$$(3.9) \qquad ((D - D')\dot{\varepsilon},\dot{\varepsilon}) \leqslant |(\dot{b},\dot{u})| \leqslant \|\dot{b}\| \cdot \|\dot{u}\| \leqslant c\,\|\dot{b}\| \cdot \|\dot{\varepsilon}\|.$$

The first three inequalities of (3.8) thus follows from the positivity of the matrix D - D'. To prove the last inequality of (3.8), use the following relation which holds for any t in I' at which the derivatives exists.

$$\dot{\varepsilon} - D^{-1}\dot{\sigma} = \frac{1}{\eta} S\dot{\alpha},$$

where $f(\sigma)\partial f(\sigma) = S\sigma$.

<u>Remark.</u> Since we assumed that the function b is piecewise analytic, the number of the changes of the state will be finite in any finite time interval. If this is the case, the continuation of the solution is obvious. However,

this is not yet proved. Hence we have to consider the case that such points accumulate to a certain $t_0 < T$. In any case, however, the value of (u,σ,α) at $t = t_0-0$ becomes definite by the uniform boundedness of the solution, and our procedure to continuate the solution is completely valid to such t_0 too. It is clear that there is no bound beyond which this continuation is impossible.

The conclusion is

Theorem 3.6. There is a unique absolutely continuous function $(u,\varepsilon,\sigma,\alpha)$ which satisfies (1.1) except at most countable $t \in I$.

Proof. The existence of the solution is already proved. Also, the above discussion shows the uniqueness too. However we shall prove the uniqueness by another method. Let $K=K_\alpha$ be defined by

$$K = \{\tau \text{ ; absolutely continuous on } t\varepsilon I \text{ and } f(\tau-\alpha) \leqslant \bar{\sigma} \}.$$

If $(u,\varepsilon,\sigma,\alpha)$ satisfies (1.1), then $\sigma \in K$ and the followings hold.

$$\sum_j (\sigma_{ij}, \phi_{p,j}) = (b_i, \phi_p) \qquad p \in P,$$

$$\int_I (\dot{\varepsilon} - C\dot{\sigma}, \tau - \sigma)dt \leqslant 0 \qquad \text{for all } \tau \in K,$$

$$\dot{\alpha} = \eta S^{-1}(\dot{\varepsilon} - C\dot{\sigma}) \qquad \text{a.e. } I \qquad (C=D^{-1}).$$

Assume that $(u,\varepsilon,\sigma,\alpha)_*$ satisfies (1.1) too. Since $\sigma \in K$ can be written as $\sigma = \alpha + \theta\bar{\sigma}$ where $f(\theta) \leqslant 1$, we have

$$\int_I (\dot{\varepsilon} - C\dot{\sigma}, \alpha + \sigma_* - \alpha_* - \sigma)dt \leqslant 0$$

$$\int_I (\dot{\varepsilon}_* - C\dot{\sigma}_*, \alpha_* + \sigma - \alpha - \sigma_*)dt \leqslant 0.$$

Define $(U,E,\Sigma,A) = (u,\varepsilon,\sigma,\alpha) - (u,\varepsilon,\sigma,\alpha)_*$. Adding these inequalities, we have

$$0 \geq \int_I (\dot{E} - C\dot{\Sigma}, A - \Sigma)dt$$

$$= \frac{1}{\eta}\int_I (S\dot{A}, A)dt + \int_I (C\dot{\Sigma}, \Sigma)dt,$$

from which the uniqueness follows.

Now as is already seen in the existence proof of the solution, the exceptional t is always an end of a time interval of positive length. But the number of such time intervals of length $\geq 1/n$ ($n \geq 1$) is finite. Hence the countability of the exceptional t follows. This completes the proof.

Remark. In this paper we considered only the *kinematic hardening* problem. The *isotropic hardening* can also be treated in the same way [3].

References

[1] T. Miyoshi, *On the existence proof in plasticity theory*, Kumamoto J. Sci. 14 (1980), 18 - 33.

[2] T. Miyoshi, *A note on the classical solutions of semi-discrete quasi-static plasticity problems*, Kumamoto J. Sci. 15 (1982), 7 - 10.

[3] T. Miyoshi, *On the convergence of the finite element solutions for plasticity problems* (in Japanese), in RIMS reports No. 430 (1981), 3 - 12.

Lecture Notes in Num. Appl. Anal., **5**, 205–231 (1982)
Nonlinear PDE in Applied Science. U.S.-Japan Seminar, Tokyo, 1982

Two Dimensional Convection Patterns in Large Aspect Ratio Systems*

by

Alan C. Newell

Program in Applied Mathematics

University of Arizona, Tucson 85721

This work was supported in part by DOA Contract DAAG29-82-C-0068.

Figure 1 below is taken from a visualization of an experiment of Gollub and McCarriar [1] in which they follow the time development of a convecting horizontal layer of fluid of depth h through the use of laser Doppler techniques. The dots are points where, at a depth of h/4 from the top of the layer, the velocity component parallel to the larger side of the box is zero and mark the boundaries of the convective rolls. The experiment is conducted in a range of Rayleigh number where nonlinear stability theory guarantees that in a horizontal layer of infinite extent, straight parallel rolls both exist and are stable.

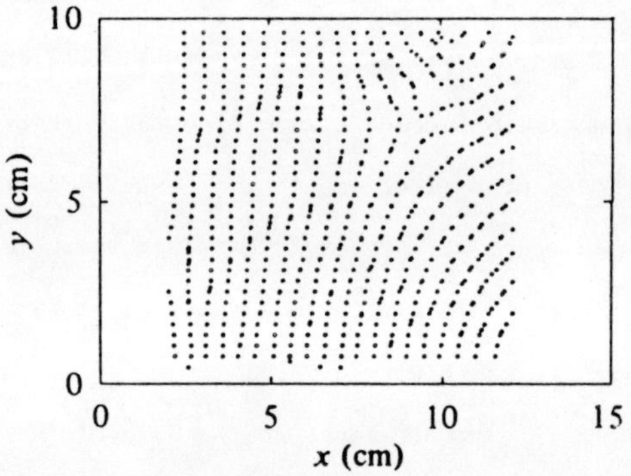

Figure 1: Doppler Map Of The Velocity Field For A Stable Convective Flow 256 Hrs. After Rayleigh Number Was Increased To 2.05 Rc. (Horizontal Diffusion Time, 40 Hrs.)

However, the existence of orientational degeneracy and a band of stable roll wavenumbers together with the fact that the rolls align themselves perpendicular to all lateral boundaries makes these solutions unattainable. The patterns that are seen are much more complicated not only containing curved rolls but also exhibiting many dislocations. Furthermore, it is not clear whether and, if so, under what circumstances the pattern achieves a time independent equilibrium. Indeed, in some cases, Gollub and McCarriar have seen patterns which are slowly time dependent over many horizontal diffusion times. The failure to reach a steady state was also noted earlier by Ahlers and Walden [2] who observed that the effective thermal conductivity of a layer of convecting fluid helium remained noisy for all Rayleigh numbers above R_C for sufficiently large aspect ratios.

Our goal in this paper is to develop a theory to describe these patterns. We start with the observation that almost everywhere in the convection field a local wavevector is defined and varies slowly over the box. At the boundaries, the wavevector is tangent. Therefore we expect that there exists locally periodic solutions defined by $f(\Theta; A, R)$ when f is 2π-periodic in Θ and

$$\nabla\Theta = \vec{k}(X, Y, T) \qquad (1)$$

$\left(\nabla = (\partial/\partial X, \partial/\partial Y)\right)$ is a slowly varying function of X, Y, T, the horizontal position coordinates and time. Indeed we know from the work of Busse [3] that such solutions as functions of Θ exist and, as we describe later, have certain stability properties. In other words, while the field variable f varies over distances of the order of the roll size, the parameters A (amplitude) and $\vec{k}$ (wavevector), which together with a knowledge of f as function of Θ describe the pattern, vary over distances of the size of the box. The inverse aspect ratio ϵ^2, the ratio of the roll wavelength d ($\simeq h$) to the linear dimension L of the box, is the only small parameter which enters the theory. The Rayleigh number R can be an order one amount above its critical value R_C, the value at which the purely conductive state becomes unstable, although it must be less than the Rayleigh number at which the straight parallel roll solution becomes unstable. The ideas we are about to describe are closely related to those of Whitham [4] who in the late sixties developed a theory to describe fully nonlinear, almost periodic wavetrains.

As a first attempt we shall use model equations, the use of which facilitates the analysis by making calculations explicit but which capture what we believe are some of the essential features of the Overbeck-Boussinesq equations. We derive equations for the slow variables $\vec{k}$, A which (i) show that all the stability criteria derived by Busse and his colleagues [3] for straight parallel rolls hold locally and (ii) reduce in the limit of small $R-R_C$ to the Newell-Whitehead-Segel [5] equations. In addition we show that the effect of curvature is to drive the roll patterns toward a state in which the local wavenumber assumes a constant value almost everywhere, a result consistent with a recent result of Pomeau and Manneville [6] for axially symmetric rolls. This is done by (i) proving that under certain natural boundary conditions, the macroscopic equations for the phase, valid for times up to the horizontal diffusion time, are deriveable from a Lyapunov functional even though the full microscopic equations for the model need not be and (ii) examining the nature of the stationary states which would be reached on this time.

But this is not the whole story as such patterns in and of themselves cannot satisfy all the boundary conditions. In between the patterns, dislocations are formed and solutions describing these structures are given. In looking at these solutions, it is clear that the time scale which the total system would need to relax to equilibrium is not simply the horizontal diffusion time L^2/ν but is this time scale multiplied by the aspect ratio L/d. Finally, we include in the model a term which takes account of the "mean drift" which occurs in systems of low to moderate Prandtl number. The presence of this effect was first pointed out by Siggia and Zippeleus [7], and can cause marked change in the behavior of the system.

I should point out that the ideas we discuss here can be extended to include situations where the basic flow f is multiperiodic. In that case f is considered to be a 2π periodic function in each of n phases Θ_i where $\nabla\Theta_i = \vec{k}_i$ and which also depends on n amplitudes A_i and a parameter (or set of parameters) R.

A more comprehensive paper is being written jointly with my colleague Mike Cross who started me thinking about these problems again during a most pleasant leave spent at the Institute of Theoretical Physics at the University of California in Santa Barbara. I am most grateful fo their hospitality and to the energetic leadership of Pierre Hohenberg who was a constant stimulus.

2. The predictions of present theories:

Consider $w(X, Y, T)$ given by

$$\frac{\partial}{\partial T}w + (\nabla^2 + 1)^2 w - Rw + w^2 w^* = 0 \qquad (2)$$

on $-\infty < X, Y < \infty$. The principal reason for choosing this model is that it possesses a simple fully nonlinear periodic solution $e^{i\Theta}$, $\Theta = \vec{k} \cdot \vec{X}$, $\vec{X} = (X, Y)$. The "conduction" solution $w = 0$ is unstable to modes of the form

$$w(X, Y, T) = We^{i\vec{k} \cdot \vec{X}}, \qquad (3)$$

whenever $R > \underset{k}{\mathrm{Min}}(k^2-1)^2 = 0$, which value is realized when $|\vec{k}| = k = 1$. The neutral stability curve in the (R, k) plane is given by $R = (k^2-1)^2$. We observe the system is degenerate in the sense that any mode of the form (3) with $|\vec{k}| = 1$ will grow at the same rate when $R > 0$. The nonlinear saturation of the linearly unstable modes is described by the Stuart-Watson [8] equation

$$W_T = RW - W^2W^* \qquad (4)$$

which in this case is an exact solution for (2) but which generally is only true for values of R near its critical value $R_C = 0$ and for amplitudes W proportional to $\sqrt{R - R_C}$. Because of the orientational degeneracy, it is indeed natural to look for solutions of the form

$$w(X, Y, T) = \sum_{|\vec{k}_j| = 1} W_j(T)e^{i\vec{k}_j \cdot \vec{X}}$$

and, in the small R limit, we can show

$$\frac{dW_j}{dT} = RW_j - W_j^2 W_j - 2 \sum_{i \neq j} W_r W_r^* W_j \quad . \qquad (5)$$

It is an easy matter to see that the only stable solution of this set is the single roll solution

$$W_1 = \sqrt{R}\; e^{i\vec{k}_1 \cdot \vec{X}}, \quad W_j = 0, \; j \neq 1. \qquad (6)$$

So, just as in the case of the Rayleigh-Benard problem under a vertically symmetric, applied temperature field with moderate Prandtl number $\nu/K \geqslant 0(1)$, single, parallel rolls are preferred. But, as we have pointed out, no direction is chosen a priori. Therefore when R is suddenly raised from subcritical to supercritical values, the system acts as a noise amplifier. At any particular location in a large horizontal layer, it will choose among the wavelengths of the noise for one close to $k_C = 1$ but it will not make any choice among directions. Therefore unless the experiment is carefully controlled (as in the case of the Busse-Whitehead [9] and Whitehead-Chen [10] experiments), rolls of approximately the critical wavelength but of different directions will spring up in different places. This directional diversity is even more accentuated due to the influence of a closed boundary. At each boundary loction, $\vec{k} \cdot \hat{n} = 0$ ($\hat{n}$ is the unit normal to the boundary) and therefore if $\hat{n}$ is continuous, rolls of all directions $\hat{k}$ are excited. Now, once the roll direction is chosen at a particular location, it becomes more stable against linear disturbances of rolls of other directions the more it grows towards its saturation amplitude. So, for early times, one finds a fluid layer resembling a sea of quasi stable patches and somehow the fluid has to find a way to resolve the incompatibilities between them.

There is also a bandwidth degeneracy. It is evident that for $R > 0$, a finite bandwidth of wavenumbers k can be excited. Indeed for $|\vec{k}| \neq 1$,

$$W_T = \left(R - (k^2 - 1)^2\right)W - W^2W^*$$

and $W \to \sqrt{R - (k^2 - 1)^2}$ asymptotically in time. We may test the (linear) stability of these solutions by setting

$$w = e^{ikX}\left(W + b_1 e^{iLX + iMY} + b_2 e^{-iLX - iMY}\right)$$

whereupon we find after some calculation that this solution is unstable if

$$2A^2(A^2)'(L^2+M^2)+2A^2(A^2)''(2k^2L^2)+4k^2L^2(A^2)'^2 > 0 \qquad (5)$$

where $A^2 = R - (k^2 - 1)^2$ and $(A^2)' = dA^2/dk^2$. The first class of instabilities occurs for

$$L = 0, \quad M \neq 0 \quad \text{and} \quad B = (A^2)(A^2)' > 0. \tag{6}$$

These correspond to the zig-zag instabilities discovered by Busse and arise when a wavevector with wavenumber $k < 1$ (in which case $(A^2)' > 0$) interacts with and gives its energy to either one of two modes $\left(k, \pm\sqrt{1 - k^2}\right)$ lying on the unit circle. The second class of instabilities occur for $M = 0$ and

$$A^2(A^2)' + 2A^2(A^2)''k^2 + k^2(A^2)'^2 > 0,$$

or equivalently when

$$\frac{d}{dk} kB > 0 \tag{7}$$

where $B = A^2\, dA^2/dk^2 = -2(k^2 - 1)\left(R - (k^2 - 1)^2\right)$. This is the Eckhaus instability and occurs when a roll has too small a wavelength. It is useful to summarize these results by way of figure 2.

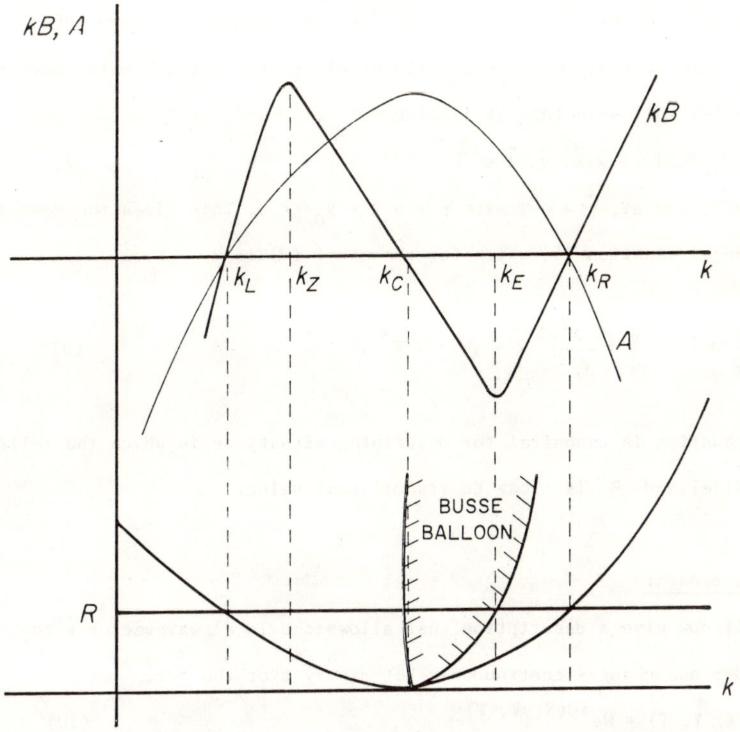

Figure 2: Graphs of A, kB and R vs. k. and the Busse Balloon

Solutions can exist for $k_L < k < k_R$ but are stable only when $k_C < k < k_E$. The shaded area is known as the Busse balloon. When derived in the context of the full Overbeck-Boussinesq equations it is somewhat more complicated. The right hand curve which is the boundary with the Eckhaus instability is replaced for large Prandtl number by a boundary to an instability to rolls in the perpendicular direction. On the other hand, for smaller Prandtl numbers, the Eckhaus stability boundary becomes linked with the skew-varicose instability which is a variant of the former when mean drift effects are included. We will show how to include these in the model. We also remark that the left hand boundary need not remain at $k = k_C = 1$ but can bend leftward depending on Prandtl number. The Busse balloon (or Busse windsock), the region of stable parallel rolls in the R, $p = \nu/K$, k plane is given in reference [3].

Since for $R > R_C$, there is an order $\sqrt{R - R_C}$ band of allowable wavenumbers in a direction parallel to $\vec{k}$ and an order $\sqrt[4]{R - R_C}$ band in the perpendicular direction, it is natural to include a richer class of solutions which have an almost parallel roll structure by letting

$$w(X, Y, T) = W(\tilde{x}, \tilde{y}, \tilde{t})e^{iX} \tag{8}$$

where $\tilde{x} = \mu^2 X$, $\tilde{y} = \mu Y$, $\tilde{t} = \mu^4 T$ with $\mu^4\chi = R - R_C \ll 1$. This gives the Newell-Whitehead-Segel equation [5] which for the model (2) is

$$\frac{\partial W}{\partial \tilde{t}} + \left(2i \frac{\partial}{\partial \tilde{x}} + \frac{\partial^2}{\partial \tilde{y}^2}\right)^2 W = \chi W - W^2 W^* . \tag{9}$$

This equation is canonical for describing situations in which the rolls are almost parallel and R is close to its critical value.

3. A new approach:

We will now give a description that allows the local wavevector $\vec{k}$ to undergo order one changes continuously but slowly over the box. Let

$$w(X, Y, T) = We^{i\Theta(X, Y, T)} \tag{10}$$

where W and $\vec{k} = \nabla\Theta$ are functions of the slow variables

$$x = \varepsilon^2 X, \; y = \varepsilon^2 Y, \; t = \varepsilon^4 T$$

and $1/\varepsilon^2$ is the aspect ratio. It is useful to write $\theta = \frac{1}{\varepsilon^2}\Theta(x, y, t)$ whence

$$\begin{aligned} &\nabla_{\vec{X}}\Theta = \nabla_{\vec{x}}\theta = \vec{k} = (m, n) = (k \cos\psi, k \sin\psi) \\ &\Theta_T = \varepsilon^2\theta_t = \varepsilon^2\sigma \end{aligned} \tag{11}$$

We will also find it useful to introduce new coordinates

$$\theta = \alpha(x, y), \quad \beta = \beta(x, y) \tag{12}$$

defined by

$$\begin{aligned} \alpha x &= k \operatorname{Cos}\psi & \beta x &= -\ell \operatorname{Sin}\psi \\ \alpha y &= k \operatorname{Sin}\psi & \beta y &= \ell \operatorname{Cos}\psi \end{aligned} \tag{13}$$

The Jacobian of the transformation from (α, β) to (x, y) is $k\ell$ and

$$k\frac{\partial}{\partial\alpha} = \operatorname{Cos}\psi\frac{\partial}{\partial x} + \operatorname{Sin}\psi\frac{\partial}{\partial y}, \quad \ell\frac{\partial}{\partial\beta} = -\operatorname{Sin}\psi\frac{\partial}{\partial x} + \operatorname{Cos}\psi\frac{\partial}{\partial y}. \tag{14}$$

The curves $\alpha(x, y)$ = constant are, of course, the loci of constant phase θ while the β coordinates are the orthogonal trajectories and measure distance along the rolls. The curvature K_α of the $\alpha(x, y)$ = constant curves is given by $\ell\frac{\partial\psi}{\partial\beta}$. The curvature of the constant curves β is $K_\beta = -k\frac{\partial\psi}{\partial\alpha}$.

In these coordinates, the compatibility conditions (11) give us that

$$K_\alpha = \ell\frac{\partial\psi}{\partial\beta} = -\frac{k}{\ell}\frac{\partial\ell}{\partial\alpha}, \quad K_\beta = -k\frac{\partial\psi}{\partial\alpha} = -\frac{\ell}{k}\frac{\partial k}{\partial\beta}, \tag{15a}$$

and

$$k\frac{\partial}{\partial\alpha}\cdot\frac{\partial\theta}{\partial t} = \frac{\partial k}{\partial t}, \quad \ell\frac{\partial}{\partial\beta}\cdot\frac{\partial\theta}{\partial t} = k\frac{\partial\psi}{\partial t}. \tag{15b}$$

Note that

$$\frac{\partial}{\partial T}Ae^{i\theta} = e^{i\theta}\left(i\varepsilon^2\sigma + \varepsilon^4\frac{\partial}{\partial t}\right)A, \tag{16}$$

$$\nabla^2 Ae^{i\theta} = e^{i\theta}\left(-k^2 + i\varepsilon^2 D_1 + \varepsilon^4 D_2\right)A, \tag{17}$$

where

$$D_1 A = 2m \frac{\partial}{\partial x}A + 2n \frac{\partial}{\partial y}A + \left(\frac{\partial m}{\partial x} + \frac{\partial n}{\partial y}\right)A \,, \tag{18a}$$

$$= 2k^2 \frac{\partial A}{\partial \alpha} + Ak \frac{\partial k}{\partial \alpha} + Ak\ell \frac{\partial \psi}{\partial \beta} \,, \tag{18b}$$

$$= 2k^2 \frac{\partial A}{\partial \alpha} + Ak \frac{\partial k}{\partial \alpha} - Ak^2/\ell \frac{\partial \ell}{\partial \alpha} \,, \tag{18c}$$

$$= \frac{k\ell}{A^2} \frac{\partial}{\partial \alpha} \frac{kA^2}{\ell} \,, \tag{18d}$$

and

$$D_2 A = \frac{\partial^2}{\partial x^2} + \frac{\partial^2}{\partial y^2} A$$

$$= \left\{\left(k \frac{\partial}{\partial \alpha}\right)^2 + \left(\ell \frac{\partial}{\partial \beta}\right)^2 + K_\alpha k \frac{\partial}{\partial \alpha} + K_\beta\, \ell \frac{\partial}{\partial \beta}\right\}A. \tag{19}$$

We now proceed to determine the equations satisfied by the slow variables $\vec{k}$ and A, the latter being the leading approximation to W which, without loss of generality, can be taken to be real. Let

$$W = A + \varepsilon^2 W_2 + \dots \,, \tag{20}$$

$$\sigma = \sigma_0 + \varepsilon^2 \sigma + \dots \,, \tag{21}$$

$$R - A^2 = R_0 + \varepsilon^2 R_2 + \dots \,. \tag{22}$$

We choose the sequences $\{\sigma_n\}$, $\{R_n\}$ in a manner so as to eliminate secular terms appearing in the expansion for W. In this case, secular means that no solution exists to the algebraic equations for W_2, W_4 etc. If we had left in the Θ variable and expanded $w = Ae^{i\Theta} + \varepsilon^2 w_2(\Theta) + \dots$, then, unless the secular terms were removed, no 2Π periodic solution for w_2 would exist. Substituting (20), (21), (22), (16), (17) into (2) gives

$$\{i\varepsilon^2\sigma + \varepsilon^4 \frac{\partial}{\partial t} + (k^2-1)^2 - i\varepsilon^2\left(D_1(k^2-1)\cdot + (k^2-1)D_1\cdot\right)$$

$$- \varepsilon^4\left(D_2(k^2-1)\cdot + (k^2-1)D_2\cdot + D_1{}^2\right) + i\varepsilon^6\left(D_1\cdot D_2 + D_2\cdot D_1\right) + \varepsilon^8 D_2{}^2$$

$$-R_0 - \varepsilon^2 R_2 - \varepsilon^4 R_4 \ldots + 2\varepsilon^2 AW_2 + \varepsilon^4(2AW_4 + W_2{}^2)\ldots\}(A + \varepsilon^2 W_2 + \ldots) = 0 \; . \tag{23}$$

What does this equation tell us? At $0(1)$

$$R_0 = (k^2-1)^2 \tag{24}$$

and so, to leading order, the amplitude A is determined from the "eikonal" equation

$$A^2 = R - (k^2-1)^2 \; . \tag{25}$$

At order ε^2 we have that

$$\left(-R_0 + (k^2-1)^2\right)W_2 + 2A^2W_2 = \text{R.H.S.}$$

Note the following interesting feature. For $R - R_C = 0(1)$, A^2 is finite and therefore, even though $R_0 = (k^2 - 1)^2$, the only terms on the RHS which give rise to secular behavior are those which are purely imaginary. On the other hand, if A^2 were small, we would have to remove all the terms on the RHS. In other words, the null space of the linearized equation is cut in half when $A = 0(1)$. What this means is that instead of having two equations, one for the amplitude A and the other for the phase of the convective pattern, we simply have a single equation for the phase. The amplitude is determined algebraically from (25). This also means, of course, that the limit to the case of small $R - R_C$ and small amplitude A is fairly subtle. [For workers in nonlinear wave theory, there is a direct analogue between this limit process and the limit process one encounters when one attempts to obtain the nonlinear Schrodinger equation from Whitham's theory.] We will carry out the perturbation analysis in such a way so as to facilitate this limit process. What we do is to use the expansion (22) which is also an amplitude expansion to reexpand A if necessary so that we can simply set $W_{2j} = 0$, $j \geqslant 1$. We find directly from (23) the following equations,

$$\sigma A - D_1(k^2-1)A - (k^2-1)D_1A + \varepsilon^4(D_1 \cdot D_2 + D_2 \cdot D_1)A = 0, \tag{26}$$

$$R-A^2 - (k^2-1)^2 = \frac{\varepsilon^4}{A}\left(A_t - (k^2-1)D_2A - D_2(k^2-1)A - D_1^2A + \varepsilon^4 D_2^2 A\right) . \tag{27}$$

Because of the simplicity of this model, these equations are exact. We will first examine these equations with a veiw to making contact with known results and then we will discuss some new consequences.

4. Connections with previous theories.

For values of R of order unity, we can neglect the RHS of equation (27) and then A is given as function of k by (25). Equation (26) tells us about the phase $\sigma = \theta_t$ and ignoring the $O(\varepsilon^4)$ terms can be written as

$$A^2 \frac{\partial\theta}{\partial t} + \frac{\partial}{\partial x} m B + \frac{\partial}{\partial y} n B = 0 , \tag{28}$$

or

$$A^2 \frac{\partial\theta}{\partial t} + \left(B + \frac{m^2}{k}\frac{dB}{dk}\right)\frac{\partial^2\theta}{\partial x^2} + \frac{2mn}{k}\frac{dB}{dk}\frac{\partial^2\theta}{\partial x \partial y} + \left(B + \frac{n^2}{k}\frac{dB}{dk}\right)\frac{\partial^2\theta}{\partial y^2} = 0 \tag{29}$$

where

$$B(k) = A^2(k)\, dA^2/dk^2 . \tag{30}$$

I want to remark at this point that the fact that the spatial terms have conservation form is not a consequence of this particular model nor the fact that it can be derived from a Lyapunov functional.

(a) The Busse Balloon holds locally.

Equation (29) is elliptic stable or unstable (in time) or hyperbolic unstable depending on which one of the following four cases obtains:

1) $B < 0, \frac{d}{dk}(kB) < 0$; Elliptic stable
2) $B > 0, \frac{d}{dk}(kB) > 0$; Elliptic unstable
3) $B > 0, \frac{d}{dk}(kB) < 0$; Hyperbolic unstable
4) $B < 0, \frac{d}{dk}(kB) > 0$; Hyperbolic unstable. (31)

These results are simply the same results which are displayed in Figure 2, except now all the variables are functions of x, y and t. Therefore we can say that all the stability features we had found when looking at the stability of straight parallel rolls continue to hold locally. Case (1) above is the Busse balloon; case (2) involves instabilities which have wavenumber dependence in both the along and perpendicular to the roll directions; case (3) is the zig-zag instability and case (4) is the Eckhaus intability. This can be easily seen by taking the local roll wavevector to be (k, 0) in which case (29) becomes

$$A\frac{\partial\theta}{\partial t} + \frac{d}{dk}(kB)\frac{\partial^2\theta}{\partial x^2} + B\,\frac{\partial^2\theta}{\partial y^2} = 0 \quad . \tag{32}$$

Hence for $B > 0$, $\frac{d}{dk}(kB) > 0$, the instability has a wavevector perpendicular to $(k, 0)$; for $B < 0$, $\frac{d}{dk}\,kB > 0$ the unstable modes are parallel to $(k, 0)$. The addition of the ε^4 term in (26) which involves higher derivatives only serves to control the growth of the instabilities after they begin. It does not inhibit them altogether nor does it of itself trigger any new instability.

The reader might like to compare this result with what happens in nonlinear wavetrains. There, the analogue of equation (26) is a second order system in x and t and so it is the ellipticity or hyperbolicity of the second order operator which determines instability or (neutral) stability of the wavetrain. For example, for a train of gravity waves on the sea surface, the hyperbolic nature of (26) changes to elliptic when the ratio of depth to wavelength is less than 1.36.

(b) <u>The Newell-Whitehead-Segel limit.</u>

To this point, we have taken variations in the directives parallel to and perpendicular to the local roll to be of the same order of magnitude. It is clear that if for some reason the local wavenumber is forced to stay approximately constant, the variations in wavenumber of order μ parallel to $\vec{k}$ are accompanied by variations of order $\sqrt{\mu}$ in the perpendicular direction (e.g. $(k_c + \mu L)^2 + (\sqrt{\mu}\,M)^2 \simeq k_c^2$) . Near $k = 1$, we find that variations perpendicular to the roll are of an order of magnitude greater than those parallel to the roll and this leads to a balance between the term $\frac{1}{A}\nabla \cdot \vec{k}\,B$ and some of the ε^4 terms in the phase equation (26).

This situation certainly obtains when R is sufficiently small, for then (see Figure 2) the bandwidth of wavenumbers parallel to the roll is $O(\sqrt{R})$ and the bandwidth perpendicular to the roll is $O(\sqrt[4]{R})$. As we have mentioned, in this limit the amplitude no longer follows the phase gradient as in (25) but the terms on the RHS of the amplitude equation (27) became equally important to these on the L.H.S. This balance is achieved when $R = \varepsilon^4\chi$. For rolls which are almost parallel

$$w(X, Y, T) = A(x, \tilde{y}, t)e^{i(X + \phi(x, \tilde{y}, t))} \tag{33}$$

and

$$\alpha = \theta = x + \varepsilon^2\phi(x, \tilde{y}) \tag{34}$$

where $x = \varepsilon^2 X$ as before and $\tilde{y} = y/\varepsilon = \varepsilon Y$, the new scaling in the perpendicular direction. It is now easy to show from (11) that

$$k = 1 + \varepsilon^2\left(\phi_x + \frac{1}{2}\phi_{\tilde{y}}^2\right), \; \psi = \varepsilon\phi_{\tilde{y}}$$

$$k\partial_\alpha = \partial_x + \phi_{\tilde{y}}\partial_{\tilde{y}}, \; \ell\partial\beta = 1/\varepsilon\; \partial_{\tilde{y}}, \; \sigma = \alpha_t = \varepsilon^2\phi_t$$

$$D_1 = 2(\partial_x + \phi_{\tilde{y}}\partial_{\tilde{y}}) + \phi_{\tilde{y}\tilde{y}}, \quad D_2 = 1/\varepsilon^2\,\partial_{\tilde{y}}^2,$$

$$K_\alpha = \ell\psi_\beta = \phi_{\tilde{y}\tilde{y}} \text{ and } K_\beta = -k\psi_\alpha = -\varepsilon\phi_{x\tilde{y}} - \varepsilon\phi_{\tilde{y}}\phi_{\tilde{y}\tilde{y}}, \tag{35}$$

where we have used subscripts in order to denote partial derivatives. Substitution of (35) into (26) and dividing by ε^2 gives (we drop the tilde on y)

$$\begin{aligned} &A\phi_t - 2\left(\phi_x + \tfrac{1}{2}\phi_y^2\right)\left(2\partial_x + 2\phi_y\partial_y + \phi_{yy}\right)A - 2\left(2\partial_x + 2\phi_y\partial_y + \phi_{yy}\right)\left(\phi_x + \tfrac{1}{2}\phi_y^2\right)A \\ &+\left(2\partial_x + 2\phi_y\partial_y + \phi_{yy}\right)A_{yy} + \partial_y^2\left(2\partial_x + 2\phi_y\partial_y + \phi_{yy}\right)A = 0 \;. \end{aligned} \tag{36}$$

It is readily shown that, if $W = Ae^{i\phi}$ in (9), equation (36) is precisely the imaginary part of equation (9). Carrying out the same calculation on (27) (recall $A \to \varepsilon^2 A$) gives the real part of equation (9).

Therefore the equations (26), (27) contain all that was previously known about roll solutions. They also contain some new information.

5. New results; some answers, more questions.

In what follows we shall take R to be of order one and therefore (27) can be replaced by (25) almost everywhere. The exceptions are those regions where $\nabla = 0(\varepsilon^{-2})$ but these points are isolated. We will concentrate on the phase equation (26),

$$A \frac{\partial\theta}{\partial t} + \frac{1}{A} \nabla (\vec{k}B) + \varepsilon^4 (D_1 \cdot D_2 + D_2 \cdot D_1) A = 0 \quad , \tag{37}$$

which may be rewritten in a variety of ways. In particular we may write

$$\nabla(\vec{k}B) = k \frac{\partial}{\partial\alpha} kB + kB\ell \frac{\partial\psi}{\partial\beta} \tag{38}$$

or in a more revealing way as

$$\nabla(\vec{k}B) = k\ell \frac{\partial}{\partial\alpha} \frac{kB}{\ell} \quad . \tag{39}$$

applying $k \frac{\partial}{\partial\alpha} \frac{1}{A}$, $\ell \frac{\partial}{\partial\beta} \frac{1}{A}$ to (37) gives us two equations for k and ψ,

$$\frac{\partial k}{\partial t} + k \frac{\partial}{\partial\alpha}\Big(\frac{1}{A^2} k \frac{\partial}{\partial\alpha} kB + \frac{kB}{A^2} \ell \frac{\partial\psi}{\partial\beta}\Big) + \varepsilon^4 k \frac{\partial}{\partial\alpha} \frac{1}{A}(D_1 \cdot D_2 + D_2 \cdot D_1)A = 0 \tag{40}$$

and

$$k\frac{\partial\psi}{\partial t} + \ell\frac{\partial}{\partial\beta}\Big(\frac{1}{A^2} k\frac{\partial}{\partial\alpha}kB + \frac{kB}{A^2} \ell \frac{\partial\psi}{\partial\beta}\Big) + \varepsilon^4 \ell \frac{\partial}{\partial\beta} \frac{1}{A}(D_1 \cdot D_2 + D_2 \cdot D_1)A = 0 \ . \tag{41}$$

For the first step, let us assume that all derivatives are of order one and consequently ignore the ε^4 terms. We will assume that everywhere k belongs to the Busse balloon $1 = k_C < k < k_E(R)$ and prove that in a region R with certain conditions on the boundary ∂R, the system relaxes to a stationary state with wavenumber k taking on the value which makes $B = 0$. This result does not depend critically on the fact that the present model is deriveable from a Lyapunov function; indeed it is also valid for systems which do not have this property. The reader might like to verify that the model

$$\Big(\frac{\partial}{\partial T} - (\nabla^2-1)\Big)(\nabla^2-1)^2 w + \big(R - ww^* - ww^*\nabla^2\big)\nabla^2 w = 0 \ , \tag{42}$$

which does not derive from Lyapunov functional, gives the phase equation

$$\big(k^2 + 1\big)^2 A^2 \theta_t k^2\big(1 - \nu k^2\big) + \nabla\cdot\vec{k}B + 0\big(\varepsilon^4\big) = 0 \tag{43}$$

where $B = A^2 \frac{dA^2}{dk^2} k^4(1-\nu k^2)^2$ and $A^2 = R - \frac{(k^2+1)^3}{k^2}$. It is crucial, however, that the order one spatial derivative terms in the phase equation have conservation form and although I have not yet categorized the class, this happens for a large class of problems. We now prove our result. For positive J^2, let

$$J^2\theta_t + \nabla\cdot(\vec{k}B) = 0 \tag{44}$$

and consider

$$F = \iint_R G(\vec{k})\, dxdy \tag{45}$$

where

$$G = -1/2\int^{k^2} B(k^2)dk^2 > 0 \tag{46}$$

is positive as we insist k belongs to the Busse balloon in which $B < 0$. Then,

$$\frac{dF}{dt} = -\int_{\partial R} \theta_t\, \vec{k}B.n\, ds - \iint_R J^2\theta_t^{\,2}\, dxdy, \tag{47}$$

n the outward unit normal to the boundary ∂R, where we have used the fact that $\nabla_{\vec{k}} G = -\vec{k}B$. Therefore, if on ∂R either (i) $\vec{k}\cdot n = 0$ (the roll axes are perpendicular to the fixed boundary) or (ii) $B = 0$ (a portion of ∂R may be a fluid boundary where $B = 0$),

$$\frac{dF}{dt} < 0 ,$$

and G decreases. But G is minimum only when $B = 0$ or $k = k_C$ the value of k for which $dA^2/dk^2 = 0$. For model (2) this is the point $k = 1$; for model (42), k_C lies to the left of $k = 1$ by an amount depending on ν.

The fact that $k \to k_C$ on a free boundary on portions of ∂R is consistent with the analysis of the stationary equation,

$$\frac{\partial}{\partial\alpha}\frac{kB}{\ell} = 0 . \tag{48}$$

This means that the quantity

$$kB/\ell = -H^2(\beta) \tag{49}$$

is constant along the orthogonal trajectories of the constant phase contours. Recall in the interval $1 < k < k_E$, $kB < 0$. This in turn means that if the β contours converge, which they will do in patches where the curvature of the phase contours increases towards a center, ℓ increases. It also means that the flux of $\vec{k}B$ between two β contours is independent of α. It may be useful for the reader to keep in mind the axially symmetric case where $\alpha = \int k(r)dr$, r the radial coordinate, $\beta = \psi = \phi$, ϕ the azimuthal coordinate, whence $\ell = \frac{1}{r}$. Now in order for the solution to remain stable we must have that

$$1 < k < k_E$$

which implies that

$$0 < |kB| < |kB|_E$$

where $|kB|_E$ is the absolute value of $k_E\, B(k_E)$. Therefore we must have that

$$0 < \ell H^2(\beta) < |kB|_E \tag{50}$$

and therefore as $\ell \to \infty$ along a β contour, $H^2(\beta)$ must tend to zero. Since it is constant along constant β contours, it must become as small as it can before the ε^4 terms in (37) enter the picture, which they will do when $\frac{\partial}{\partial\alpha} = 0\,(\varepsilon^{-2})$ and $\ell = 0(\varepsilon^{-2})$. Thus, in order that the inequality (50) holds, $H^2(\beta) = \varepsilon^2 K(\beta)$ and hence $kB = 0(\varepsilon^2)$ everywhere that $\ell = 0(1)$. But, since kB is small only near $k_c = 1$, we must have $k = 1 + 0(\varepsilon^2)$. Near the sink, we can find solutions of (37) iteratively in the form

$$k = 1 + \frac{\varepsilon^2\gamma^2}{r + \varepsilon^2\rho} + \cdots, \quad \rho = \frac{3\gamma^2 + 1}{4\gamma^2} \tag{51}$$

where $\gamma^2 = \frac{1}{4R} K^2(\beta)$, which indicates that as $r \to \varepsilon^2$, k goes from $k_C = 1$ to a value somewhere between $k_C = 1$ and k_E.

Thus our first prediction is that on the time scale ε^{-4}, the horizontal diffusion time, patches form which satisfy the boundary conditions $\vec{k} \cdot n = 0$ along portions of the box boundary, and in which $k \to k_c$ almost everywhere. Examine the numerical experiments of Greenside, Coughran and Schryer [11] (figure 3) carried out on equation (2) for real ψ and the real experiments of Bergé [12] (figure 4).

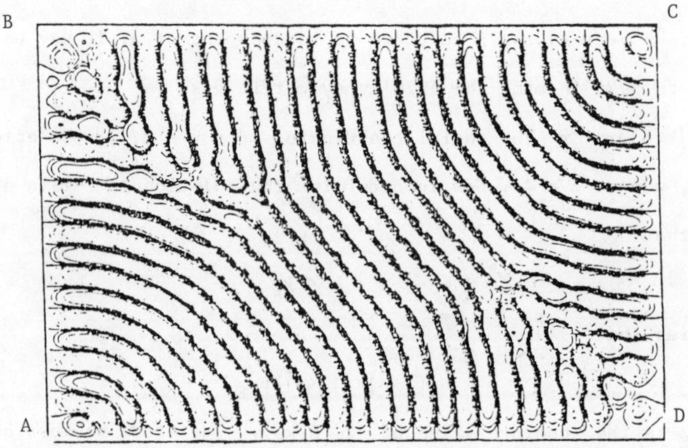

Fig. 3: Numerical Integration of (2), ψ Real
For Time $\sim$ Horizontal Diffusion Time

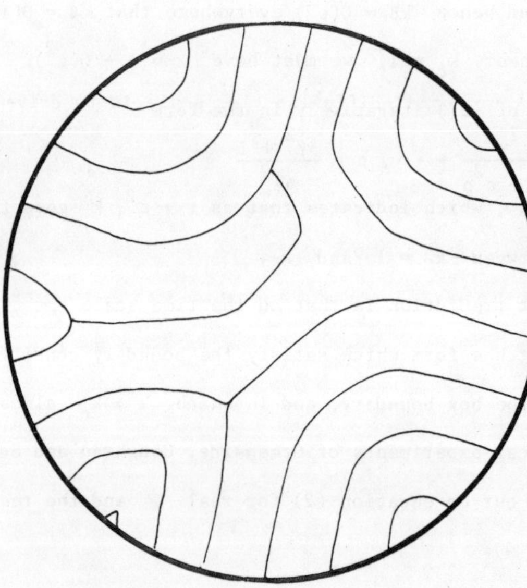

Fig. 4: From an experiment of P. Bergé. Contours of Constant Downward Velocity. Aspect Ratio of 16. $R \sim 2R_c$.

Note in the rectangular geometry of figure 3, that patches with circular symmetry form about the corners A and C. In the circular geometry of Bergé'sreal experiment, one again sees circular patches forming about sinks which are attached to the boundary. Moreover, it is abundantly clear from these figures that the box cannot be tiled with these patches. Certain areas, for example the corner B and D in figure 3, are quite incompatible. In order to compensate for these mismatches, the ε^4 terms of the phase equation (37) must be incorporated in the analysis.

One way is to take $\partial/\partial\alpha$, $\partial/\partial\beta$ to be $O(\varepsilon^{-2})$ but this simply brings us back to the microscopic theory. Another way, which retains the fundamental idea that the convection field can be described by a slowly varying wavevector $\vec{k}$, is to recognize that the terms

$$\frac{1}{A}\nabla(\vec{k}B) \quad \text{and} \quad \varepsilon^4(D_1 \cdot D_2 + D_2 \cdot D_1)\, A \tag{52}$$

can balance when the β derivatives are $O(\varepsilon^{-1})$ and $k = 1 + O(\varepsilon^2)$. This reduces both terms in (53) to $O(\varepsilon^2)$ and since we saw that the main effect of the dynamics on the horizontal diffusion time scale is to drive k towards k_c, this approximation is not at all unreasonable. A little algebra shows that if $R = O(1)$, $k = 1 + O(\varepsilon^2)$, then A is $\sqrt{R}$ to within $O(\varepsilon^4)$ and the stationary patterns which one might expect to reach on the ε^{-6} time scale (which is the horizontal diffusion time scale multiplied by the aspect ratio ε^{-2}) are given by

$$-2\nabla \cdot \vec{k}(k^2-1) + \varepsilon^4\left(\ell \frac{\partial}{\partial\beta}\right)^3\psi = 0 \tag{53}$$

and the compatibility conditions (11) are

$$\frac{\partial}{\partial\alpha}(\ell^{-1}) = \frac{\partial\psi}{\partial\beta} \tag{54}$$

$$\frac{\ell\partial}{\partial\beta}(k^{-1}) = \frac{\partial\psi}{\partial\alpha}. \tag{55}$$

Equations (54), (55) show us that $\frac{\partial\psi}{\partial\beta}$ is at most order one and therefore ψ itself is of order ε. For these solutions, then, the rolls are <u>locally</u> almost straight. Since we are now working in distances of order ε (as measured in box units; $1/\varepsilon$ in roll wavelength units) we can make the following local approximations, which are very similar to those made when we were deriving the Newell-Whitehead-Segel equations (36). Let (ξ,η) be locally the 'across and along' roll cordinates and

$$\theta = \xi + \varepsilon^2\phi(\xi,\zeta),\ \zeta = \eta/\varepsilon. \tag{56}$$

Using (56) together with (12), (13) and (14) we find

$$\ell\frac{\partial}{\partial\beta} = \frac{1}{\varepsilon}\frac{\partial}{\partial\zeta},$$

$$K_\alpha = \ell\frac{\partial\psi}{\partial\beta} = \frac{\partial^2\phi}{\partial\zeta^2}$$

$$k^2 = 1 + \varepsilon^2\left(\frac{\partial\phi}{\partial\xi} + \frac{1}{2}\left(\frac{\partial\phi}{\partial\zeta}\right)^2\right), \tag{57}$$

$$k\frac{\partial}{\partial\alpha} = \frac{\partial}{\partial\xi} + \frac{\partial\phi}{\partial\zeta}\frac{\partial}{\partial\zeta},$$

and equation (53) is (using subscripts for partial derivatives)

$$-4\phi_{\xi\xi} - 8\phi_\zeta\phi_{\xi\zeta} - 4\phi_\xi\phi_{\zeta\zeta} - 6\phi_\zeta^2\phi_{\zeta\zeta} + \phi_{\zeta\zeta\zeta\zeta} = 0, \tag{58}$$

which is precisely the Newell-Whitehead-Segel equation (36) and (9) with amplitude A held constant. We are now going to discuss solutions of this equation which lend some insight into Figure 5 which is the a sequel to Figure 3

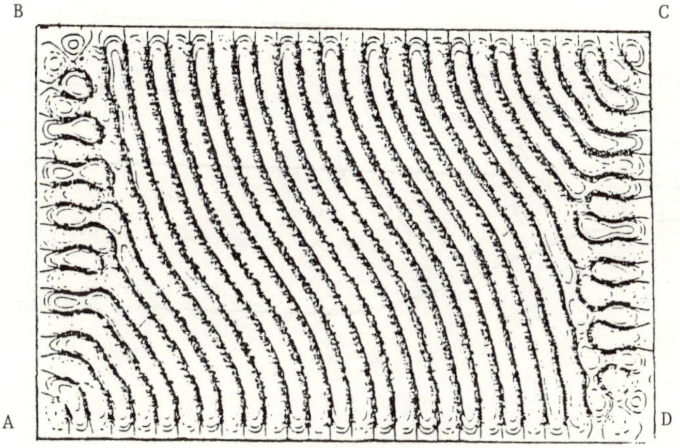

Figure 5: Numerical Integration of (2), ψ real
For Time >> Horizontal Diffusion Time.

Notice that in order to compensate for the incompatibilities in the corner B of Figure 3, the circular rolls emanating from AD have undergone a change and have introduced dislocations along the wall AB. Roll number 7, counting from A along AD, doubles in width as it approaches the side AB and undergoes a dislocation. Roll number 9 detaches from AB altogether and forms a series of dislocations along AB (which we call a grain boundary) and then attaches itself to the upper wall BC. Rolls 10 through 25 take on an S shape in which the approximate distance over which significant changes occur is the square root of the box dimension, or the 'along the roll' scaling in equation (58).

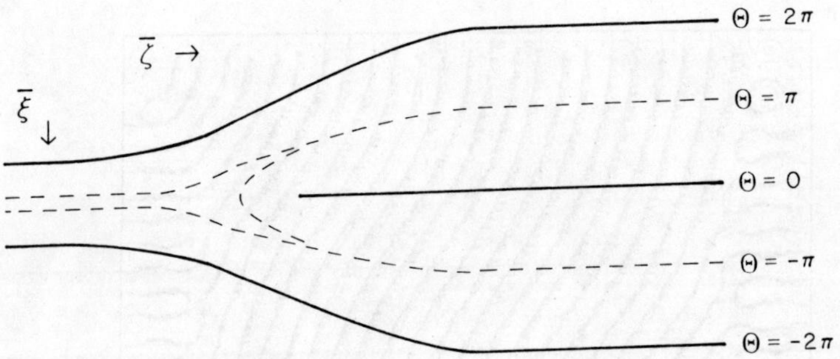

Figure 6: Dislocation

We first note that a property of this equation is that if

$$\phi(\xi,\zeta) \text{ solves } (58), \text{ so does } -\phi(-\xi,\zeta). \tag{59}$$

Also observe that $\Theta = \varepsilon^{-2}\theta$ is given by

$$\Theta = \bar{\xi} + \theta(\xi,\zeta), \ \xi = \varepsilon^2\bar{\xi}. \tag{60}$$

The shapes of the phase contours near dislocations suggest that we search for self similar solutions of the form

$$\phi(\xi,\zeta) = F(z), \ z = \zeta/\sqrt{2\xi}, \ \xi > 0. \tag{61}$$

Notice that in cell units $z = \zeta/\sqrt{2\xi} = \bar{\zeta}/\sqrt{2\bar{\xi}}$ where $\zeta = \varepsilon\bar{\zeta}$, $\xi = \varepsilon^2\bar{\xi}$ and

$$\Theta = \overline{\xi} + F(\overline{\xi}/\sqrt{2\overline{\xi}}) \, . \tag{62}$$

$F(z)$ satisfies the equation, $F' = \frac{dF}{dz}$,

$$F'''' = 4z^2F'' + 12zF' - 12zF'F'' - 8F'^2 + 6F'^2F'' \tag{63}$$

which is really an equation for $G = F'$. It has the symmetry property that if

$(F(z), G(z))$ solves (63), so does

$$(\overline{F}(z), \overline{G}(z)) = (F(-z), -G(-z)). \tag{64}$$

Equation (63) has a one parameter family of solutions $F(C;z)$ with derivatives $G(C,z)$ which decay as Ce^{-z^2} as $z \to -\infty$. This can be seen by linearizing (63) which then has error function solutions. For C very small, these solutions behave very much like the error function solutions; they are symmetric about $z = 0$ and lead to a jump in F of $\Delta F = F(\infty) - F(-\infty) = \sqrt{\pi}C$. However (63) is nonlinear and the bigger C gets the closer F approaches its pole solutions (actually G has the pole; F has a logarithmic singularity)

$$F(z) \sim \ell n \frac{1}{z-z_0} \tag{65}$$

representing a balance between F'''' and $6F'^2F''$. Thus there is a critical value C_0 of C, which from numerical calculations is approximately .564, above which the solutions do not exist over the line $-\infty < z < \infty$. As C approaches C_0 from below, ΔF is very sensitive to changes in C, going from a value of 3.14 ($\simeq \pi$) at $C = .54$ to 7.93 at $C = .565$. In figure 7, we graph $F(z)$, $\overline{F}(z)$ for $C = \pi$ and in Figure 8, we draw the contour of constant phase Θ

$$\Theta = \overline{\xi} + \overline{F}\left(\zeta/\sqrt{2\overline{\xi}}\right) \, . \tag{66}$$

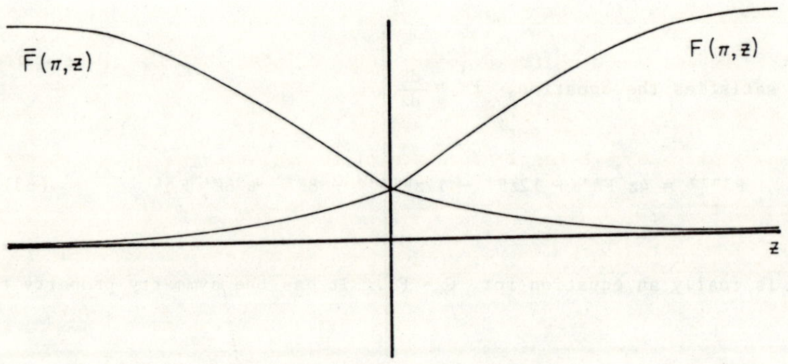

Figure 7: Graphs of $\overline{F}(\overline{\pi},z)$, $F(\pi,z)$.

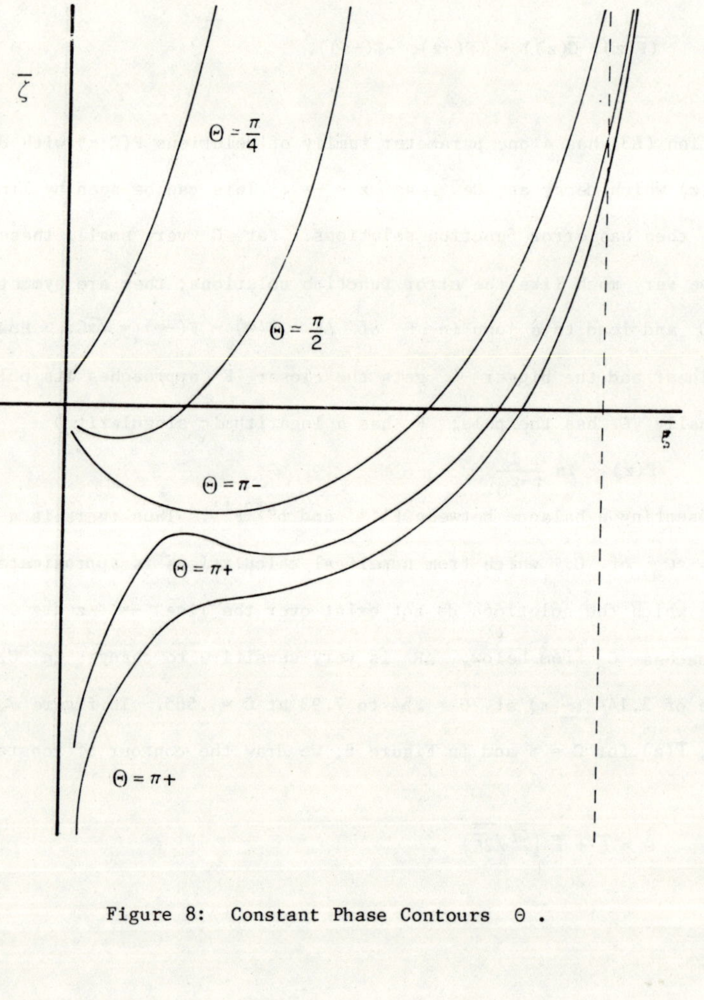

Figure 8: Constant Phase Contours Θ .

For values of Θ slightly greater than π, the contours are defined for all $\overline{\zeta}$. For values less than π, the phase contours intersect the $\overline{\xi} = 0$ axis at the origin. For $\overline{\xi} < 0$, we use the symmetry property (60) to infer that the phase contours in this region are simply a reflection of those for $\overline{\xi} > 0$ in the $\overline{\xi} = 0$ axis. These solutions seem to give a fairly accurate picture of the real dislocations seen in experiments.

Finally, we indicate how to include mean drift terms in the model. Consider

$$\frac{\partial w}{\partial T} + (\nabla^2+1)^2 w - Rw + w^2 w^* + u \cdot \nabla w = 0 \tag{67}$$

where $u = \nabla x\ \tau z$ (z is the unit vector perpendicular to X,Y) and

$$\nabla^2 \tau = 1/2p \,\{\nabla w^* \,.\, \nabla(\nabla^2 w) + (*)\} \tag{68}$$

Following the previous analysis, we find that the slow equation for the phase is

$$A\theta_t + \frac{k\ell}{A}\frac{\partial}{\partial\alpha}\frac{kB}{\ell} + Ak\ell\,\frac{\partial\tau}{\partial\beta} + 0(\varepsilon^4) = 0 \tag{69}$$

where

$$p\,\nabla^2\tau = k\ell\,\frac{\partial}{\partial\beta}\,k\ell\,\frac{\partial}{\partial\alpha}\,\left(\frac{kA^2}{\ell}\right). \tag{70}$$

In (68), the parameter 1/p mimics the effect of low Prandtl number situations where mean drift is caused by the nonlinear advection terms in the momentum equations. In (70), ∇ refers to the slow derivatives with respect to $(x,y) = \varepsilon^2(X,Y)$.

6. SUMMARY.

In this paper we have presented a mathematical framework for describing convection patterns which includes all previous theories and from it we have made several predictions about the manner in which the patterns evolve. In particular, we suggest that on the horizontal diffusion time scale T_H, the convection field develops patches, often of a circular nature surrounding a sink, in which the wavenumber is constant. The incompatibility of these patches is ironed out over the longer time scale of the aspect ratio times T_H and the process involves a gliding motion (compare Figures 3 and 5) in which roll dislocations move in a direction perpendicular to the roll axis. The climb motion, where the dislocations move along the roll axis, occur on the scale T_H as their role is to adjust wavelength, although small adjustments of order ε^2 will be made on the $\varepsilon^{-2}T_H$ scale.

While we believe we have made a start, many questions still remain open. Some of these are.

1. For what class of models is the flow on the horizontal diffusion time scale a gradient one; equivalently, for which models does (44) obtain?
2. What is the effect of the mean drift term? What parallel conclusions can we draw?
3. Do the patterns ever settle down or do they always remain noisy? If the former is the case, is it a consequence of geometry where the dislocations get stuck in corners? In a circular geometry, one might argue that the glide motion never stops. If the latter is the case, does the resulting chaotic motion lie on a low dimensional strange attractor, one which, for example, mimics the very gentle heaving of the glide motion as it rotates around the box?

REFERENCES

1. Gollub, I. P. and McCarriar A. R. 1982. Phys. Rev. A 26, 3470.

2. Ahlers G. and Walden R. W. 1980 Phys Rev. Lett. 44, 445.

3. Busse F. H. 1980 Hydrodynamic instabilities and the transition to turbulence. 97-136. Eds. H. L. Swinney and J. P. Gollub. Publ. Springer-Verlag.

4. Whitham G. B. 1970. J. Fluid Mech 44, 373.

5. Newell A. C. and Whithead J. A. 1969. J. Fluid Mech. 38, 279. Segel, L. A. 1969 J. Fluid Mech. 38, 203.

6. Pomeau Y. and Manneville P. 1981. Phys. Lett 40, 1067.

7. Siggia E. and Zippelius A. 1982. J. Fluid Mech. to appear.

8. Stuart J. T. 1960. J. Fluid Mech. 9, 353-370. Watson J. 1960. J. Fluid Mech. 9, 371-389.

9. Busse F. H. and Whitehead J. A. 1971. J. Fluid Mech. 47, 305-320.

10. Chen M. M. and Whitehead J. A. 1968. J. Fluid Mech. 31, 1.

11. Greenside H. S., Coughran W. M. and Schryer N. L. 1982. Preprint.

12. Berge , P. 1980. Chaos and Order in Nature pp. 14-24. Ed. H. Haken. Publ. Springer-Verlag.

Lecture Notes in Num. Appl. Anal., **5**, 233–251(1982)
Nonlinear PDE in Applied Science. U.S.-Japan Seminar, Tokyo, 1982

Stationary free boundary problems for circular flows with or without surface tension

Hisashi OKAMOTO *

Department of Mathematics
Faculty of Science
University of Tokyo
Hongo Bunkyo-ku Tokyo
113 Japan

Free boundary problems for flows circulating around a circle or sphere are considered. It is revealed that the surface tension plays a crucial role concerning perturbations and bifurcations of a trivial flow. Main tools are implicit function theorems (classical or generalized) and bifurcation theory due to Sattinger or Golubitsky & Schaeffer. Therefore all the classical solution near the trivial one are dealt with.

§1. Physical meaning.

Consider a fluid around a planet. We keep a figure like the Jupiter in mind. We consider a plane perpendicular to the axis of rotation and we regard the flow as a two dimensional one. We assume that the flow is encircled with two closed Jordan curves Γ and γ. The inner curve Γ represents the surface of the planet,whence Γ is a given curve. For simplicity we assume that Γ is the unit circle in $\mathbb{R}^2$. The outer curve γ represents a free boundary to be sought. The outside of γ is assumed to be a vacuum or to be filled with a perfect fluid whose pressure is given. Hence we treat a one phase problem. The flow region is denoted by Ω_γ ,i.e., we denote by Ω_γ the doubly connected domain between Γ and γ. Finally we assume that the fluid is incompressible,inviscid and irrotational. Then the problem is formulated by the stream function V as follows.

<u>**PROBLEM A.**</u> Find a closed Jordan curve γ outside Γ and a function V in Ω_γ such that

$$(1.1) \qquad \Delta V = 0 \qquad \text{in } \Omega_\gamma ,$$

$$(1.2) \qquad V = 0 \qquad \text{on } \Gamma ,$$

$$(1.3) \qquad V = a \qquad \text{on } \gamma ,$$

$$(1.4) \qquad \frac{1}{2}|\nabla V|^2 + Q + \sigma K_\gamma = \text{unknown constant} \qquad \text{on } \gamma ,$$

$$(1.5) \qquad |\Omega_\gamma| = \omega_0 .$$

The quantities appearing above are defined below.

a , ω_0 ; prescribed positive constants,

σ ; the surface tension coefficient ($\geqq 0$) ··· given ,

Q ; a given function defined outside Γ ,

K_γ ; the curvature of γ ,the sign of which is taken to be positive if γ is convex,

$|\Omega_\gamma|$; the area of Ω_γ.

REMARK 1.1. The equation (1.4) is a consequence of Bernoulli's law and the Laplace equation arising in the theory of surface tension. In fact,Bernoulli's law asserts that

$$(1.6) \qquad \frac{1}{2}|\nabla V|^2 + p + \Psi = \text{unknown constant} \quad \text{on } \gamma$$

where p is the pressure of the fluid and Ψ is a potential of the volume force. On the other hand,the Laplace equation is expressed as

$$(1.7) \qquad p = p_{ext} + \sigma K_\gamma ,$$

where p_{ext} is the known pressure of the external atmosphere. Putting $Q = p + p_{ext}$, we obtain (1.4) from (1.6) and (1.7). In this regard, $Q \equiv 0$ or $Q = -g/r$

(g ; a constant , $r = (x^2 + y^2)^{1/2}$) is an interesting case.

Trivial solution. If Q is radially symmetric, then there exists the following trivial solution. Take a number $r_0 > 1$ such that $\pi r_0^2 - \pi = \omega_0$. Then a circle γ_0 of radius r_0 with the origin as its center is a solution for any $\sigma \geqq 0$. In fact the corresponding stream function V is represented as

$$V = V(r) = \frac{a}{\log r_0} \log r \qquad (1 < r < r_0). \tag{1.8}$$

The unknown constant in (1.4) is $\frac{1}{2}(a/r_0 \log r_0)^2 + Q(r_0) + \sigma/r_0$.

Our aim is to study perturbations and bifurcations of this trivial solution. Our analysis is based on classical or generalized implicit function theorems and the bifurcation theory due to Sattinger [5] or Golubitsky and Schaeffer [2].

Now let us consider the case where the fluid is governed by the Navier-Stokes equation:

PROBLEM B. Find a closed Jordan curve γ and functions $\underline{V} = (V_1, V_2)$, P such that

$$-\nu \Delta \underline{V} + (\underline{V} \cdot \nabla)\underline{V} = -\nabla P + \nabla(g/r) \qquad \text{in } \Omega_\gamma , \tag{1.9}$$

$$\operatorname{div} \underline{V} = 0 \qquad \text{in } \Omega_\gamma , \tag{1.10}$$

$$\underline{V} = \underline{b} \qquad \text{on } \Gamma , \tag{1.11}$$

$$\underline{V} \cdot \underline{n} = 0 \ , \quad \underline{t}T(\underline{V})\underline{n} = 0 \qquad \text{on } \gamma , \tag{1.12}$$

$$-\underline{n}T(\underline{V})\underline{n} = \sigma K_\gamma \qquad \text{on } \gamma , \tag{1.13}$$

$$|\Omega_\gamma| = \omega_0 \ . \tag{1.14}$$

The quantities appearing above are defined below.

$\underline{V} = (V_1, V_2)$; the velocity vector , P ; the pressure ,

ν ; the kinematic viscosity , $\underline{n}$; the outward normal vector on γ ,

$\underline{t}$; a tangent vector on γ ,

$T(\underline{V})$; the stress tensor , the components of which are

$$T(\underline{V})_{ij} = \nu\left(\frac{\partial V_i}{\partial x_j} + \frac{\partial V_j}{\partial x_i} \right) - P\delta_{ij} \qquad (i , j = 1 , 2),$$

$\underline{b}$; a prescribed $\mathbb{R}^2$-valued function on Γ satisfying $\int_\Gamma \underline{b}\cdot\underline{n}\, d\Gamma = 0$.

A three dimensional analogue of PROBLEM B is also considered (see §3).

§2. Mathematical Formulation and results for PROBLEM A.

We prepare some symbols.

$\Omega = \{ (x,y) \in \mathbb{R}^2 ; 1 < x^2 + y^2 < \infty \}$, $S^1 = \{ (x,y) \in \mathbb{R}^2 ; 1 = x^2 + y^2 \}$,

$C^{m+\alpha}(\overline{\Omega})$, $C^{m+\alpha}(S^1)$ ($m = 0,1,2,\cdots$, $0 < \alpha < 1$) ; the Hölder spaces with usual norm $\| \|_{m+\alpha,\Omega}$, $\| \|_{m+\alpha,S^1}$.

We fix a number $\alpha \in (0,1)$ and a function $Q_0 \in C^{2+\alpha}([1, \infty))$. The typical case is $Q_0 = Q_0(r) = -g/r$ or $Q_0 \equiv 0$.

When a small $u \in C^{3+\alpha}(S^1)$ is given , we denote by γ_u a closed Jordan curve which is parametrized in the polar coordinates as $(r_0 + u(\theta) , \theta)$ ($0 \leqq \theta < 2\pi$). Hereafter we identify a function on S^1 with a 2π-periodic function on $\mathbb{R}$. We denote a domain between Γ and γ_u by Ω_u. The curvature of γ_u is denoted by K_u. It is represented as

$$K_u = \frac{(r_0 + u)^2 + 2(u')^2 - (r_0 + u)u''}{[(r_0 + u)^2 + (u')^2]^{3/2}} .$$

(' means the differentiation with respect to θ). $\partial/\partial\nu_u$ means the differentiation along the outward normal vector on γ_u. V_u denotes the unique solution of the Dirichlet problem

$$(2.1)\qquad \Delta V_u = 0 \qquad \text{in } \Omega_u ,$$

$$(2.2)\qquad V_u|_\Gamma = 0 \ , \ V_u|_{\gamma_u} = a .$$

For $u \in C^{3+\alpha}(S^1)$, $Q \in C^{2+\alpha}(\overline{\Omega})$ and $\xi \in \mathbb{R}$, we put

$$(2.3)\qquad F_1(a,Q;u,\xi) = \left\{ \frac{1}{2}|\nabla V_u|^2 + Q \right\}\Big|_{\gamma_u} + \sigma K_u - \xi_0 - \xi ,$$

where $\xi_0 = \frac{1}{2}(a/r_0 \log r_0)^2 + Q_0(r_0) + \sigma/r_0$,

$$(2.4)\qquad F_2(a,Q;u,\xi) = \frac{1}{2}\int_0^{2\pi}(r_0 + u(\theta))^2 d\theta - \pi - \omega_0 ,$$

$$(2.5)\qquad F(a,Q;u,\xi) = (F_1(a,Q;u,\xi),F_2(a,Q;u,\xi)).$$

Using a canonical pull-back,we regard $F_1(a,Q;u,\xi)$ as a function on S^1. Then it is easy to see that $F(a,Q_0;0,0) = (0,0)$ and that $\{\gamma_u , V_u\}$ is a solution for Q if and only if $F(a,Q;u,\xi) = (0,0)$ for some $\xi \in \mathbb{R}$. Note that $F(a,\cdot;\cdot,\cdot)$ is a continuous mapping from a neighborhood of $(Q_0;0,0)$ in $C^{2+\alpha}(\overline{\Omega}) \times C^{3+\alpha}(S^1) \times \mathbb{R}$ into $C^{1+\alpha}(S^1) \times \mathbb{R}$.

Now perturbation of the trivial solution is possible in the following sense.

<u>**THEOREM 1.**</u> *Assume that* $\sigma > 0$. *Define* a_n *by*

$$a_n = \left\{ \frac{\sigma(n^2-1)/r_0^2 + \frac{\partial Q_0}{\partial r}(r_0)}{\frac{1}{r_0} + \frac{n}{r_0}\,\frac{r_0^n + r_0^{-n}}{r_0^n - r_0^{-n}}} \right\}^{1/2} r_0 \log r_0$$

for $n \in \underline{\mathbb{N}} = \{ m \in \mathbb{N} \ ; \ \sigma(n^2-1)/r_0^2 + \frac{\partial Q_0}{\partial r}(r_0) \geqq 0 \}$. *Then, for any* $a \notin \{a_n\}_n$, *there exists a positive constant* δ *such that for any* $Q \in C^{2+\alpha}(\overline{\Omega})$ *satisfying* $\| Q - Q_0 \|_{2+\alpha,\Omega} < \delta$ *we have a solution* $\{u,\xi\}$ of the equation $F(a,Q;u,\xi) = (0,0)$. *The solution is unique in some neighborhood of the origin.*

<u>**THEOREM 2.**</u> *Assume that* $\sigma = 0$. *Let* $Q_0 \in C^{19+\alpha}(\overline{\Omega})$ *satisfy*

(2.6) $$\frac{\partial Q_0}{\partial r}(r_0) < a^2/r_0^3(\log r_0)^2.$$

Then there exists a positive constant ε *such that for any* $Q \in C^{10+(1/2)+\alpha}(\overline{\Omega})$ *satisfying* $\| Q - Q_0 \|_{10+(1/2)+\alpha,\Omega} < \varepsilon$ *we have a solution* $\{u,\xi\}$ *of the equation* $F(a,Q;u,\xi) = (0,0)$.

The next two theorems state uniqueness or nonuniqueness of the solution. We put $G(a;u,\xi) = F(a,Q_0;u,\xi)$.

<u>THEOREM 3</u>. *Fix a nutural number* n. *Assume that* $\sigma > 0$. *Assume also that*

(2.7) $$a_m \neq a_n \qquad \textit{for all} \quad m \neq n.$$

Then there exists a branch of nontrivial solutions of $G(a;u,\xi) = (0,0)$ *through* $(a_n;0,0)$. *If* n *is sufficiently large, then the bifurcation occurs subcritically.*

<u>THEOREM 4</u>. *Assume that* $\sigma = 0$ *and* (2.6). *Then, in some* $C^{2+\alpha}$*-neighborhood of* γ_0, *there exists no solution other than* γ_0.

We prove THEOREM 1 by means of a classical implicit function theorem. On the contrary we use a generalized implicit function theorem due to Zehnder [8] in order to prove THEOREM 2. THEOREM 3 is proved by a bifurcation theory due to [1,2,5]. THEOREM 4 is a consequence of the maximum principle. In §4 we will give outlines of the proofs. For the details, see Okamoto [3,4].

§3. Results for the Navier-Stokes problem.

Using the notation in the preceding section, we formulate PROBLEM B as follows. Firstly, for a given $u \in C^{3+\alpha}(S^1)$, we consider the solution of

(3.1) $$-\nu\Delta\underline{U} + (\underline{U}\cdot\nabla)\underline{U} = -\nabla q + \nabla(g/r) \qquad \text{in } \Omega_u,$$

(3.2) $$\operatorname{div}\underline{U} = 0 \qquad \text{in } \Omega_u,$$

(3.3) $$\underline{U} = \underline{b} \qquad \text{on } \Gamma,$$

(3.4) $$\underline{U}\cdot\underline{n} = 0 \quad , \quad \underline{t}T(\underline{U})\underline{n} = 0 \qquad \text{on} \quad \gamma_u ,$$

(3.5) $$\int_{\Omega_u} (q - g/r)dx = 0,$$

where $\underline{b} \in X \equiv \{ \underline{\beta} \in C^{3+\alpha}(\Gamma)^2 \ ; \ \int_\Gamma \underline{\beta}\cdot\underline{n}\, d\Gamma = 0 \}$. The boundary condition (3.4) constitutes a complementary condition in the sense of Agmon,Douglis and Nirenberg. Therefore,for sufficiently small $u \in C^{3+\alpha}(S^1)$ and $\underline{b} \in X$,such a solution $\underline{U}$,q is determined uniquely and continuously from u. We denote it by $\underline{V}_u$,P_u. Hence we can define a mapping H by the equalities below.

(3.6) $$H_1(\underline{b};u,\xi) = \underline{n}T(\underline{V}_u)\underline{n}|_{\gamma_u} + \sigma K_u - \xi_0 - \xi \ ,$$

(3.7) $$H_2(\underline{b};u,\xi) = \frac{1}{2}\int_0^{2\pi} (r_0 + u(\theta))^2 d\theta - \pi - \omega_0 \ ,$$

(3.8) $$H(\underline{b};u,\xi) = (H_1(\underline{b};u,\xi), H_2(\underline{b};u,\xi)) \ ,$$

where $u \in C^{3+\alpha}(S^1)$ ($\|u\|_{3+\alpha} \ll 1$) , $\xi \in \mathbb{R}$, $\underline{b} \in X$ ($\|\underline{b}\|_{3+\alpha} \ll 1$) , $\xi_0 \equiv (\sigma - g)/r_0$.

Obviously $H(0;0,0) = (0,0)$. Furthermore $\{\gamma_u, \underline{V}_u, P_u\}$ is a solution for $\underline{b} \in X$ if and only if $H(\underline{b};u,\xi) = (0,0)$ for some $\xi \in \mathbb{R}$. Similarly to THEOREM 1 we obtain the following

<u>THEOREM 5</u>. *We can choose a positive constant* η *such that, for any* $\underline{b} \in X$ *satisfying* $\|\underline{b}\|_{3+\alpha} < \eta$ *, there exists a* $\{u,\xi\} \in C^{3+\alpha}(S^1)\times\mathbb{R}$ *which solves* $H(\underline{b};u,\xi) = (0,0)$. *The solution is unique in some neighborhood of* $(0,0)$ *in* $C^{3+\alpha}(S^1)\times\mathbb{R}$.

In order to treat the three dimensional version of PROBLEM B,we employ polar coordinates r , θ , ϕ determined by

$$x = r\cos\theta\cos\phi , \quad y = r\cos\theta\sin\phi , \quad z = r\sin\theta . \qquad \left(1 < r \ , \ -\frac{\pi}{2} < \theta < \frac{\pi}{2} \ , \quad 0 \leq \phi < 2\pi \right)$$

The problem to be considered is written as

PROBLEM B'. Find a S^2-like surface γ and functions $\underline{V} = (V_r, V_\theta, V_\phi), P$ such that

$$(3.9) \qquad - \nu\Delta\underline{V} + (\underline{V}\cdot\nabla)\underline{V} = - \nabla P + \nabla(g/r) \qquad \text{in } \Omega_\gamma ,$$

$$(3.10) \qquad \operatorname{div} \underline{V} = 0 \qquad \text{in } \Omega_\gamma ,$$

$$(3.11) \qquad \underline{V} = \underline{b} \qquad \text{on } r = 1 ,$$

$$(3.12) \qquad \underline{V}\cdot\underline{n} = 0 \quad , \quad \underline{t}T(\underline{V})\underline{n} = 0 \qquad \text{on } \gamma ,$$

$$(3.13) \qquad - \underline{n}T(\underline{V})\underline{n} = \sigma H_\gamma \qquad \text{on } \gamma ,$$

$$(3.14) \qquad |\Omega_\gamma| = \omega_0 ,$$

where Ω_γ is a domain between S^2 and γ , $|\Omega_\gamma|$ is the volume of Ω_γ. H_γ is the mean curvature of γ. The second equation in (3.12) should be interpreted as $T(\underline{V})\underline{n} = [\underline{n}T(\underline{V})\underline{n}]\underline{n}$.

NOTATION.

$$X = \{ u \in C^{3+\alpha}[-\tfrac{\pi}{2},\tfrac{\pi}{2}] \ ; \ u'(\pm \tfrac{\pi}{2}) = 0 \ , u'''(\pm \tfrac{\pi}{2}) = 0 \} ,$$

$$Y = \{ u \in C^{1+\alpha}[-\tfrac{\pi}{2},\tfrac{\pi}{2}] \ ; \ u'(\pm \tfrac{\pi}{2}) = 0 \} ,$$

$$Z = \{ b \in C^{3+\alpha}[-\tfrac{\pi}{2},\tfrac{\pi}{2}] \ ; \ b(\pm \tfrac{\pi}{2}) = b'(\pm \tfrac{\pi}{2}) = b''(\pm \tfrac{\pi}{2}) = b'''(\pm\tfrac{\pi}{2}) = 0 \}.$$

If a small $u \in X$ is given,we denote by γ_u a surface parametrized as

$$x = (r_0+u(\theta))\cos\theta\cos\phi \quad , \ y = (r_0 + u(\theta))\cos\theta\sin\phi \quad , \ z = (r_0 + u(\theta))\sin\theta.$$

Here $r_0 > 1$ is determined by $\frac{4\pi}{3}r_0^3 - \frac{4\pi}{3} = \omega_0$.

For sufficiently small $b \in Z$ and $u \in X$ we solve (3.9) through (3.12) for $\underline{b} = (0,0,b)$ ($b \in Z$). This is uniquely determined for small $b \in Z$ and $u \in X$. Then we put it $\underline{V_u}$, P_u. Note that $\underline{V_u}$ and P_u are independent of ϕ.

Now we define a mapping $\hat{H}$ in a way similar to the case of H.

$$\hat{H}(b;u,\xi) = (\hat{H}_1(b;u,\xi),\hat{H}_2(b;u,\xi)) , \tag{3.15}$$

$$\hat{H}_1(b;u,\xi) = \underline{n}T(V_{\underline{u}})\underline{n}|_{\gamma_u} + \sigma H_u - \xi_0 - \xi , \tag{3.16}$$

$$\hat{H}_2(b;u,\xi) = \pi\int_{-\pi/2}^{\pi/2}(r_0+u)^2\cos^2\theta\,\{u'\sin\theta + (r_0+u)\cos\theta\}d\theta - \frac{4\pi}{3} - \omega_0. \tag{3.17}$$

__REMARK__. The mean curvature H_u of γ_u is represented as

$$2H_u = \frac{u'\tan\theta}{(r_0+u)[(r_0+u)^2+(u')^2]^{1/2}} + [(r_0+u)^2+(u')^2]^{-1/2} + \frac{(r_0+u)^2+2(u')^2-(r_0+u)u''}{[(r_0+u)^2+(u')^2]^{3/2}} . \tag{3.18}$$

$\hat{H}$ is a continuous mapping from a neighborhood of $(0;0,0)$ in $Z\times X\times \mathbb{R}$ into $Y\times\mathbb{R}$. Then we have the following

__THEOREM 6__. *If* $b\in Z$ *is sufficiently small, then there exists* $\{u,\xi\}\in X\times\mathbb{R}$ *such that* $\hat{H}(b;u,\xi) = (0,0)$. *The solution is unique in some neighborhood of the origin.*

§4. Outline of the proof.

4.1. THEOREM 1 is proved if we have shown that the Fréchet derivative of F with respect to $\{u,\xi\}$ is an isomorphism from $C^{3+\alpha}(S^1)\times\mathbb{R}$ onto $C^{1+\alpha}(S^1)\times\mathbb{R}$ for $a\notin\{a_n\}_n$. To show this we have to calculate the derivative of F explicitly:

__Claim__. F is a C^1-mapping and its derivative is given by

$$D_{u,\xi}F(a,Q;u,\xi) = \begin{pmatrix} D_uF_1(a,Q;u,\xi) & D_\xi F_1(a,Q;u,\xi) \\ D_uF_2(a,Q;u,\xi) & D_\xi F_2(a,Q;u,\xi)\end{pmatrix} ,$$

$$D_uF_1(a,Q;u,\xi)w = \frac{\partial V_u}{\partial r}[\,\Phi_u w + \Sigma(V_u)w\,] + \frac{\partial Q}{\partial r}\Big|_{\gamma_u} w \tag{4.1}$$

$$+ \sigma\{f_0(u)w + f_1(u)w + f_2(u)w\} \qquad (\, w \in C^{3+\alpha}(S^1)),$$

$$D_\xi F_1(a,Q;u,\xi)\lambda = -\,\lambda \qquad (\,\lambda \in \mathbb{R}\,)\,, \tag{4.2}$$

$$D_uF_2(a,Q;u,\xi)w = \int_0^{2\pi}(r_0 + u(\theta))w(\theta)d\theta \qquad (\, w \in C^{3+\alpha}(S^1)\,), \tag{4.3}$$

$$D_\xi F_2(a,Q;u,\xi)\lambda = 0 \qquad (\,\lambda \in \mathbb{R}\,). \tag{4.4}$$

Here we have put

$$f_0(u) = \frac{\partial K_u}{\partial u}\;,\quad f_1(u) = \frac{\partial K_u}{\partial u'}\;,\quad f_2(u) = \frac{\partial K_u}{\partial u''}\,.$$

(' means the differentiation with respect to θ .)

The function $\Sigma(V_u)$ is defined by

$$\Sigma(V_u) = [(r_0 + u\,)^2 + (u')^2]^{-1/2}\left\{ r\frac{\partial^2 V_u}{\partial r^2} - u'\frac{\partial}{\partial r}\Big[\,\frac{1}{r}\frac{\partial V_u}{\partial \theta}\,\Big]\right\}\Big|_{\gamma_u}\,. \tag{4.5}$$

The operator Φ_u is defined by $\Phi_u w = \dfrac{\partial U}{\partial \nu_u}$,using the solution U of

$$\Delta U = 0 \qquad \text{in } \Omega_u\,,$$

$$U|_\Gamma = 0\;\;,\;\; U|_{\gamma_u} = -\frac{\partial V_u}{\partial r}\Big|_{\gamma_u} w\,.$$

The proof of (4.2,3,4) is straightforward. To show (4.1) it is sufficient to prove that

$$D_uT(u)w = \Phi_u w + \Sigma(V_u)w \;\;, \tag{4.6}$$

where $T(u) = |\nabla V_u||_{\gamma_u}$. The formula above is proved in [3]. Here we only give a formal calculation to derive (4.6). Firstly we extend V_u to a $C^{3+\alpha}$-class function on some neighborhood of $\overline{\Omega}_u$. Secondly note that $|\nabla V_u||_{\gamma_u} = \nabla V_u \cdot \vec{\nu}_u$ since V_u is a constant on γ_u. Then we have

$$T(u+w) - T(u) = [\nabla V_{u+w}|_{\gamma_{u+w}} - \nabla V_u|_{\gamma_{u+w}}]\cdot\vec{\nu}_{u+w}$$

$$+ [\nabla V_u|_{\gamma_{u+w}} - \nabla V_u|_{\gamma_u}]\cdot\vec{\nu}_{u+w}$$

$$+ \nabla V_u|_{\gamma_u}\cdot[\vec{\nu}_{u+w} - \vec{\nu}_u]$$

$$\equiv I_1 + I_2 + I_3.$$

Putting $\bar{U} = V_{u+w} - V_u$, we obtain $\Delta\bar{U} = 0$ in $\Omega_{u+w} \cap \Omega_u$, $\bar{U} = 0$ on Γ and

$$\bar{U}|_{\gamma_{u+w}} = a - V_u|_{\gamma_{u+w}} = V_u|_{\gamma_u} - V_u|_{\gamma_{u+w}} \simeq -\frac{\partial V_u}{\partial r} w .$$

Hence $I_1 = \Phi_u w$ modulo $o(\|w\|_{3+\alpha})$. It is easy to see

$$I_2 \simeq (\nabla V_u|_{\gamma_{u+w}} - \nabla V_u|_{\gamma_u})\cdot\vec{\nu}_u \simeq \Sigma(V_u)w.$$

Since $\vec{\nu}_{u+w} - \vec{\nu}_u = \vec{t} + o(\|w\|_{3+\alpha})$ with a tangent vector $\vec{t}$ on γ_u, we obtain $I_3 \simeq 0$. From these considerations we find (4.6).

Now we show that $A(a) \equiv D_{u,\xi}F(a,Q_0;0,0)$ is an isomorphism for $a \notin \{a_n\}_n$.

<u>Claim</u>. If $a \notin \{a_n\}_n$, then $A(a)$ is injective.

<u>Proof</u>. Assume that $A(a)(w,\lambda) = (0,0)$. We represent w by the Fourier series:

$$w = \sum_{n=1}^{\infty} b_n \sin n\theta + \sum_{n=0}^{\infty} c_n \cos n\theta .$$

Then we have $b_n S(a,n) = c_n S(a,n) = 0$ $(n \geqq 1)$, $\lambda = c_0 \times\{\text{something}\}$ and $c_0 = 0$, where we have put

$$S(a,n) = \frac{\sigma(n^2-1)}{r_0^2} - \left(\frac{a}{r_0 \log r_0}\right)^2 \left(\frac{n}{r_0}\frac{r_0^n + r_0^{-n}}{r_0^n - r_0^{-n}} + \frac{1}{r_0}\right) + \frac{\partial Q_0}{\partial r}(r_0).$$

Since $S(a,n)$ vanishes if and only if $a = a_n$, we see that $A(a)$ is injective if $a \notin \{a_n\}_n$. Q.E.D.

On the other hand, it holds that $A(a)$ = " an isomorphism " + " a compact operator ". Using the claim above and the Riesz-Schauder theory we can conclude that $A(a)$ is an isomorphism for $a \notin \{a_n\}_n$.

4.2. Proof of THEOREM 2. When $\sigma = 0$, $D_{u,\xi}F(a,Q_0;0,0)$ is no longer an isomorphism from $C^{3+\alpha}(S^1)\times\mathbb{R}$ onto $C^{1+\alpha}(S^1)\times\mathbb{R}$. However, it is an isomorphism from $C^{3+\alpha}(S^1)\times\mathbb{R}$ onto $C^{2+\alpha}(S^1)\times\mathbb{R}$. From this fact one observes that we are in a position to use a generalized implicit function theorem. Among others we use a one due to Zehnder [8]. In verifying several assumptions of the generalized implicit function theorem, we use a priori estimates of Schauder type which are borrowed from Schaeffer [7]. For the details, see Okamoto [3] in which the proof of THEOREM 4 is included.

4.3. Proof of THEOREM 3. From now on we put $G(a;u,\xi) = F(a,Q_0;u,\xi)$. In the proof of THEOREM 1 we have shown that $A(a) \equiv D_{u,\xi}F(a,Q_0;0,0)$ is an isomorphism for $a \notin \{a_n\}_n$ and that $A(a_n)$ has a null-space spanned by $(\cos n\theta\,,0)$ and $(\sin n\theta\,,0)$. (Here we have used (2.7).) In order to use a theory of bifurcation from simple eigenvalue we use tha following Banach space:

$$X^{m+\alpha} = \{\, u \in C^{m+\alpha}(S^1)\ ;\ u(\theta) = u(2\pi-\theta) \quad (0 \leqq \theta < 2\pi) \,\}$$

with the norm $\| \ \|_{m+\alpha}$. Let G^* denote the restriction of G on $X^{3+\alpha}\times\mathbb{R}$. Then it holds that the range of $G^*(a;\cdot,\cdot)$ is included in $X^{1+\alpha}\times\mathbb{R}$ and the null-space of $D_{u,\xi}G^*(a_n;0,0)$ is spanned only by $(\cos n\theta\,,0)$. Consequently we can apply THEOREM 1.7 of Crandall and Rabinowitz [1]. The details are in [4].

To see whether the bifurcation occurs supercritically or subcritically, we proceed as follows. (The details are also in [4].) Let a two dimensional subspace of $C^{3+\alpha}(S^1)$ spanned by $\cos n\theta$ and $\sin n\theta$ be $<\cos n\theta\,,\sin n\theta>$. We denote a canonical projection from $C^{3+\alpha}(S^1)$ onto $<\cos n\theta\,,\sin n\theta>$ by P. Then we define functions ϕ and ξ by the equations below.

$$(4.7) \qquad (I-P)G_1(a;x\cos n\theta + y\sin n\theta + \phi(a;x,y)\,,\xi(a;x,y)) = 0\ ,$$

(4.8) $$G_2(a\,;x\cos n\theta + y\sin n\theta + \phi(a;x,y)\,,\xi(a;x,y)) = 0.$$

The assumption (2.7) and the classical implicit function theorem ensures that ϕ and ξ are well-defined in some neighborhood of $(a_n;0,0)$ in $\mathbb{R}^3$, and that their ranges are in $(I-P)C^{3+\alpha}(S^1)$, $\mathbb{R}$, respectively. Then the equation

(4.9) $$F(a;x,y) \equiv PG_1(a\,;x\cos n\theta + y\sin n\theta + \phi(a;x,y)\,,\xi(a;x,y)) = 0$$

is a bifurcation equation. If we write

$$F(a;x,y) = F_1(a;x,y)\cos n\theta + F_2(a;x,y)\sin n\theta\ ,$$

then the solution set near $(a_n;0,0)$ is in a one-to-one correspondence with

$$\{\ (a;x,y)\ ;\ F_1(a;x,y) = F_2(a;x,y) = 0\ \}\ .$$

Since the original problem is O(2)-covariant,we have

<u>**PROPOSITION 4.1.**</u> *The bifurcation function* F *is a* C^∞*-mapping. There exists a* C^∞*-mapping* F^* *defined in some neighborhood of* $(a_n;0)$ *in* $\mathbb{R}^2$ *such that*

(4.10) $$F_1(a;x,y) = xF^*(a\,;x^2+y^2)\ ,$$

(4.11) $$F_2(a;x,y) = yF^*(a\,;x^2+y^2).$$

By this proposition we observe that the solution set is composed of $\{\ x = y = 0\ \}$ and $\{\ F^*(a;\ x^2+y^2) = 0\ \}$. Of course the former corresponds to the trivial solution. To deal with the nontrivial ones we expand F^* as

$$F^*(a\,;x^2+y^2) = A_n(a-a_n) + B_n(x^2+y^2) + \text{higher order terms}.$$

By the result of Golubitsky and Schaeffer [2],it holds that $F = 0$ is O(2)-equivalent to

$$\begin{pmatrix} x\{A_n(a-a_n) + B_n(x^2+y^2)\} \\ y\{A_n(a-a_n) + B_n(x^2+y^2)\} \end{pmatrix} = \begin{pmatrix} 0 \\ 0 \end{pmatrix},$$

if $A_n \neq 0$ and $B_n \neq 0$. Therefore the direction of the bifurcating branch is determined by the sign of A_n and B_n. Since A_n and B_n are given by

$$A_n \cos n\theta = \frac{\partial^2 F}{\partial a \partial x}(a_n;0,0) \quad , \quad 6B_n \cos n\theta = \frac{\partial^3 F}{\partial x^3}(a_n;0,0) \ ,$$

we have to compute the third order derivative of F, hence of G_1. To this end we prepare some symbols.

<u>NOTATION</u>. For $w \in C^{3+\alpha}(S^1)$ we denote by $U(w)$ the solution of

$$\Delta U = 0 \qquad \text{in } 1 < r < r_0 \ ,$$

$$U|_\Gamma = 0 \ , \quad U|_{r=r_0} = - \frac{aw}{r_0 \log r_0} \ ,$$

For $w, z \in C^{4+\alpha}(S^1)$ the symbol $Y(w,z)$ denotes the solution of

$$\Delta Y = 0 \qquad \text{in } 1 < r < r_0 \ ,$$

$$Y = 0 \qquad \text{on } \Gamma \ ,$$

$$Y|_{r=r_0} = - \frac{\partial U(w)}{\partial r} z - \frac{\partial U(z)}{\partial r} w + \frac{awz}{r_0^2 \log r_0} \ .$$

For $w_1, w_2, w_3 \in C^{5+\alpha}(S^1)$ the symbol $X(w_1,w_2,w_3)$ denotes the solution of

$$\Delta X = 0 \qquad \text{in } 1 < r < r_0 \ ,$$

$$X = 0 \qquad \text{on } \Gamma \ ,$$

$$X|_{r=r_0} = - \sum_{i=1}^{3} \frac{\partial Y(w_i, w_{i+1})}{\partial r} w_{i+2} - \sum_{i=1}^{3} \frac{\partial^2 U(w_i)}{\partial r^2} w_{i+1} w_{i+2} - \frac{2aw_1 w_2 w_3}{r_0^3 \log r_0} \ ,$$

where we put $w_{i+3} = w_i$.

NOTATION'.

$$Lw = \frac{\partial U(w)}{\partial r} - \frac{aw}{r_0^2 \log r_0} \qquad (w \in C^{3+\alpha}(S^1)),$$

$$B(w,z) = \frac{\partial Y(w,z)}{\partial r} + \frac{\partial^2 U(w)}{\partial r^2} z + \frac{\partial^2 U(z)}{\partial r^2} w + \frac{2awz}{r_0^3 \log r_0} + \frac{aw'z'}{r_0^3 \log r_0}$$

$$(w , z \in C^{4+\alpha}(S^1)).$$

Now the third order derivative of G_1 is given by

(4.12)
$$D_u^3 G_1(a;0,0)(w_1,w_2,w_3)$$
$$= \sum_{i=1}^{3} B(w_i,w_{i+1})Lw_{i+2} + \frac{\partial^3 Q_0}{\partial r^3}(r_0)w_1 w_2 w_3$$
$$+ \frac{a}{r_0 \log r_0}\left(\frac{\partial}{\partial r} X(w_1,w_2,w_3) + \sum_{i=1}^{3} \frac{\partial^2 Y(w_i,w_{i+2})}{\partial r^2} w_{i+2} \right)$$
$$+ \frac{a}{r_0 \log r_0} \sum_{i=1}^{3} \frac{\partial^3 U(w_i)}{\partial r^3} w_{i+1} w_{i+2}$$
$$- \frac{3a^2}{r_0^3 (\log r_0)^2}\left(2w_1 w_2 w_3 + \sum_{i=1}^{3} w_i w'_{i+1} w'_{i+2} \right)$$
$$+ \frac{a}{r_0^3 \log r_0} \sum_{i=1}^{3} \frac{\partial U(w_i)}{\partial r} w'_{i+1} w'_{i+2}$$
$$+ \frac{3\sigma}{r_0^4}\left(-w_1 w_2 w_3 - 2\sum_{i=1}^{3} w''_i w_{i+1} w_{i+2} - \sum_{i=1}^{3} w_i w'_{i+1} w'_{i+2} + \sum_{i=1}^{3} w''_i w'_{i+1} w'_{i+2} \right) .$$

On the other hand,we have

$$6B_n \cos n\theta = PD_u^3 G_1(a_n;0,0)(\cos n\theta ,\cos n\theta ,\cos n\theta) .$$

Therefore,in principle,we can compute B_n by the formula above. However,it is very difficult to decide its sign,since it is very complicated. But we have

$$B_n \sim - \sigma n^4/4r_0^4 \qquad (n \to \infty).$$

Hence B_n is negative for a large n. On the other hand, A_n is negative for any n. In fact we have

$$A_n = -\frac{2a_n}{r_0 \log r_0}\left(\frac{n}{r_0^2 \log r_0}\frac{r_0^n + r_0^{-n}}{r_0^n - r_0^{-n}} + \frac{1}{r_0^2 \log r_0} \right).$$

Thus we see that the bifurcation is subcritical for a large n.

4.4. Proof of THEOREMS 5 and 6. It is not so hard to verify that H and $\hat{H}$ is a C^1-mapping. Hence the proof of THEOREM 5 or 6 are completed by checking that $D_{u,\xi}H(0;0,0)$ or $D_{u,\xi}\hat{H}(0;0,0)$ is an isomorphism,respectively. The derivatives are given below.

$$D_u H_1(0;0,0)w = -\frac{\sigma}{r_0^2}(w + w'') + \frac{gw}{r_0^2} \qquad (w \in C^{3+\alpha}(S^1)),$$

$$D_\xi H_1(0;0,0)\lambda = -\lambda \qquad (\lambda \in \mathbb{R}),$$

$$D_u H_2(0;0,0)w = r_0\int_0^{2\pi} w(\theta)d\theta \qquad (w \in C^{3+\alpha}(S^1)),$$

$$D_\xi H_2(0;0,0)\lambda = 0 \qquad (\lambda \in \mathbb{R}).$$

$$D_u \hat{H}_1(0;0,0)w = \frac{\sigma}{2r_0^2}(-w'' + w'\tan\theta - 2w) + \frac{gw}{r_0^2} \qquad (w \in X),$$

$$D_\xi \hat{H}_1(0;0,0)\lambda = -\lambda \qquad (\lambda \in \mathbb{R}),$$

$$D_u \hat{H}_2(0;0,0)w = 3\pi r_0^2\int_{-\pi/2}^{\pi/2} w(\theta)\cos^3\theta\, d\theta + \pi r_0^2\int_{-\pi/2}^{\pi/2} w'(\theta)\cos^2\theta\sin\theta\, d\theta$$

$$(w \in X),$$

$$D_\xi \hat{H}_2(0;0,0)\lambda = 0 \qquad (\lambda \in \mathbb{R}).$$

In a way similar to the proof of THEOREM 1 we can prove that $D_{u,\xi}H(0;0,0)$ is an isomorphism from $C^{3+\alpha}(S^1)\times\mathbb{R}$ onto $C^{1+\alpha}(S^1)\times\mathbb{R}$. To treat $D_{u,\xi}\hat{H}(0;0,0)$, we first show that it is injective.

If $D_{u,\xi}\hat{H}(0;0,0)(w,\lambda) = (0,0)$, then w and λ satisfy

(4.13) $$-w'' + w'\tan\theta - 2w + \frac{2g}{\sigma}w - \frac{2r_0^2}{\sigma}\lambda = 0 ,$$

(4.14) $$3\int_{-\pi/2}^{\pi/2} w(\theta)\cos^3\theta\, d\theta + \int_{-\pi/2}^{\pi/2} w'(\theta)\cos^2\theta\sin\theta\, d\theta = 0.$$

We change the variable from θ to $s \equiv \sin\theta$. Then $\hat{w}(s) \equiv w(\theta)$ satisfies

$$(s^2-1)\frac{d^2\hat{w}}{ds^2} + 2s\frac{d\hat{w}}{ds} + 2(\frac{g}{\sigma} - 1)\hat{w} - \frac{2r_0^2}{\sigma}\lambda = 0 .$$

Expanding $\hat{w}$ by the series of the Legendre polynomials,we see that $\hat{w}$ must be a constant. Then (4.14) implies that $w \equiv 0$. Consequently $(w,\lambda) = (0,0)$.

To show that Range $D_{u,\xi}\hat{H}(0;0,0) = Y \times \mathbb{R}$,we do as follows. Firstly we define operators A_μ and B_μ by the equalities below.

$$D_{u,\xi}\hat{H}(0;0,0) = A_\mu + B_\mu ,$$

$$A_\mu(w,\lambda) = (v,\zeta) ,$$

(4.15) $$v = \frac{\sigma}{2r_0^2}(-w'' + w'\tan\theta + \mu w) - \lambda \equiv \frac{\sigma}{2r_0^2}\Psi_\mu w - \lambda , \qquad (w \in X),$$

(4.16) $$\zeta = 3\pi r_0^2\int_{-\pi/2}^{\pi/2} w(\theta)\cos^3\theta\, d\theta + \pi r_0^2\int_{-\pi/2}^{\pi/2} w'(\theta)\cos^2\theta\sin\theta\, d\theta$$

(μ is a positive parameter). Then $B_\mu \in L(X\times\mathbb{R}, Y\times\mathbb{R})$ is a compact operator. By the Riesz-Schauder theory it is sufficient to show that A_μ is an isomorphism from $X\times\mathbb{R}$ onto $Y\times\mathbb{R}$ for some μ. Therefore,for a given $(v,\zeta) \in Y\times\mathbb{R}$, we have to find a $(w,\lambda) \in X\times\mathbb{R}$ satisfying (4.15) and (4.16). Observe that we have only to show the surjectivity of Ψ_μ. To show the surjectivity of Ψ_μ , we employ the following

NOTATION.

$$H = L^2(-\frac{\pi}{2}, \frac{\pi}{2}; \cos\theta\, d\theta) ,$$

$$V = \{ f \in H \; ; \; f' \in H \},$$

$$a(w,v) = \int_{-\pi/2}^{\pi/2} w'(\theta)v'(\theta)\cos\theta \, d\theta + \mu\int_{-\pi/2}^{\pi/2} w(\theta)v(\theta)\cos\theta \, d\theta .$$

In virtue of the Lax-Milgram theorem there exists, for a given $f \in H$, a unique $w \in V$ such that

$$a(w,v) = (f,v)_H \qquad (v \in V).$$

This equality formally implies $\Psi_\mu w = f$. Then we show that $w \in X$ if $f \in Y$. This is shown in an elementary way. Hence we omit the proof. By the procedure above we arrive at THEOREM 6.

Acknowledgment. The author wishes to express his deep gratitude to Professor H. Fujita for his unceasing encouragement. He is also grateful to Professor H. Weinberger for his important advice and helpful discussion during his stay in University of Tokyo.

* Partially supported by the Fujukai.

REFERENCES

[1] M.G. Crandall and P.H. Rabinowitz, Bifurcation from simple eigenvalue, J. Func. Anal., 8 (1971) 321 - 340.

[2] M. Golubitsky and D. Schaeffer, Imperfect bifurcation in the presence of symmetry, Comm. Math. Phys., 67 (1979) 205 - 232.

[3] H. Okamoto, On the stationary free boundary problem for a circulating flow without surface tension, (preprint).

[4] H. Okamoto, Bifurcation phenomena in a free boundary problem for a circulating flow with surface tension, (preprint).

[5] D.H. Sattinger, Group Theoretic Methods in Bifurcation Theory, Springer

Lecture Notes in Math., No. 762 (1979).

[6] D.H. Sattinger, On the free surface of a viscous fluid motion, Proc. R. Soc. London A, 349 (1976) 183-204.

[7] D. Schaeffer, The capacitor problem, Indiana Univ. Math. J., 24 (1975) 1143-1167.

[8] E. Zehnder, Generalized implicit function theorems with applications to some small divisor problems, I, Comm. Pure Appl. Math., 28 (1975) 91-140.

Lecture Notes in Num. Appl. Anal., **5**, 253–257 (1982)
Nonlinear PDE in Applied Science. U.S.-Japan Seminar, Tokyo, 1982

FOCUSING SINGULARITY FOR THE NONLINEAR SCHROEDINGER EQUATION

G. Papanicolaou,*
Courant Institute, New York University

D. McLaughlin,*
University of Arizona

M. Weinstein,*
Stanford University

We summarize recent results on the focusing singularity of the nonlinear Schroedinger equation.

In this note we shall give a brief account of some recent work that we carried out motivated by the observations and calculations of Zakharov and Synakh [1]. A detailed exposition is given in [2] and in [3]. In [4] the results of careful numerical computation are reported.

The nonlinear Schroedinger equation

$$(1) \qquad 2i\phi_t + \Delta\phi + |\phi|^{2\sigma}\phi = 0\,, \qquad t > 0\,, \quad x \in \mathbb{R}^N$$

$$\phi(0,x) = \phi_0(x)$$

arises as a canonical problem in which focusing (the nonlinearity with the plus sign in (1)) competes with dispersion (the Laplacian in (1)). Among the many specific contexts where this occurs we mention nonlinear optics, where $N = 2$ and $\sigma = 1$, plasma problems, $N = 3$, $\sigma = 1$, water waves, etc. The case $N = 1$, $\sigma = 1$ has been studied extensively and was first shown to be solvable by the inverse scattering method by Zakharov and Shabat [5].

In the analysis of (1) three cases with distinct behavior arise. The subcritical case $\sigma < 2/N$ where dispersion dominates and a global solution in $C([0,\infty);H^1(\mathbb{R}^N))$ exists. Here $H^1(\mathbb{R}^N)$ denotes the usual Sobolev space of functions with square integrable derivatives. This result is proved in detail in [6]. In the critical case $\sigma = \frac{2}{N}$ and in the supercritical case $\sigma > \frac{2}{N}$ it is known that solutions of (1) will blow up in a finite time, i.e. their H^1 norm will become infinite [7].

Based on numerical evidence and some heuristic calculations, Zakharov and Synakh [1] conclude that in the case $N = 2$, $\sigma = 1$ (critical case) if an axially symmetric solution becomes singular at $t = t^*$ then near t^* it has the form

$$(2) \qquad |\phi(t,x)| \sim (t^*-t)^{-2/3} R(|x|(t^*-t)^{-2/3})$$

where $R(r)$ is the "ground state" solution of

$$(3) \qquad \frac{d^2R}{dr^2} + \frac{1}{r}\frac{dR}{dr} - R + R^3 = 0 , \qquad R > 0 , \quad r > 0 ,$$

$$\frac{dR}{dr}(0) = 0 , \qquad R(\infty) = 0 .$$

Careful numerical computations [4] indicate that indeed the blowup occurs with the power 2/3 as obtained by Zakharov and Synakh.

Concerning the nature of the singularity in the supercritical case little seems to be known.

We have looked in detail into the problem of understanding the form (2) of the blowing up solution in the critical case. For technical reasons we have so far restricted attention to the case $N = 1$, $\sigma = 2$. We have shown that in this case there is a function $z(t,x)$ in $H^1(\mathbb{R}^1)$ for $-\varepsilon_0 \le t < 0$, $0 < \varepsilon_0$ sufficiently small, such that for each $\lambda_0 \neq 0$

$$(3) \qquad \phi(t,x) = \frac{\lambda_0}{(-t)^{2/7}} R\left(\frac{x}{(-t)^{4/7}}\right) e^{-i7\lambda_0(-t)^{-1/7}} + z(t,x)$$

is a solution of (1) in $-\varepsilon_0 \le t < 0$ and

(4) $$(-t)^{2/7} \sup_x |z(t,x)| \to 0 \quad \text{as} \quad t \to 0 .$$

In other words we have shown that singular solutions of the form (3) exist with z being a lower order correction in view of (4). We do not know why solutions of the form (3) arise as singular solutions for a broad class of initial data as has been observed numerically.

The main tool in the analysis is the study of the linearized problem about R Schroedinger equation

(5) $$2iw_t + \Delta w - w + (\sigma+1)R^{2\sigma}w + \sigma R^{2\sigma}\bar{w} = 0 .$$

If $w = u + iv$ then we may rewrite (5) in system form

(6) $$2\binom{u}{v}_t = L\binom{u}{v} \qquad L = \begin{pmatrix} 0 & L_- \\ -L_+ & 0 \end{pmatrix}$$

$$L_+ = -\Delta + 1 - (2\sigma+1)R^{2\sigma}$$
$$L_- = -\Delta + 1 - R^{2\sigma} .$$

On pairs of functions $\binom{f}{g}$ in $H^1 \times H^1$ define the bilinear form

(7) $$B\left(\binom{f}{g},\binom{p}{q}\right) = (f, L_+p) + (g, L_-q) .$$

One verifies easily that this bilinear form is invariant for solutions of (6). However, B is not an inner product in $H^1 \times H^1$ because it is not positive definite owing to the nullspace that L has.

One easily finds that the function

(8) $$n_1 = \binom{0}{-R} , \quad n_2 = \binom{-R'}{0} , \quad n_3 = \binom{0}{xR} , \quad n_4 = \begin{pmatrix} \frac{1}{2}R + xR' \\ 0 \end{pmatrix}$$

satisfy

(9) $$Ln_1 = Ln_2 = 0\,, \qquad L^2 n_3 = L^2 n_4 = 0\,.$$

Moreover these null vectors are associated with the classical symmetries of our problem that take a solution $\phi(t,x)$ into

$$\lambda^{1/\sigma}\phi\left(\lambda^2(t-t_0)\,,\ \lambda(x-\xi t-x_0)\right)\, e^{i\left[\xi(x-\xi t-x_0)+\frac{\lambda^2+\xi^2}{2}t\right]}$$

where (λ,ξ,x_0,t_0) are four parameters.

One might expect that the bilinear form B restricted to functions in $H^1\times H^1$ that are orthogonal to four function pairs η_1, η_2, η_3, η_4 (the biorthogonal basis for example) is positive. This is true in the subcritical case $\sigma < \frac{2}{N}$ ($N=1$ in the present discussion) and in fact B becomes then equivalent to the standard inner product in $H^1\times H^1$. But this is not true in the critical case!

In the critical case there is one more symmetry to the problem ($N=1$, $\sigma=2$).

$$\phi(t,x) \to \lambda\,\phi(\tau,\theta)e^{i\frac{\dot{a}}{a}\frac{x^2}{2}}$$

where $\lambda = a^{-1}$, $\tau = \int^t \lambda^2 ds$, $\theta = \lambda x$ and $\ddot{a} = 0$, i.e. a is a linear function. (Notice that this transformation leads to singular solutions with $\sqrt{t}$ singularity; they have never been observed in numerical experiments.) Therefore, in addition to (8) we have

(10) $$n_5 = \begin{pmatrix} 0 \\ x^2 R\end{pmatrix} \qquad \text{with} \qquad L^3 n_5 = 0\,.$$

But, without having another classical symmetry, we also have

(11) $$n_6 = \binom{\rho}{0}\,, \quad L_+\rho = -x^2 R\,, \qquad \text{with} \qquad L^4 n_6 = 0\,.$$

It can be shown that $n_1, n_2, \ldots, n_6$ span now the (generalized) null-space of L and that B restricted to functions orthogonal to six function pairs $\eta_1,\ldots,\eta_6$ is an inner product equivalent to $H^1 \times H^1$.

One now looks for solutions of the form (3) and one must show a $z(t,x)$ with the correct properties exists. The power 2/7 emerges as the only suitable candidate for this purpose and the structure of the nullspace discussed above is essential.

*Supported by Air Force grant AFOSR-80-0228.

REFERENCES

[1] Zakharov, V.E. and Synakh, V.S., The nature of the self-focusing singularity, JETP 41 (1976) 465-468.

[2] McLaughlin, D., Papanicolaou, G. and Weinstein, M., Focusing and saturation in nonlinear beams, to appear.

[3] Weinstein, M., N.Y.U. Dissertation, 1982.

[4] Sulem, P.L., Sulem, C. and Patera, K., Numerical investigation of focusing singularities, to appear.

[5] Zakharov, V.E. and Shabat, A.B., Exact theory of two-dimensional self focusing and one-dimensional self modulation of waves in nonlinear media, JETP 34 (1972) 62-69.

[6] Ginibre, J. and Velo, G., On a class of nonlinear Schrödinger equations I, II, J. Funct. Anal. 32 (1979) 1-32, 33-71.

[7] Glassey, R.T., On the blowing-up of solutions to the Cauchy problem for the nonlinear Schrödinger equation, J. Math. Phys. 18 (1977) 1794-1797.

Lecture Notes in Num. Appl. Anal., **5**, 259–271 (1982)
Nonlinear PDE in Applied Science. U.S.-Japan Seminar, Tokyo, 1982

Soliton Equations as Dynamical Systems on Infinite Dimensional Grassmann Manifold

Mikio Sato

RIMS, Kyoto University, Kyoto 606

Yasuko Satc

Mathematics Department, Ryukyu University, Okinawa 903-01

In the winter of 1980-81 it was found that the totality of solutions of the Kadomtsev - Petviashvili equation as well as of its multi-component generalization forms an infinite dimensional Grassmann manifold [1]. In this picture the time evolution of a solution is interpreted as the dynamical motion of a point on this manifold. A generic solution corresponds to a generic point whose orbit (in the infinitely many time variables) is dense in the manifold, whereas degenerate solutions corresponding to points bound on those closed submanifolds which are stable under the time evolution describe the solutions to various specialized equations such as KdV, Boussinesq, nonlinear Schrödinger, sine-Gordon, etc.

We foresee that a similar structural theory should hold also for multi-dimensional 'integrable' systems.

§1. The universal Grassmann manifold

For a vector space $V=V(N)$ (say, over $\mathbb{C}$) of dimension N $(=m+n)$ the Grassmann manifold $GM(m,V)$ $(=GM(m,n))$ is by definition the parameter space for the totality of m-dimensional subspaces in V. We can write

$$GM(m,V) = \{m\text{-frames in } V\} / GL(m)$$

where an m-frame means an m-tuple of linearly independent vectors. $GM(m,V)$ is a homogeneous space of the general linear group $GL(V)$.

Further, it is viewed as an algebraic submanifold (of dimension mn) of the $\binom{N}{m}-1$ dimensional projective space $\mathbb{P}(\wedge^m V)$ by letting an m-frame $(\xi^{(0)},\cdots,\xi^{(m-1)})$ correspond to the exterior product $\xi^{(0)}\wedge\cdots\wedge\xi^{(m-1)}\in\wedge^m V$ (the canonical projective embedding). If $\xi^{(i)}=\xi_{0i}e_0+\cdots+\xi_{N-1,i}e_{N-1}$ where $e_0,\cdots,e_{N-1}$ denote a basis of V, then $\xi^{(0)}\wedge\cdots\wedge\xi^{(m-1)}=\sum_{0\leq\ell_0<\cdots<\ell_{m-1}<N}\xi_{\ell_0\cdots\ell_{m-1}}e_{\ell_0}\wedge\cdots\wedge e_{\ell_{m-1}}$ with $\xi_{\ell_0\cdots\ell_{m-1}}=\det(\xi_{\ell_i j})_{i,j=0,\cdots,m-1}$. These $\xi_{\ell_0\cdots\ell_{m-1}}$, $0\leq\ell_i<N$, (which are antisymmetric in suffixes) satisfy the Plücker's relations:

$$\sum_{i=0}^{m}(-)^i\xi_{k_0\cdots k_{m-2}\ell_i}\xi_{\ell_0\cdots\hat{\ell}_i\cdots\ell_m}=0$$

and vice versa; i.e. a point in the ambient $\mathbb{P}(\wedge^m V)$ lies in the embedded GM(m,V) if and only if its projective coordinates $\xi_{\ell_0\cdots\ell_{m-1}}$, $0\leq\ell_i<N$, satisfy the Plücker's relations (i.e. are Plücker coordinates).

To each set of suffixes $(\ell_0,\cdots,\ell_{m-1})$, $0\leq\ell_0<\cdots<\ell_{m-1}<N$, we associate a Young diagram Y consisting of rows of length $\ell_{m-1}-(m-1),\cdots,\ell_1-1,\ell_0$, respectively (cf. H. Weyl, The Classical Groups, Princeton, 1939) and often identify them; e.g. Plücker coordinates are also written ξ_Y, the diagrams Y being those contained in the $m\times n$ rectangular diagram Δ_{mn}.

After Weyl's celebrated work Young diagrams (of vertical size $\leq$ N) classify irreducible tensor representations of GL(V). Denoting by R_{ij} the contragredient of the irreducible representation space labeled by the $i\times j$ rectangular diagram Δ_{ij}, our GM(m,V) is the projective algebraic manifold corresponding to the graded algebra $\bigoplus_{j=0}^{\infty}R_{mj}$. (Here multiplication is unambiguously defined because $R_{mi}\otimes R_{mj}$ containes $R_{m,i+j}$ exactly once.) We can also write:

$$GM(m,V)=(\widetilde{GM}(m,V)-\{0\})\,/\,GL(1),$$

where $\widetilde{GM}(m,V)=\{(\xi_Y)_{Y\subset\Delta_{mn}}\,|\,\xi_Y \text{ satisfy the Plücker's relations}\}\subset\wedge^m V$.

Let $m\leq m'$ and $n\leq n'$. Then: (i) if $(\xi'_Y)_{Y\subset\Delta_{m'n'}}$ satisfies the Plücker's relations, so does its restriction to Y's within Δ_{mn} (whence $\widetilde{GM}(m',n')\to\widetilde{GM}(m,n)$). On the other hand, (ii) $(\xi_Y)_{Y\subset\Delta_{mn}}$ satisfies the Plücker's relations if and only if $(\xi'_Y)_{Y\subset\Delta_{m'n'}}$ does, ξ'_Y being defined by $\xi'_Y = \xi_Y$ or $= 0$ according as $Y\subset\Delta_{mn}$ or not (whence $\widetilde{GM}(m,n) \hookrightarrow \widetilde{GM}(m',n')$. (i) and (ii) combined give the commutative diagram

$$\begin{array}{ccc} \widetilde{GM}(m',n') & \longrightarrow & \widetilde{GM}(m,n) \quad \text{(restriction)} \\ \text{id}\uparrow\wr & & \text{id}\uparrow\wr \\ \widetilde{GM}(m',n') & \hookleftarrow & \widetilde{GM}(m,n) \quad \text{(embedding)}. \end{array}$$

Hence, defining the universal Grassmann manifold $GM = (\widetilde{GM}-\{0\}) / GL(1)$ and its dense submanifold $GM^{fin} = (\widetilde{GM}^{fin} - \{0\}) / GL(1)$ by

$$\widetilde{GM} = \{(\xi_Y)_{Y:\text{all diagrams}} \mid \xi_Y \text{ satisfy all the Plücker's relations}\},$$

$$\widetilde{GM}^{fin} = \{(\xi_Y)_Y \in \widetilde{GM} \mid \xi_Y = 0 \text{ for almost all } Y\}$$

respectively, we have

$$\widetilde{GM} = \{(\xi_Y)_{Y:\text{all diagrams}} \mid (\xi_Y)_{Y\subset\Delta_{mn}} \in \widetilde{GM}(m,n) \text{ for any } m \text{ and } n\},$$

$$\widetilde{GM}^{fin} = \bigcup_{m,n} \widetilde{GM}(m,n), \text{ and}$$

$$\begin{array}{ccc} \widetilde{GM} & \xrightarrow{\text{surjective}} & \widetilde{GM}(m,n) \\ \text{dense}\uparrow\cup & & \text{id}\uparrow\wr \\ \widetilde{GM}^{fin} & \hookleftarrow & \widetilde{GM}(m,n). \end{array}$$

To each $\xi \in GM(m,n)$ (resp. $\in GM$) uniquely corresponds a diagram $Y\subset\Delta_{mn}$ (resp. an unrestricted Y) in such a way that, for the Plücker coordinates of ξ, $\xi_Y \neq 0$ while $\xi_{Y'} = 0$ unless $Y'\supset Y$; and, denoting by $GM^Y(m,n)$ those points to which the given Y corresponds, we have a cellular decomposition $GM(m,n) = \bigsqcup_{Y\subset\Delta_{mn}} GM^Y(m,n)$, with $GM^Y(m,n) \simeq \mathbb{C}^{mn-|Y|}$, $|Y|$ = size of $Y = \ell_0+\cdots+\ell_{m-1}-\frac{1}{2}m(m-1)$ (resp. $GM = \bigsqcup_Y GM^Y$).

Consider the infinite dimensional vector space $\mathbf{V}$ (resp. $\dot{\mathbf{V}}$) consisting of elements $\xi = (\xi_\nu)_{\nu\in\mathbb{Z}}$, with $\xi_\nu \in \mathbb{C}$, $\xi_\nu = 0$ for $\nu \ll 0$ (resp. for $\nu \gg 0$). (Setting $e_\mu = (\delta_{\mu\nu})_{\nu\in\mathbb{Z}} \in \mathbb{V}$ one also writes $\xi = \sum_{-\infty\ll\nu<\infty} \xi_\nu e_\nu$ (resp. $\xi = \sum_{-\infty<\nu\ll\infty} \xi_\nu e_\nu$).) Further, by introducing the dual (or contragredient) basis $(e^*_\mu)_{\mu\in\mathbb{Z}}$ to $(e_\mu)_{\mu\in\mathbb{Z}}$ and the dual space $\mathbb{V}^* = \{\xi^* = \sum_{-\infty<\nu\ll\infty} \xi^*_\nu e^*_\nu \mid \xi^*_\nu \in \mathbb{C}\}$ (resp. $\dot{\mathbb{V}}^* = \{\xi^* = \sum_{-\infty\ll\nu<\infty} \xi^*_\nu e^*_\nu \mid \xi^*_\nu \in \mathbb{C}\}$) to $\mathbb{V}$ (resp. to $\dot{\mathbb{V}}$) so that their pairing is given by the effectively finite sum: $\langle\xi^*,\xi\rangle = \sum \xi^*_\nu \xi_\nu$, our vector space naturally acquires the weak topology (or rather, S. Lefschetz's linear topology, in which our space is locally linearly compact). (Any locally convex topology on a vector space induces via its dual a linear topology there, and its subsapce is closed by the latter if and only if it is so by the former.)

Define subspaces $\mathbb{V}^{(m)}$ of $\mathbb{V}$ (resp. subspaces $\dot{\mathbb{V}}^{(m)}$ of $\dot{\mathbb{V}}$), $m \in \mathbb{Z}$, by $\mathbb{V}^{(m)}$ (resp. $\dot{\mathbb{V}}^{(m)}$) $= \{(\xi_\nu)_{\nu\in\mathbb{Z}} \in \mathbb{V}$ (resp. $\dot{\mathbb{V}}$) $\mid \xi_\nu = 0$ for $\nu<m\}$. Then we have

GM(resp. $\mathrm{GM}^{\mathrm{fin}}$) = {closed subspaces V of $\mathbb{V}$ (resp. $\dot{\mathbb{V}}$) | The dimensions of Ker and Coker of the natural map $V \to \mathbb{V}/\mathbb{V}^{(0)}$ (resp. $\to \dot{\mathbb{V}}/\dot{\mathbb{V}}^{(0)}$) are both finite and coincide.}
= {closed subspaces V of $\mathbb{V}$ (resp. $\dot{\mathbb{V}}$) | $\dim V\cap\mathbb{V}^{(\nu)}$ (resp. $\dim V\cap\dot{\mathbb{V}}^{(\nu)}$) $= |\nu|$ for $\nu \ll 0$},

where the closedness of V is a consequence of the other conditions and the qualifier is dispensable for $V\subset\mathbb{V}$, while it is not for $V\subset\dot{\mathbb{V}}$. Also we have, for any diagram Y parametrized by $(\ell_0,\cdots,\ell_{m-1})$,

$$\mathrm{GM}^Y = \{V\subset\mathbb{V} \mid \dim V\cap\mathbb{V}^{(\nu)} \leqq k \text{ if and only if } \nu > -m+\ell_{m-1-k} \ (\nu\in\mathbb{Z},\ k\in\mathbb{N})\},$$

understanding that $\ell_\nu = \nu$ for $\nu < 0$.

Between these extremes, GM and $\mathrm{GM}^{\mathrm{fin}}$, come various intermediates. For example we define, for $r=1,2,\cdots$,

$$\mathrm{GM}^{\mathrm{ana}(r)} = \{(\xi_Y)_Y \in \mathrm{GM} \mid \sqrt[|Y|]{|\xi_Y|/(|Y|/r)!} \text{ are bounded as } |Y| \to \infty\}$$

and for $0 \leqq a < \infty$,

$$GM^{\exp(a)} = \{(\xi_Y)_Y \in GM \mid \overline{\lim_{|Y|\to\infty}} \sqrt[|Y|]{|\xi_Y|} \leqq a\},$$

so that we have

$$GM \supset GM^{ana(1)} \supset GM^{ana(2)} \supset \cdots \supset GM^{\exp(a)} \supset GM^{\exp(0)} \supset GM^{fin}.$$

Then

$GM^{ana(r)}$ (resp. $GM^{\exp(a)}$) = {closed subspaces V of $\mathbb{V}^{ana(r)}$ (resp. of $\mathbb{V}^{\exp(a)}$) | The dimensions of Ker and Coker of the natural map $V \to \mathbb{V}^{ana(r)}/(\mathbb{V}^{ana(r)})^{(0)}$ (resp. $\to \mathbb{V}^{\exp(a)}/(\mathbb{V}^{\exp(a)})^{(0)}$) are both finite and coincide.}

where

$\mathbb{V}^{ana(r)}$ (resp. $\mathbb{V}^{\exp(a)}$) = $\{(\xi_\nu)_{\nu\in\mathbf{Z}} \mid \sqrt[\nu]{|\xi_\nu|/(\nu/r)!}$ are bounded and $\sqrt[\nu]{|\xi_{-\nu-1}|(\nu/r)!}$ tend to 0 as $\nu\to\infty$ (resp. $\overline{\lim_{\nu\to\infty}} \sqrt[\nu]{|\xi_\nu|} \leqq a$, $\overline{\lim_{\nu\to\infty}} \sqrt[\nu]{|\xi_{-\nu-1}|} < a^{-1}$)$\}$,

$\mathbb{V}^{ana(r)*}$ (resp. $\mathbb{V}^{\exp(a)*}$) = $\{(\xi_\nu)_{\nu\in\mathbf{Z}} \mid (\xi_{-\nu-1})_{\nu\in\mathbf{Z}} \in \mathbb{V}^{ana(r)}$ (resp. $\mathbb{V}^{\exp(a)}$)$\}$

and

$(\mathbb{V}^{ana(r)})^{(m)}$ (resp. $(\mathbb{V}^{\exp(a)})^{(m)}$) = $\{(\xi_\nu)_{\nu\in\mathbf{Z}} \subset \mathbb{V}^{ana(r)}$ (resp. $\mathbb{V}^{\exp(a)}$) | $\xi_\nu = 0$ for $\nu < m\}$.

§2. Time evolution on GM

Denoting by Λ the shift operator:

$$\Lambda e_\nu = e_{\nu-1}, \quad \Lambda \sum \xi_\nu e_\nu = \sum \xi_\nu e_{\nu-1},$$

we define for $\xi \in \widetilde{GM}$ its evolution in time variables $t = (t_1, t_2, \cdots)$ by $\xi(t) = e^{t_1\Lambda + t_2\Lambda^2 + \cdots}\xi$.

In the case of $\xi \in \widetilde{GM}^{fin}$, $\xi(t)$ is again in $\widetilde{GM}^{fin}$ for any $t_\nu \in \mathbb{C}$, $\nu = 1, 2, \cdots$. For general $\xi \in \widetilde{GM}$, however, $\xi(t)$ should be understood as a generalized element whose components are formal power series in $(t_1, t_2, \cdots)$ rather than complex numbers. (In the case of $\xi \in \widetilde{GM}^{ana(r)}$, one has $\xi(t) \in \widetilde{GM}^{ana(r)}$ if $|t_\nu|$ is sufficiently small for $\nu = r$ and are 0 for $\nu > r$. For $\xi \in \widetilde{GM}^{\exp(a)}$, one has $\xi(t) \in \widetilde{GM}^{\exp(a)}$ for $t_\nu \in \mathbb{C}$ subject to the condition $\overline{\lim}\sqrt[\nu]{|t_\nu|} < a^{-1}$.)

In any case we have, for the Plücker coordinates $\xi_Y(t)$ of $\xi(t)$,

$$\xi_Y(t) = \chi_Y(\partial_t)\xi_\phi(t) \quad \text{and} \quad \xi_\phi(t) = \sum_Y \xi_Y \cdot \chi_Y(t),$$

where ϕ denotes the empty Young diagram, $\chi_Y(t)$ denotes the character polynomial for the general linear group, and $\chi_Y(\partial_t)$ denotes the differential operator obtained from $\chi_Y(t)$ by replacing t_ν by $\frac{1}{\nu}\frac{\partial}{\partial t_\nu}$. (After H. Weyl, $\chi_Y(t)$ admits various expressions, one of which is

$$\chi_Y(t) = \sum_{\nu_1+2\nu_2+\cdots=|Y|} \pi_Y(1^{\nu_1}2^{\nu_2}\cdots)\frac{t_1^{\nu_1}t_2^{\nu_2}\cdots}{\nu_1!\nu_2!\cdots},$$

where $\pi_Y(1^{\nu_1}2^{\nu_2}\cdots)$ is the irreducible character of the symmetric permutation group of $|Y|$ letters, labeled by the Young diagram Y and evaluated at the conjugacy class consisting of ν_1 cycles of size 1, ν_2 cycles of size 2, etc.)

We call $\xi_\phi(t)$ the τ function of ξ (Notation: $\tau(t;\xi)$ or $\tau(t)$). The above formulae show that $\tau(t)$ plays the role of generating function for Plücker coordinates:

$$\xi_Y(t) = \chi_Y(\partial_t)\tau(t;\xi), \quad \xi_Y = \chi_Y(\partial_t)\tau(t;\xi)\,|_{t\mapsto 0},$$

$$\tau(t'+t;\xi) = \tau(t';\xi(t)) = \sum_Y \xi_Y(t)\chi_Y(t'),$$

and that the Plücker's relations for $(\xi_Y(t))_Y$ assume the form of quadratic differential equations, or, what amounts to the same, the form of 'bilinear' equations of R. Hirota.

Summing up, we have

Theorem 1. Although any $f(t) \in \mathbb{C}[[t_1,t_2,\cdots]]$ admits the formal expansion of the form: $f(t) = \sum_Y c_Y\chi_Y(t)$, where the coefficients are uniquely given by $c_Y = \chi_Y(\partial_t)f(t)\,|_{t\mapsto 0}$, it represents the τ function of some $\xi \in \widetilde{GM}$ if and only if its coefficients c_Y satisfy the Plücker's relations.

Theorem 2. An $f(t) \in \mathbb{C}[[t_1,t_2,\cdots]]$ is the τ function of some $\xi \in GM$ if and only if it satisfies the Hirota bilinear equations of the form

$$\sum_{i=0}^{m} (-)^i \chi_{k_0\cdots k_{m-2}\ell_i}\left(\frac{D_t}{2}\right)\chi_{\ell_0\cdots\hat{\ell}_i\cdots\ell_m}\left(-\frac{D_t}{2}\right)\tau\cdot\tau = 0.$$

Moreover these exhaust all the Hirota equations to be satisfied by τ.

These quadratic differential equations are also equivalent to the quadratic difference equations. Namely,

Theorem 3. (Addition formulae) For any $\alpha \in \mathbb{C}$ we set $[\alpha] = (\alpha,\frac{1}{2}\alpha^2,\frac{1}{3}\alpha^3,\cdots)$ so that $t+[\alpha] = (t_1+\alpha,\ t_2+\frac{1}{2}\alpha^2,\cdots)$. Let $\alpha_i \in \mathbb{C}$ for $i = 0,\cdots,N-1$ and define

$$\zeta_{\ell_0\cdots\ell_{m-1}}(t) = \Delta(\alpha_{\ell_{m-1}},\cdots,\alpha_{\ell_0})\tau(t+[\alpha_{\ell_0}]+\cdots+[\alpha_{\ell_{m-1}}]),\ 0 \le \ell_i < N$$

with $\Delta(\alpha_{m-1},\cdots,\alpha_0) = \prod_{m>i>j\geq 0} (\alpha_i-\alpha_j)$. Then $\zeta_{\ell_0\cdots\ell_{m-1}}(t)$ satisfy the Plücker's relations for $GM(m,V(N))$. This property again characterizes the function τ.

E.g. we have

$$(\alpha_1-\alpha_0)(\alpha_3-\alpha_2)\tau(t+[\alpha_0]+[\alpha_1])\tau(t+[\alpha_2]+[\alpha_3])$$

$$-\ (\alpha_2-\alpha_0)(\alpha_3-\alpha_1)\tau(t+[\alpha_0]+[\alpha_2])\tau(t+[\alpha_1]+[\alpha_3])$$

$$+\ (\alpha_3-\alpha_0)(\alpha_2-\alpha_1)\tau(t+[\alpha_0]+[\alpha_3])\tau(t+[\alpha_1]+[\alpha_2]) = 0.$$

Denote by $E_{\nu\mu}$ the linear operator on $\mathbb{V}$ sending e_μ to e_ν and all the other e_κ, $\kappa \neq \mu$, to 0 (i.e. $E_{\nu\mu}\sum\xi_\kappa e_\kappa = \xi_\mu e_\nu$), and by $L_{\mu\nu}$ the vector field on $\widetilde{GM}$ induced by $E_{\nu\mu}$ (i.e. $(1+\varepsilon L_{\mu\nu})F(\xi) \equiv F((1+\varepsilon E_{\nu\mu})\xi)$ mod ε^2 for any function F on $\widetilde{GM}$). Since any $F(\xi)$ is a function of the Plücker coordinates ξ_Y's of ξ, $L_{\mu\nu}$ is also characterized by: $L_{\mu\nu}\xi_{\ell_0\cdots\ell_{m-1}} = \sum_{0\leq i<m} \delta_{\nu+m,\ell_i} \times \xi_{\ell_0\cdots\mu+m\cdots\ell_{m-1}}$ assuming $\nu+m$ and $\mu+m \geq 0$. (This poses no restriction on the diagram Y labelled by $(\ell_0,\cdots,\ell_{m-1})$ since $(0,1,\cdots,k-1,\ell_0+k,\cdots,\ell_{m-1}+k)$ also labels the same Y for any $k \in \mathbb{N}$.)

For the shift operator Λ we have: $\Lambda^n = \sum_{\nu\in\mathbb{Z}} E_{\nu,\nu+n}$, $n \in \mathbb{Z}$. Further, define the operator K s.t. $\Lambda K - K\Lambda = 1$ by $K\sum \xi_\nu e_\nu = \sum \nu \xi_{\nu-1} e_\nu$ to have $f(K\Lambda)\Lambda^n = \sum_\nu f(\nu) E_{\nu,\nu+n}$ for any polynomial $f(\nu)$ of ν, and in particular, $\frac{K^k}{k!}\Lambda^{k+n} = \sum_\nu \binom{\nu}{k} E_{\nu,\nu+n}$.

For $n \neq 0$, the infinitesimal operator $1+\varepsilon f(K\Lambda)\Lambda^n$, mod ε^2, induces the well-defined infinitesimal transformation on $\widetilde{GM}$, and one can write

$$(1+\varepsilon U_n^{(k)})F(\xi) \equiv F((1+\varepsilon \frac{K^k}{k!}\Lambda^{k+n})\xi) \quad \text{mod } \varepsilon^2$$

with $U_n^{(k)} = \sum_\nu \binom{\nu}{k} L_{\nu+n,\nu}$, while for $n=0$ we introduce another vector field M defined by $M\xi_Y = \xi_Y$ and set: $U_0^{(k)} = \sum_{\nu<0} \binom{\nu}{k}(L_{\nu\nu} - M) + \sum_{\nu \geqq 0} \binom{\nu}{k} L_{\nu\nu}$ to have an well-defined vector field on $\widetilde{GM}$. (Indeed, $U_0^{(k)}\xi_Y = f_k(Y)\xi_Y$ where $f_k(Y) = \sum_{0 \leqq i < m} \left(\binom{\ell_i - m}{k} - \binom{i-m}{k}\right)$. In particular $U_0^{(0)} = 0$.) M commutes $L_{\mu\nu}$ and $U_\nu^{(k)}$, and $U_\nu^{(k)}$'s satisfy the commutation relation

$$[U_\nu^{(k)}, U_\mu^{(\ell)}] = \sum_{j \geqq 0} \left(\binom{\ell+\mu}{j}\binom{k+\ell-j}{\ell} - \binom{k+\nu}{j}\binom{k+\ell-j}{k}\right) U_{\nu+\mu}^{(k+\ell-j)} + \delta_{\nu,-\mu} \cdot (-)^\ell \binom{\ell+\mu}{\ell+k+1} M.$$

<u>Theorem 4</u>. $\tau(t;\xi)$, as the function of t and $\xi \in \widetilde{GM}$, satisfies, and is characterized up to an arbitrary constant factor by, the following holonomic system of linear differential equations:

$$((L_{\mu'\nu'} - \delta_{\mu'\nu'})L_{\mu\nu} + (L_{\mu\nu'} - \delta_{\mu\nu'})L_{\mu'\nu})\tau = 0 \quad \text{for } \mu,\mu',\nu,\nu' \in \mathbf{Z},$$

$$(U_n^{(0)} - \frac{\partial}{\partial t_n})\tau = 0, \quad (U_{-n}^{(0)} - nt_n)\tau = 0 \quad \text{for } n=1,2,\cdots.$$

Indeed, the first equations (which are of the second order) restrict the solution to a linear form $\sum_Y c_Y(t)\xi_Y$ of the Plücker coordinates ξ_Y while the remaining (first order) equations fix the coefficients $c_Y(t)$ to $c \cdot \chi_Y(t)$.

Here we see that the holonomic system of these linear equations on $\{t\} \times \widetilde{GM}$ produces no linear equation but the system of non-linear (quadratic or Hirota) equations of Theorem 2, upon elimination of the variables $\xi \in \widetilde{GM}$ (i.e. upon taking the direct image by the projection $\{t\} \times \widetilde{GM} \to \{t\}$), in a sharp contrast to the finite dimensional case [2].

Also remarkable is the close resemblance between this holonomic system and the system characterizing theta functions [3]. (Theorem 3 also suggests analogy between τ and θ.)

The holonomic system generated by these equations in Theorem 4 contains also the equations of the form: $(U_\nu^{(k)}-T_\nu^{(k)})\tau = 0$, $k\in\mathbb{N}$, $\nu\in\mathbb{Z}$, where $T_\nu^{(k)}$ is a differential operator in t of the form: $T_\nu^{(k)} = \frac{1}{k!}\sum_{\nu_0,\dots,\nu_{k-1}\in\mathbb{Z},\nu_0+\dots+\nu_{k-1}=\nu} s_{\nu_0}s_{\nu_1}\cdots s_{\nu_{k-1}}$ + terms of lower degree, with $s_\nu = \frac{\partial}{\partial t_\nu}$, 0, or $|\nu|t_{|\nu|}$ according as $\nu>0$, $=0$, or <0. ($T_0^{(0)} = 0$, $T_\nu^{(0)} = s_\nu$.)

§3. Soliton equations and their solutions

Consider the totality $\mathcal{E}_{\mathcal{R}}$ of the microdifferential operators in the formal category $P = \sum_{-\infty<\nu\ll\infty} a_\nu(x)(\frac{d}{dx})^\nu$, where the coefficients $a_\nu(x)$ are taken from a given differential ring $\mathcal{R}$ (i.e. an associative algebra endowed with the derivation $\frac{d}{dx}:\mathcal{R}\to\mathcal{R}$). If $a_\nu(x) = 0$ for $\nu>m$ we write $P\in\mathcal{E}_{\mathcal{R}}^{(m)}$. Together with P its adjoint $P^* = \sum(-\frac{d}{dx})^\nu a_\nu(x)$ is again in $\mathcal{E}_{\mathcal{R}}$, and for $P,Q\in\mathcal{E}_{\mathcal{R}}$ their product $PQ\in\mathcal{E}_{\mathcal{R}}$ is well-defined by employing the Leibniz rule $(\frac{d}{dx})^\nu a(x) = \sum_{k\geq 0}\binom{\nu}{k}a^{(k)}(x)(\frac{d}{dx})^{\nu-k}$ for $\nu\in\mathbb{Z}$. Setting $a_{-1}(x) = \mathrm{Res}\,P\,dx$, we have $\mathrm{Res}\,P\,dx = -\mathrm{Res}\,P^*dx$. Thus $\mathcal{E}_{\mathcal{R}}$ constitutes a (non-commutative) ring including $\mathcal{D}_{\mathcal{R}} = \{$differential oprators$\}$ as a subring. We have: $P = P_+ + P_-$ with $P_+ = \sum_{0\leq\nu\ll\infty} a_\nu(x)(\frac{d}{dx})^\nu\in\mathcal{D}_{\mathcal{R}}$, $P_- = \sum_{\nu<0} a_\nu(x)(\frac{d}{dx})^\nu\in\mathcal{E}_{\mathcal{R}}^{(-1)}$, yielding the decomposition

$$\mathcal{E}_{\mathcal{R}} = \mathcal{D}_{\mathcal{R}}\oplus\mathcal{E}_{\mathcal{R}}^{(-1)}.$$

In the following we choose $\mathcal{R} = \mathbb{C}[[x]]$, the ring of formal power series in x, and simply write $\mathcal{E}_{\mathbb{C}[[x]]} = \mathcal{E}$; similarly with $\mathcal{E}^{(m)}$ and $\mathcal{D}$. Then $\mathbb{V}$ of §1 is canonically isomorphic to the quotient module of $\mathcal{E}$ by its maximal left ideal $\mathcal{E}x$ as left $\mathcal{E}$ modules, by letting $\xi = \sum_{-\infty\ll\nu<\infty}\xi_\nu e_\nu\in\mathbb{V}$ correspond to the residue class of $\sum\xi_\nu(\frac{d}{dx})^{-\nu-1}$ mod $\mathcal{E}x$ and the action of $P(x,\frac{d}{dx})\in\mathcal{E}$ on ξ be defined by $\xi \mapsto P(K,\Lambda)\xi$. Hereafter we identify them: $\mathbb{V} = \mathcal{E}/\mathcal{E}x$. Further we write $\mathbb{V} = \mathbb{V}^*$ by identifying $\sum \xi_\nu e_\nu \in \mathbb{V}$ with $\sum (-)^\nu \xi_{-\nu-1} e^*_\nu \in \mathbb{V}^*$, so that we have: $\langle \xi', \xi \rangle = - \langle \xi, \xi' \rangle$, $\langle \xi', P\xi \rangle = \langle P^*\xi', \xi \rangle$.

We also set

$$\mathcal{E}^{ana} = \{\sum_\nu a_\nu(x)(\tfrac{d}{dx})^\nu \,|\, \exists \delta > 0 \text{ s.t. } a_\nu(x) \text{ are holomorphic in } |x| < \delta,$$

$$\sqrt[\nu]{|a_\nu(x)|/\nu!} \to 0 \text{ and } \sqrt[\nu]{\nu!|a_{-\nu-1}(x)|} \text{ are bounded as } \nu \to \infty,$$

$$\text{both uniformly in } |x| < \delta\},$$

$$\dot{\mathcal{E}} = \{ \sum_{-\infty \ll \nu < \infty} a_\nu(x)(\tfrac{d}{dx})^\nu \,|\, \exists k \in \mathbb{N} \text{ s.t. } a_\nu(x) \text{ are polynomials of } x \text{ of degree} \leqq k\},$$

and get: $\mathbb{V}^{ana(1)} = \mathcal{E}^{ana}/\mathcal{E}^{ana}x$, $\dot{\mathbb{V}} = \dot{\mathcal{E}}/\dot{\mathcal{E}}x$.

Consider the operator $W = \sum_{\nu \geqq 0} w_\nu \cdot (\frac{d}{dx})^{-\nu} \in \mathcal{E}^{(0)}_{\mathcal{R}}$ which is monic (i.e. $w_0=1$) so that W^{-1} is again an operator of the same kind which we shall write $W^{-1} = \sum_{\nu \geqq 0} (\frac{d}{dx})^{-\nu} \cdot w^*_\nu$ with $w^*_0 = 1$. Let $\mathcal{W}$ denote the totality of such monic operators $W \in \mathcal{E}^{(0)}_{\mathcal{R}}$ with $\mathcal{R} = \mathbb{C}((x))$ (= the field of formal Laurent series in x, which is the field of quotients of $\mathbb{C}[[x]]$), satisfying the additional condition that there exists $m, n \in \mathbb{N}$ s.t. $x^m W$ and $W^{-1}x^n$ both $\in \mathcal{E}^{(0)}_{\mathbb{C}[[x]]}$ (i.e. $x^m w_\nu$ and $x^n w^*_\nu \in \mathbb{C}[[x]]$ for $\nu=1,2,\ldots$). Set $V^\phi = \bigoplus_{\nu<0} \mathbb{C}e_\nu \subset \mathbb{V}$; V^ϕ is also characterized by the property that its Plücker coordinates $\xi_Y = 1$ or 0 according as $Y=\phi$ or not. For $W \in \mathcal{W}$ we set $\gamma(W) = (W^{-1}x^n)V^\phi$, where n is so chosen that $W^{-1}x^n \in \mathcal{E}^{(0)}_{\mathbb{C}[[x]]}$. This definition of $\gamma(W)$ does not depend on the choice of such n. (This is because $xV^\phi = V^\phi$.)

Theorem 5. For $W \in \mathcal{W}$, $\gamma(W) \in GM$ and this map is bijective, namely

$$\gamma : \mathcal{W} \overset{\sim}{\to} GM.$$

In this correspondence, the inverse images of GM^{fin} and $GM^{ana(1)}$ are given by $\mathcal{W}^{fin}$ and $\mathcal{W}^{ana(1)}$, respectively, where $\mathcal{W}^{ana(1)} = \mathcal{W} \cap \mathcal{E}^{ana(1)}$ and

$$\mathcal{W}^{fin} = \mathcal{W} \cap \dot{\mathcal{E}} = \{W \in \mathcal{W} \,|\, \exists m, n \in \mathbb{N}, \text{ s.t. } W(\tfrac{d}{dx})^m \text{ and } (\tfrac{d}{dx})^n W^{-1} \in \mathcal{D}_{\mathbb{C}((x))}\}.$$

Theorem 6. Let $\xi(t) = e^{t_1\Lambda+t_2\Lambda^2+\cdots}\xi$ as in §2, and let $W = W(t) = \gamma^{-1}(\xi(t)) \in \mathcal{W}$ be the corresponding microdifferential operator. Then the evolution of $W(t)$ in t is given by

$$[W] \qquad \frac{\partial W}{\partial t_n} = B_n W - W\left(\frac{d}{dx}\right)^n, \text{ with } B_n = \left(W\left(\frac{d}{dx}\right)^n W^{-1}\right)_+ .$$

(B_n is a differential operator of the n-th order.)

Theorem 5 tells that conversely any solution of [W] is given in the above form in a unique way. More explicitly we have

$$w_\nu = \left.\frac{p_\nu(-\partial_t)\tau(t;\xi)}{\tau(t;\xi)}\right|_{t_1\mapsto x+t_1}, \quad w_\nu^* = \left.\frac{p_\nu(\partial_t)\tau(t;\xi)}{\tau(t;\xi)}\right|_{t_1\mapsto x+t_1},$$

where p_ν represents the character polynomial χ_Y for $Y = \Delta_{1,\nu}$.

Put $L = W\frac{d}{dx}W^{-1}$. Then $L = \sum_{\nu\geq 0} u_\nu\left(\frac{d}{dx}\right)^{1-\nu}$ with $u_0=1$, $u_1=0$, and u_ν are differential polynomials of $w_1,\cdots,w_{\nu-1}$, and the above system of evolution equations for W immediately implies that for L as follows:

$$[L] \qquad \frac{\partial L}{\partial t_n} = B_n L - L B_n, \text{ with } B_n = (L^n)_+$$

which is also equivalent to the following system:

$$[B] \qquad \frac{\partial B_m}{\partial t_n} - \frac{\partial B_n}{\partial t_m} + B_m B_n - B_n B_m = 0,$$

which constitutes the integrability condition for

$$[\psi] \qquad \frac{\partial \psi}{\partial t_n} = B_n \psi.$$

(Incidentally, the explicit solution for ψ is given by $\psi = \left.\frac{\tau(t;\xi')}{\tau(t;\xi)}\right|_{t_1\mapsto t_1+x}$, where ξ' is any element of GM containing $\Lambda\xi$ (as subspaces in $\mathbb{V}$).)

[L] (or [B]) gives infinite number of non-linear equations for $u_2, u_3, \cdots$, known as the equations of Kadomtsev - Petviashvili hierarchy (e.g. $3u_{2,t_2t_2} + (-4u_{2,t_3} + u_{2,t_1t_1t_1} + 12u_2u_{2,t_1})_{t_1} = 0$).

Explicit forms of the equations are easier to obtain for $v_2, v_3, \cdots$ than for

$u_2, u_3, \cdots$ where v_n's are defined as coefficients of $\frac{d}{dx} = L + v_2 L^{-1} + v_3 L^{-2} + \cdots$.
(v_n is a differential polynomial of $u_2, \cdots, u_n$ and conversely u_n is that of $v_2, \cdots, v_n$; e.g. $v_2 = -u_2$, $v_3 = -u_3$, etc. and $u_2 = -v_2$, $u_3 = -v_3$, etc.) Namely we have

$$[\psi'] \qquad p_n(-\partial_t)\psi = v_n \psi, \text{ for } n \geqq 2,$$

and its integrability condition

$$[v] \qquad p_n(-\partial_t) v_{m+1} + (J_{n,m}[v])_x = 0, \quad \text{for } m,n \geqq 1,$$

with

$$J_{n,m}[v] = v_{m+n} + \frac{1}{2} \sum_{\substack{i,i',j,j' \geqq 1 \\ i+i'=m,\, j+j'=n}} v_{i+j} v_{i'+j'}$$

$$+ \frac{1}{3} \sum_{\substack{i,i',i'',j,j',j'' \geqq 1 \\ i+i'+i''=m,\, j+j'+j''=n}} v_{i+j} v_{i'+j'} v_{i''+j''} + \cdots$$

as the equivalents of $[\psi]$ and $[L]$, respectively.

Again, v_n is explicitly given by $v_n = \frac{\partial}{\partial t_1}(p_{n-1}(-\partial_t)\log\tau)\Big|_{t_1 \mapsto t_1 + x}$, for $n \geqq 2$.

so far, accounts are given for the 1-component case. To generalize it to the r-component case we shall modify the notations as follows. For $\nu \in \mathbf{Z}$ and $0 \leqq i < r$ the basis element $e_{r\nu+i} \in \mathbf{V}$ is rewritten as $e_\nu^{(i)}$ and operators Λ^r and $\sum_{\nu \in \mathbf{Z}} E_{r\nu+i, r\nu+i}$ as Λ and E_{ii}, respectively, so that we now have

$$\Lambda e_\nu^{(i)} = e_{\nu+1}^{(i)}, \quad E_{ii} e_\nu^{(j)} = \delta_{ij} e_\nu^{(j)}.$$

For $\xi \in \widetilde{GM}$ we define its evolution in the new set of time variables $t = (t_\nu^{(i)})_{0 \leqq i < r, \nu = 1,2,\cdots}$ by $\xi(t) = e^{\sum t_\nu^{(i)} E_{ii} \Lambda^\nu} \xi$.

Let the Young diagram Y be labelled by $(\ell_0, \cdots, \ell_{mr-1})$, and for each $i = 0, \cdots, r-1$, suppose that there are m_i of ℓ_ν's s.t. $\ell_\nu \equiv i \pmod r$, whom we rewrite as $(\ell_\nu^{(i)} r + i)_{\nu = 0, \cdots, m_i - 1}$. Set $m_i' = m_i - m$ to have $\sum m_i' = 0$, and call Y_i the Young diagram labelled by $(\ell_0^{(i)}, \cdots, \ell_{m_i-1}^{(i)})$. Then we see that the single diagram Y and the composite object $((Y_0, m_0'), \cdots, (Y_{r-1}, m_{r-1}'))$ correspond to each other in 1 to 1 manner. Accordingly, we rewrite ξ_Y as $\pm\xi_{(Y_0,m'_0),\cdots,(Y_{r-1},m'_{r-1})}$, with the possible change of sign caused by rearrangement of the suffixes ℓ_ν. If $m'_i = 0$ for $i=0,\cdots,r-1$, it is simply written as $\pm\xi_{Y_0,\cdots,Y_{r-1}}$.

All the results for the 1-component case are, mutatis mutandis, generalized to the r-component case. For example,

$$\tau(t;\xi) = \xi_{\phi,\cdots,\phi}(t) = \sum_{Y_0,\cdots,Y_{r-1}} \xi_{Y_0\cdots Y_{r-1}} \chi_{Y_0}(t^{(0)})\cdots\chi_{Y_{r-1}}(t^{(r-1)})$$

with $t^{(i)} = (t_1^{(i)}, t_2^{(i)},\cdots)$, and, as for $W = \gamma^{-1}\xi(t)$,

$$\frac{\partial W}{\partial t_n^{(i)}} = B_n^{(i)} W - W E_{ii}\left(\frac{d}{dx}\right)^n \quad \text{with} \quad B_n^{(i)} = \left(W E_{ii}\left(\frac{d}{dx}\right)^n W^{-1}\right)_+ .$$

References

[1] M. Sato, Soliton equations as dynamical systems on an infinite dimensional Grassmann manifold, RIMS Kokyuroku 439 (1981), 30-46.

[2] M. Sato, T. Kawai and M. Kashiwara, Microfunctions and pseudo-differential equations, Lecture Notes in Math. 287, Springer (1973), 265-529.

[3] M. Sato, Pseudo-differential equations and theta functions, Astérisque 2 et 3 (1973), 286-291.

Lecture Notes in Num. Appl. Anal., **5**, 273–288 (1982)
Nonlinear PDE in Applied Science. U.S.-Japan Seminar, Tokyo, 1982

L^1 and L^∞ Approximation of Vector Fields in the Plane

Gilbert Strang

Massachusetts Institute of Technology

ABSTRACT

We study four problems, two in L^1 and two in L^∞, whose analogues in L^2 are the familiar minimum principles which lead to the Laplace equation. One possibility is to be given the boundary value $\psi = g$ and to minimize $\|\nabla\psi\|_1$ or $\|\nabla\psi\|_\infty$; the gradient at a point (x,y) in Ω is measured by $|\nabla\psi|^2 = \psi_x^2 + \psi_y^2$. In the other problems we are given a vector field $v: \Omega \to R^2$, and minimize either $\|\nabla w - v\|_1$ or $\|\nabla w - v\|_\infty$. In each case we use the duality theory of convex analysis to give equivalent statements of the problem, often with an interpretation in mechanics and often partly solved. Nevertheless some questions still remain open.

Approximation in the L^2 norm leads to linear equations (which are forbidden at this conference). In L^p, $1 < p < \infty$, the equations become nonlinear but much of the analysis continues to apply. In L^1 and L^∞, however, the situation is entirely different: it is not the differential equation but the underlying variational principle that leads to an existence theory, and suggests how to construct the optimal solution.

This note studies four typical problems for scalar valued functions on a simply connected domain $\Omega \subset R^2$ with sufficiently smooth boundary Γ. Each of the problems has a dual--a maximization instead of a minimization--and in the applications the dual is of equal importance. Where one variational statement is the "static" formulation of a problem in mechanics, with stresses as the unknown, the other is the "kinematic" form in terms of velocities. We will study several combinations of boundary conditions and inhomogeneous terms, but not every possible combination, because already we ask the reader's consent about one more thing. In addition to the dual of each problem, there is another pair of optimizations (equivalent to the given one) created by a special situation in R^2, that the general solution of $\operatorname{div} \sigma = 0$ is $\sigma = (\psi_y, -\psi_x)$ for some ψ. Therefore it will happen that each of our problems has four equivalent forms, and that one of them is simpler to solve than the others. (For the others we may learn only the value of the maximum or minimum, by duality, without finding the function which achieves that value.) Some questions will remain unsolved even with four alternatives.

The problems arise in the study of plasticity and optimal plastic design, and elsewhere; we will give references rather than a complete description of these applications. And we do the same for the proofs of duality; in our problems they come directly from the techniques of Ekeland-Temam [1], who applied the Moreau-Rockafellar theory to a sequence of important examples in partial differential equations. Our chief purpose is to contribute some additional examples, and they have developed from our joint work with Robert Kohn and Roger Temam. We mention that optimal design leads to more complicated variational problems ([2-3] is part of a large literature), and also that perfect plasticity in R^3 has required a new space of vector-valued displacements and a corresponding analysis [4-5]. The problems in this note are easier, but they have natural interest and it may be useful to organize them more systematically.

At the end we discuss some applications in optimal design and the nonconvex problems to which they lead.

1. The minimum of $\|\nabla\psi\|_\infty$ with Dirichlet data

It is this form which can be solved directly, but we begin with the four equivalent problems:

1A. MIN $\|\sigma\|_\infty$ subject to $\operatorname{div}\sigma = 0$ and $\sigma\cdot n = f$ on Γ

1B. MAX $\int_\Gamma uf$ subject to $\iint_\Omega |\nabla u| = 1$

1C. MIN $\|\nabla\psi\|_\infty$ subject to $\psi = g$ on Γ

1D. MAX $\int_\Gamma g\tau\cdot n$ subject to $\operatorname{div}\tau = 0$, $\iint|\tau| = 1$.

The conversion between 1A and 1C is by $\sigma = (\psi_y, -\psi_x)$, a rotation through $\pi/2$ of $\nabla\psi$. It follows that $f = \sigma \cdot n = \nabla\psi \cdot t = \partial\psi/\partial s$, the tangential derivative of ψ. Therefore $\psi = g = \int f\, ds$ up to a constant on Γ; it is assumed that $\int f\, ds = 0$ on the closed curve Γ.

Similarly 1B and 1D are connected by $\tau = (u_y, -u_x)$, and integration by parts on Γ. 1B is dual to 1A, and 1D to 1C.

The solution of 1C comes from an extension lemma proved independently by McShane and Kirszbraun for Lipschitz functions on a metric space. Their construction applies equally to Hölder continuous functions, and was followed by deeper results of Whitney and Calderon. In our case the metric is the shortest distance within Ω; it is Euclidean distance, if Ω is convex. Then the quantity $\|\nabla\psi\|_\infty$ to be minimized is the Lipschitz constant for ψ. The lemma extends g on Γ to ψ on Ω with <u>no increase</u> in the Lipschitz constant, by the simple construction

$$\psi(P) = \max_{Q \in \Gamma} [g(Q) - d(P,Q)].$$

Therefore the minimum in 1C is immediate: it equals the Lipschitz constant for g, and Problem 1 is solved.

This provides a natural analogue for continuous flows of the <u>max flow-min cut theorem</u> of Ford and Fulkerson. Instead of a finite network with capacity constraints on the edges, the flow through Ω is described by a vector field σ and its capacity by $|\sigma| \leq 1$. The possibilities of varying capacity $|\sigma| \leq c(x,y)$ and nonzero sources $\operatorname{div} \sigma = F$ within Ω are included in [6]. Iri created a similar theory with a neat application to traffic in Tokyo [7].

We give a brief but very informal derivation of the optimality conditions that connect 1A to 1B, with $u(x,y)$ as Lagrange multiplier for the constraint $\operatorname{div} \sigma = 0$. The saddle-point (Lagrangian) form is

$$\min_{\sigma\cdot n=f} \max_{u} [\|\sigma\|_\infty + \iint u \operatorname{div} \sigma] = \max_{u} \min_{\sigma} [\|\sigma\|_\infty - \iint \sigma\cdot\nabla u + \int uf] .$$

The final minimum over σ is $-\infty$ if $\iint|\nabla u| > 1$, and otherwise it is $\int uf$ as in 1B. The optimality conditions are

$$\|\sigma\|_\infty = \iint \sigma\cdot\nabla u, \quad \iint|\nabla u| = 1,$$

which gives an interesting form for u. It is the characteristic function, or more exactly a multiple $1/|P-Q|$ of the characteristic function, of the set bounded by the line between P and Q on Γ --where these are the points (not necessarily unique) at which g attains its Lipschitz constant. Thus $\nabla u = 0$ except across this line; on the line, ∇u is a singular measure of mass one and σ is the normal vector of magnitude $\|\sigma\|_\infty$.

<u>Example 1</u>: $g = \sin\theta$ on the circle Ω of radius $r < 1/2$. The Lipschitz constant is $1/r$, between the points $P = (0,r)$ and $Q = (0,-r)$. An optimal u is zero in the semicircle $x > 0$ and $1/2r$ in the complement $x \leq 0$. An optimal ψ (not the same as McShane's extension) is $\psi = y/r$, with $\sigma = (\psi_y, -\psi_x) = (1/r, 0)$, and σ is normal to the diameter PQ.

A less trivial example, with more uniqueness, would start with an elliptical Ω.

2. The minimum of $\|\nabla\psi\|_1$ with Dirichlet data

Again we give the four equivalent extremal principles:

2A. MIN $\iint_\Omega |\sigma|$ subject to div $\sigma = 0$ and $\sigma \cdot n = f$

2B. MAX $\int_\Gamma uf$ subject to $\|\nabla u\|_\infty = 1$

2C. MIN $\iint |\nabla\psi|$ subject to $\psi = g$ on Γ

2D. MAX $\int g\tau \cdot n$ subject to div $\tau = 0$ and $\|\tau\|_\infty = 1$.

The connections among 2A-2D are the same as for 1A-1D.

In this case it is again 2C (the least gradient problem [8]) that can be solved most directly, using the coarea formula for a function $\psi(x,y)$ of bounded variation:

$$\iint |\nabla\psi| dxdy = \int_{-\infty}^{\infty} |\gamma_t| dt,$$

where γ_t is the level set where $\psi = t$ and $|\gamma_t|$ is its length. More precisely, since ψ could be constant over a set of positive area, we construct $E_t = \{(x,y) \in \Omega: \psi(x,y) \geq t\}$ and then $\gamma_t = \partial E_t \setminus \Gamma$. This coarea formula and its generalizations are a valuable tool in geometric measure theory [9-10]; for a smooth function ψ, or a piecewise linear function, the proof is straightforward.

To minimize $\|\nabla\psi\|_1$ is to minimize for each t the length of γ_t, subject to the requirement that γ_t connects the boundary points at which $g = t$. (Again a more precise form may be required; the set E_t must contain the set on which $g \geq t$. We note that for ψ of bounded variation, with trace g in $L^1(\Gamma)$, the function $|\gamma_t|$ is defined for almost all t and is integrable.)

Example 2. $g = \cos 2\theta$ on the unit circle Ω. The L^2 norm of ψ is smallest for the harmonic function $\psi = r^2 \cos 2\theta$, zero only on the four rays to $\theta = \pm \pi/4, \pm 3\pi/4$. But the L^1 norm is minimized by a function ψ which vanishes over the whole inscribed square of side $\sqrt{2}$ with these rays as diagonals. Between the square and the circle, the level lines ψ = constant are straight, to minimize $|\gamma_t|$. To the right of the square, for example, $\psi = 2x^2-1$: constant on vertical lines x = constant, and agreeing at the boundary $r = 1$ with $g = \cos 2\theta = 2\cos^2\theta-1$.

In the duality with 2B, the optimality condition imitates 1B above to give

$$\iint|\sigma| = \iint\sigma\cdot\nabla u, \quad \|\nabla u\|_\infty = 1.$$

Therefore ∇u is a unit vector in the direction of σ where $\sigma \neq 0$, and elsewhere $|\nabla u| \leq 1$. In our example we had $\psi = 2x^2-1$ and $\sigma = (0,-4x)$ in the section $x \geq 1/\sqrt{2}$ of the circle. Therefore $\nabla u = (0,1)$ and $u = -y$. Similarly $u = \pm x$ or $\pm y$ in the four quarters of the circle--but not uniquely so in the inscribed square where $\psi = 0$ and $\sigma = 0$.

3. The L^1 approximation of a vector field by a gradient

The given vector field is $v = (v_1(x,y), v_2(x,y))$, and it is itself a gradient if $F = \partial v_1/\partial y - \partial v_2/\partial x = 0$. Otherwise the maxima and minima will exceed zero in our four equivalent problems:

3A. $\text{MAX} -\iint_\Omega v\cdot\sigma$ subject to $\operatorname{div}\sigma = 0$, $\sigma\cdot n = 0$ on Γ,
$|\sigma| \leq 1$ in Ω

3B. $\text{MIN} \iint_\Omega |\nabla w - v|$

3C. $\text{MAX} \iint F\psi$ subject to $\psi = 0$ on Γ, $|\nabla\psi| \leq 1$ in Ω

3D. $\text{MIN} \iint |\tau|$ subject to $\text{div}\, \tau = -F$.

In 3D, τ is the rotation of $\nabla w - v$ through $\pi/2$; the divergence of $\nabla w^\perp = (-w_y, w_x)$ is automatically zero, so that $\text{div}\, \tau = -\text{div}\, v = -F$. Again we use a property special to R^2.

In this case it is 3C which can be solved at sight, provided we assume $F \geq 0$. To make $\iint F\psi$ as large as possible is to maximize ψ among functions vanishing on Γ with $|\nabla\psi| \leq 1$. The extremal function is ψ = distance to Γ, and this choice gives the optimal values in 3A-3D.

For F of varying sign the problem is much more difficult and interesting; we believe it to be unsolved. On an interval in R^1 the optimal ψ has $\psi' = \pm 1$, and the "breakpoints" can be determined.

A construction of the optimal w (assuming $F \geq 0$) was given by Mosolov [11]. The computation is more delicate than that of ψ, and is guided by the optimality condition

$$\sigma = \frac{\nabla w - v}{|\nabla w - v|} \quad \text{at every point where} \quad \nabla w \neq v.$$

Example 3. $v = (y,-x)$ and $F = 2$ on the unit circle. On this domain $\psi = 1-r$, the distance to the boundary. Therefore $\sigma = (\psi_y, -\psi_x)$ is the unit vector field tangent to the concentric circles r = constant. It happens that in this instance $w \equiv 0$. The optimality condition displayed above is satisfied by $\sigma = -v/|v|$, and therefore $v = (y,-x)$ "cannot be approximated by a gradient." The closest approximation is null.

This choice of v arises naturally in the torsion of a cylindrical rod. The vector σ gives the shearing stress and $-\iint v\cdot\sigma$ is the associated moment resisting an external force that twists the bar. The maximum in 3A, subject to the limitation $|\sigma| \leq 1$ for a plastic material, is the limit moment of a bar with cross-section Ω. This is the largest torque the bar can resist; the angle of twist approaches infinity and the bar becomes fully plastic ($|\sigma| = 1$ throughout Ω).

The dual variable $w(x,y)$ measures the "warping" of each cross-section out of its original plane--and for the circular bar of Example 3 each cross-section remains plane and $w = 0$.

The minimization 3D has a further mechanical interpretation. It again refers to a bar with cross-section Ω, but now subject to the axial force $F(x,y)$. This body force is resisted by shear stresses $\tau = (\tau_{xz}, \tau_{yz})$, and $\operatorname{div} \tau = -F$ expresses equilibrium. Then $\iint |\tau|$ is proportional to the minimum weight of a bar which can withstand the load F. It is a specialization of the Michell-Prager theory of optimal design to the case of pure shear [12].

4. The L^∞ approximation of a vector field by a gradient

For given v, with $F = -\operatorname{div} v^\perp$ and $\tau = (\nabla w - v)^\perp$ as before, the equivalent problems are

4A. MAX $-\iint \sigma\cdot v$ subject to $\operatorname{div} \sigma = 0$, $\sigma\cdot n = 0$ on Γ, $\|\sigma\|_1 = 1$

4B. MIN $\|\nabla w - v\|_\infty$

4C. MAX $\iint F\psi$ subject to $\psi = 0$ on Γ, $\|\nabla\psi\|_1 = 1$

4D. MIN $\|\tau\|_\infty$ subject to $\operatorname{div} \tau = -F$.

Our method is to apply the coarea formula to 4C. We want to show that the optimal ψ is a multiple of the characteristic function C of some subset $E \subset \Omega$. The coarea formula can be written as

$$\|\nabla\psi\|_1 = \int_{-\infty}^{\infty} \|\nabla C_t\|_1 \, dt,$$

where C_t is the characteristic function of $E_t = \{\psi \geq t\}$. It is not difficult [6] to show that also

$$\iint_\Omega F\psi = \int_{-\infty}^{\infty} (\iint_\Omega FC_t) \, dt.$$

Suppose M is the maximum in 4C, and suppose that

$$\iint_\Omega FC < (M-\epsilon)\|\nabla C\|_1$$

for every characteristic function C. Then choosing $C = C_t$ for the optimal ψ (or more precisely $C = C_{nt}$ for a maximizing sequence of functions ψ_n) we would contradict the previous equations by integrating over t. Therefore the maximum (or supremum) is attained in 4C by a normalized characteristic function $\psi = C/\|\nabla C\|_1$.

In other words, Problem 4C is equivalent to the simpler problem

$$M = \operatorname*{MAX}_C \frac{|\iint_E FC|}{\|\nabla C\|_1} = \operatorname*{MAX}_{E\subset\Omega} \frac{(\text{area of } E)}{(\text{perimeter of } E)} \text{ for } F \equiv 1.$$

Whenever F is constant, we have an <u>isoperimetric problem</u>: maximize area/perimeter within Ω. For $F > 0$ it is a weighted isoperimetric problem. And for F of varying sign, it is a generalized isoperimetric problem which is new to us.

We have not mentioned the boundary condition $\psi = 0$, which seems to be violated if $\partial C \cap \Gamma \neq 0$. Nevertheless the analysis can be justified; it is the condition $\psi = 0$ which must be relaxed [13], and the correct form of 4C is

$$\text{MAX} \iint_\Omega F\psi \quad \text{subject to} \quad \iint_\Omega |\nabla\psi| + \int_\Gamma |\psi| = 1.$$

Similarly Problem 2C can be relaxed by the boundary integral $\int|\psi-g|$. The effect in Problem 4C is that the length of $\partial C \cap \Gamma$ is included in the perimeter of C, and we reach the isoperimetric problem described above.

We show by example that the optimal ψ can be computed (at least for $F = 1$ and for simple domains). Since ψ is piecewise constant, changing only at the boundary of the optimal set C in the isoperimetric problem, $\sigma = (\psi_y, -\psi_x)$ is a singular measure supported on ∂C --a "line of δ-functions." The optimality condition connecting it to τ gives only moderate information:

$$|\tau| = \|\tau\|_\infty \quad \text{on} \quad \partial C \quad (\text{and} \quad \tau \perp \partial C).$$

We have no method for computing τ in Ω.

Example 4: $v = (y,-x)$ and $F = 2$ on the unit circle Ω. This is the oldest of all isoperimetric problems, and the subset which maximizes area/perimeter is $C = \Omega$. The vector field τ is radial, with $\text{div}\,\tau = -2$, and again the nearest gradient to v is $\nabla w = 0$.

Example 5: $v = (y,-x)$ and $F = 2$ on the unit square Ω, with vertices at $(\pm 1/2, \pm 1/2)$. The optimal subset C is neither the whole square nor the inscribed circle. Instead it is a compromise [6,14] whose boundary coincides with Γ except for quarter circles of radius $(2 + \sqrt{\pi})^{-1}$ in the four corners of the square. (They are tangent to the square, so ∂C is smooth.) We have so far been unsuccessful in determining a corresponding vector field τ with $\operatorname{div} \tau = (2 + \sqrt{\pi})\|\tau\|_\infty$. It exists, by duality theory, and we have offered a modest prize (10,000 Yen at the U.S.-Japan Seminar) for its discovery.

It would be interesting to compare these L^1 and L^∞ optimizations with the corresponding discrete problems in ℓ^1 and ℓ^∞. There the optimality conditions are classical, and the best approximations must approach our solutions (including the characteristic functions) in an irregular but consistent way. The rate of convergence, and the pattern of error in the discrete problems, should be visible from numerical experiments--since this is a class of discrete problems in which the continuous limit can be solved.

We hesitate to propose a more complete list of dual variational problems of the same type. Harmonic functions will appear for optimization in L^2, and the special property of two dimensions (which produced all the second pairs of equivalent problems) introduces the conjugate harmonic. Within the list above there are combinations of conditions that earlier entered only separately, for example the combination

$$\text{MAX} \iint F\psi \quad \text{subject to} \quad \psi = g \quad \text{on} \quad \Gamma \quad \text{and} \quad \|\nabla\psi\|_\infty = 1.$$

This corresponds to applying both shear and torsion to a cylindrical rod, and it is solved (if the constraints are compatible and $F > 0$) by the maximal function $\psi(P) = \min\,[g(Q) + d(P,Q)]$.

A related problem mixes L^2 and L^∞, and has become a fundamental example in the theory of variational inequalities:

$$\text{MIN} \iint \tfrac{1}{2}|\nabla\psi|^2 - \psi F \quad \text{subject to} \quad \psi = 0 \quad \text{on} \quad \Gamma, \; \|\nabla\psi\|_\infty \leq 1.$$

The dual minimizes a combined norm

$$\|\nabla w - v\|_{1,2} = \iint \min\left(\tfrac{1}{2}|\nabla w - v|^2, |\nabla w - v| - \tfrac{1}{2}\right) dxdy.$$

This seems appropriate also for "robust statistics," in which the least squares model (Gauss-Markov linear regression) is natural--except that it assigns too much weight to observations that lie far outside the normal range. These outliers are less significant in L^1, and a mixed norm is more realistic.

Finally we mention <u>optimal design</u>, which is subject to all these constraints and one more: it begins as a nonconvex problem, to minimize the support of σ. A typical case, for longitudinal shear in a plastic cylinder, is

$$\text{INF} \iint 1_{\{\sigma \neq 0\}} \quad \text{subject to} \quad \text{div}\, \sigma = 0, \; \sigma \cdot n = f, \; \|\sigma\|_\infty \leq 1.$$

The integrand jumps from 0 to 1 at $\sigma = 0$, and from 1 to ∞ at $|\sigma| = 1$ (since $|\sigma| > 1$ is inadmissible). The equivalent "relaxed problem" replaces this integrand by the largest convex function which does not exceed it: it equals $|\sigma|$ for $|\sigma| \leq 1$ and ∞ for $|\sigma| > 1$. In this new problem

$$\text{MIN} \iint |\sigma| \quad \text{subject to} \quad \text{div}\, \sigma = 0, \; \sigma \cdot n = f, \; \|\sigma\|_\infty \leq 1$$

there exists an optimal solution (which is a weak limit of the highly oscillatory minimizing sequences in the original problem). The solution can actually be computed by modifying the construction in 2C to account for the new constraint $|\sigma| = |\nabla\psi| \leq 1$. It gives the admissible structure of minimum weight.

There is also a more delicate class of nonconvex problems,

whose relaxed form fails to be convex: instead it is polyconvex. In a forthcoming paper with Kohn we study elastic design subject to two loads, leading to the new minimum principle

$$\mathrm{INF} \iint 1_{\{|\sigma|+|\tau|\neq 0\}} + |\sigma|^2 + |\tau|^2 \quad \text{subject to } \operatorname{div}\sigma = \operatorname{div}\tau = 0,$$
$$\sigma\cdot n = f, \quad \tau\cdot n = g.$$

In this case σ,τ represents a 2 by 2 matrix; with n loads it would be 2 by n. Only the case of a single load leads to an equivalent convex problem; for $n = 2$ the relaxed integrand is polyconvex--a convex function of σ,τ, and their determinant $D = \sigma^{\perp}\cdot\tau$. The underlying theory was developed abstractly by Morrey [15], and more recently by Ball [16] and Dacorogna [17]. Perhaps our example is the first involving all three arguments in which this polyconvexification has been found.

ACKNOWLEDGEMENT

We gratefully acknowledge the support of the National Science Foundation (MCS 81-02371 and INT 81-00464) and the Army Research Office (DAAG 29-80-K0033).

REFERENCES

[1] Ekeland, I. and Temam, R., Convex Analysis and Variational Problems (North-Holland, Amsterdam, 1976).

[2] Kohn, R. and Strang, G., Structural design optimization, homogenization, and relaxation of variational problems, in: Papanicolaou, G. (ed.), Disordered Media, Lecture Notes in Physics 154 (Springer-Verlag, New York, 1982).

[3] Rozvany, G.I.N., Optimal Design of Flexural Systems (Pergamon, Oxford, 1976).

[4] Matthies, H., Strang, G. and Christiansen, E., The saddle point of a differential program, in: Glowinski, R., Rodin, E., and Zienkiewicz, O.C. (eds.), Energy Methods in Finite Element Analysis (John Wiley, New York, 1979).

[5] Temam, R. and Strang, G., Functions of bounded deformation, Arch. Rat. Mech. Anal. 75 (1980) 7-21.

[6] Strang, G., Maximal flow through a domain, Mathematical Programming (to appear).

[7] Iri, M., Theory of flows in continua as approximation to flows on networks, in: Prekopa, A. (ed.) Survey of Mathematical Programming (North-Holland, Amsterdam, 1978).

[8] Bombieri, E., DeGiorgi, E. and Giusti, E., Minimal cones and the Bernstein problem, Inventiones Math. 7 (1969) 243-268.

[9] Fleming, W. and Rishel, R., An integral formula for total gradient variation, Archiv der Mathematik 11 (1960) 218-222.

[10] Federer, H., Geometric Measure Theory (Springer-Verlag, New York, 1969).

[11] Mosolov, P.P., On the torsion of a rigid-plastic cylinder, PMM 41 (1977) 344-353.

[12] Strang, G. and Kohn, R., Optimal design of cylinders in shear, in: Whiteman, J. (ed.), The Mathematics of Finite Elements and Applications IV (Academic Press, London, 1982).

[13] Strang, G., A family of model problems in plasticity, in: Glowinski, R., Lions, J. L. (eds.), Proceedings of the Symposium on Computing Methods in Applied Sciences, Lecture Notes in Mathematics 704 (Springer-Verlag, New York, 1979).

[14] Strang, G., A minimax problem in plasticity theory, in: Nashed, Z. (ed.), Functional Analysis Methods in Numerical Analysis, Lecture Notes in Mathematics 701 (Springer-Verlag, New York, 1979).

[15] Morrey, C.B., Multiple Integrals in the Calculus of Variations (Springer-Verlag, Berlin, 1966).

[16] Ball, J.M., Convexity conditions and existence theorems in nonlinear elasticity, Arch. Rat. Mech. Anal. 63 (1977) 337-403.

[17] Dacorogna, B., Weak Continuity and Weak Lower Semi-continuity of Nonlinear Functionals, Springer Lecture Notes in Mathematics 922 (Springer-Verlag, Berlin, 1982).

Lecture Notes in Num. Appl. Anal., **5**, 289–311 (1982)
Nonlinear PDE in Applied Science. U.S.-Japan Seminar, Tokyo, 1982

Deformation formulas and their applications to spectral and evolutional inverse problems*

Takashi SUZUKI

Department of Mathematics
Faculty of Science
University of Tokyo

We describe our recent results on spectral and evolutional inverse problems. Our main interest lies in the uniqueness of the problems and 24 theorems will be stated.

§1. Summary.

In this article, two topics are taken up: inverse problems for evolution equations and inverse problems for spectral theories. These two are associated with each other, and our methodology of the study is the same. That is, the deformation formulas, which are simple and are easily proven.

Our first topic, the inverse theory for evolution equations, studies the following problem: By observing within a ceatain area the values of the solution of an evolution equation, can we determine the coefficients or the initial value of the equation ? Many authors have been interested in such a problem, and some of their works are referred to in Suzuki [29,30,31]. We here refer to Lavrentiev [13,14], Nakagiri [25] and Kohn-Vogelius [42], and also Kitamura-Nakagiri [12], Seidman [28], Pierce [27], Suzuki-Murayama [41] and Murayama [23]. Actually, the latters are related to the problems which we study here. In this article, we consider parabolic equations on compact intervals and on circles, and give conditions for the equations to be determined through various observations. Details will be described in §§ 2 and 3.

Our second topic, the inverse spectral theory, determines a differential operator by its spectrum. This kind of problem has been investigated for Sturm-

Liouville's operator by V. Ambarzumian, G. Borg, N. Levinson, I.M. Gel'fand, B.M. Levitan, M.G. Gasymov, V.A. Marchenko, B.Y. Levin, H. Hochstadt and B. Lieberman ([1], [2], [17], [5], [18], [20], [16], [10], [11]), and for Hill's equation by G. Borg, H. Hochstadt, P.D. Lax, W. Goldberg, H. Flashka, B.A. Dubrovin, V.A. Mateev, S.P. Novikov, H.P. Mckean, P. van Morebeke, E. Trubowitz and others ([2], [9], [1 5], [6,7], [4], [3], [21], [22]). In §§4 and 5 of the present article, we reexamine these studies from the viewpoint of our deformation formulas, and come up with some new results, among which are involved crucial improvements on their theorems.

Throughout the present article, the "deformation formulas" play an important role. The first and the second ones will be stated in §§ 2 and 3, respectively. These connect eigenfunctions of two separate differential operators by integral transformations, of which kernels satisfy certain hyperbolic equations. We are inspired by the Gel'fand-Levitan theory [5] in deriving these formulas. By using these deformation formulas, the answers to our inverse problems are sometimes arrived at a certain non-linear equation, which we call the "G - equation". A typical example of this can be seen in Theorem 2 of §2.

This article is made up of five sections. As mentioned above, we show in §2 some theorems on inverse problems for parabolic equations which are proven by the first deformation formula. Theorems on similar problems whose proofs are based on the second deformation formula will be stated in §3. §§ 4 and 5 are devoted to the study of inverse spectral problems for Sturm-Liouville's operator and for Hill's equation, respectively.

In this article, proofs of theorems are not explicitly stated, except for a few allusions to them. See the papers referred to there, for the proofs.

§2. Inverse problems for evolution equations (I).

For $p \in C^1[0,1]$, $h \in R$, $H \in R$ and $a \in L^2(0,1)$, Let $(E_{p,h,H,a})$ be the parabolic equation

(2.1) $$\frac{\partial u}{\partial t} + (p(x) - \frac{\partial^2}{\partial x^2})u = 0 \qquad (0<t<\infty,\ 0<x<1),$$

with the boundary condition

(2.2) $$\frac{\partial u}{\partial x} - hu\big|_{x=0} = \frac{\partial u}{\partial x} + Hu\big|_{x=1} = 0 \qquad (0<t<\infty),$$

and with the initial condition

(2.3) $$u\big|_{t=0} = a(x) \qquad (0<x<1).$$

As is well-known, if the initial value a and the coefficients (p,h,H) are given, there exists a unique solution u=u(t,x) of $(E_{p,h,H,a})$. In particular, the values of the solution on the boundaries $\xi=0,1$ are determined. Hence we obtain the mapping

$$F^1 = F^1_{T_1,T_2}: \ (p,h,H,a) \mapsto \{\langle u(t,0),\ u(t,1)\rangle \mid T_1 \leqq t \leqq T_2\},$$

for some T_1,T_2 in $0 \leqq T_1 < T_2 < \infty$.

Let $(p,h,H,a) \in C^1[0,1] \times R \times R \times L^2(0,1)$ be given with the solution u=u(t,x) of $(E_{p,h,H,a})$, and consider the set

$$M^1_{p,h,H,a} \equiv F^1_{T_1,T_2}{}^{-1}(F^1_{T_1,T_2}(p,h,H,a)).$$

It denotes the totality of equations $(E_{q,j,J,b})$ whose solutions v=v(t,x) have the same boundary values as those of u:

$$M^1_{p,h,H,a} = \{(q,j,J,b) \in C^1[0,1] \times R \times R \times L^2(0,1) \mid \text{the solution } v=v(t,x) \text{ of the equation } (E_{q,j,J,b}) \text{ satisfies}$$

(2.4) $$v(t,\xi) = u(t,\xi) \qquad (T_1 \leqq t \leqq T_2;\ \xi=0,1)\}.$$

Since u and v are analytic in $t \in (0,\infty)$, (2.4) is equivalent to

(2.4') $$v(t,\xi) = u(t,\xi) \qquad (0<t<\xi;\ \xi=0,1),$$

so that $M^1_{p,h,H,a}$ is independent of T_1 or T_2.

It is obvious that $(p,h,H,a)\in M^1_{p,h,H,a}$ holds. In the case of

$$M^1_{p,h,H,a} = \{(p,h,H,a)\}, \tag{2.5}$$

on the other hand, these values $\{u(t,\xi) \mid T_1\leq t\leq T_2;\ \xi=0,1\}$ determine the equation $(E_{p,h,H,a})$. However, (2.5) does not always hold. For instance $u\equiv 0$ follows from $a\equiv 0$ for any (p,h,H), hence

$$M^1_{p,h,H,0} \supseteq \{(q,j,J,0) \mid q\in C^1[0,1],\ j\in R,\ J\in R\}$$

for each (p,h,H). We wish to give a condition on (p,h,H,a) for (2.5) to hold. To this end we introduce the following

Notation. The realization in $L^2(0,1)$ of the differential operator $p(x)\frac{\partial^2}{\partial x^2}$ with the boundary condition (2.2) is denoted by $A_{p,h,H}$. The eigenvalues and the eigenfunctions of $A_{p,h,H}$ are denoted by $\{\lambda_n \mid n=0,1,2,\cdots\}$ and $\{\phi(\cdot,\lambda_n) \mid n=0,1,2,\cdots\}$, respectively, the latter being normalized by $\phi(0,\lambda_n)=1$. □

Furthermore, noting that each λ_n is simple $(-\infty<\lambda_0<\lambda_1<\cdots\to\infty)$, we give

Definition. For each $a\in L^2(0,1)$, we call

$$N = \#\{\ \lambda_n \mid (a,\phi(\cdot,\lambda_n))_{L^2(0,1)} = 0\ \}$$

the "degenerate number" of a with respect to $A_{p,h,H}$, where $(\ ,\ )_{L^2(0,1)}$ denotes the L^2-inner product. □

Then we have

Theorem 1 (Murayama [23], Suzuki [32]). (2.5) holds if and only if $N=0$, where N is the degenerate number of a with respect to $A_{p,h,H}$. □

This theorem is shown by the following assertion: The set $\{\lambda_n,\ \phi(1,\lambda_n) \mid n=0,1,2,\cdots\}$ characterizes the operator $A_{p,h,H}$. Murayama [23] found that $\{\lambda_n,\ \phi(1,\lambda_n) \mid n=0,1,2,\cdots\}$ corresponds one to one to the spectral function of $A_{p,h,H}$, and showed the assertion by the Gel'fand-Levitan theory [5]. Suzuki [32] proved the assertion directly by the following first deformation formula: Let $\phi=\phi(x,\lambda)\varepsilon C^2[0,1]$ be the solution of

$$(2.6)\qquad (-\frac{d^2}{dx^2}+p(x))\phi=\lambda\phi \quad (0\leq x\leq 1), \qquad \phi(0,\lambda)=1, \qquad \phi'(0,\lambda)=h$$

for each $\lambda\varepsilon R$. (This notation is compatible to that of $\phi(\cdot,\lambda_n)$.) Set $D=\{(x,y)\mid 0<y<x<1\}$. Then,

Lemma 1. For each $p,q\varepsilon C^1[0,1]$ and $h,j\varepsilon R$, there exists a unique $K=K(x,y)\varepsilon C^2(\overline{D})$ such that

$$(2.7.1)\qquad K_{xx}-K_{yy}+p(y)K=q(x)K \qquad (\text{on } \overline{D})$$

$$(2.7.2)\qquad K(x,x)=(j-h)+\frac{1}{2}\int_0^x (q(s)-p(s))ds \qquad (0\leq x\leq 1)$$

$$(2.7.3)\qquad K_y(x,0)=hK(x,0) \qquad (0\leq x\leq 1). \quad \square$$

Lemma 2 (first deformation formula, Suzuki [32]). For K in Lemma 1,

$$(2.8)\qquad \psi(x,\lambda)=\phi(x,\lambda)+\int_0^x K(x,y)\phi(y,\lambda)dy \qquad (0\leq x\leq 1)$$

satisfies

$$(2.9)\qquad (-\frac{d^2}{dx^2}+q(x))\psi=\lambda\psi \quad (0\leq x\leq 1), \qquad \psi(0,\lambda)=1, \qquad \psi'(0,\lambda)=j,$$

for each $\lambda\varepsilon R$. $\square$

The point is that K is indepenent of λ. On account of it, conditions on $\{\lambda_n, \phi(1,\lambda_n)\mid n=0,1,2,\cdots\}$ can be concentrated into those on K through (2.8). [32] proved the assertion mentioned above by the study of these conditions on K, together with (2.7). On the other hand, the proofs of Lemmas 1 and 2 are elementary. In fact, Lemma 1 is obtained by the method of Picard [26] (that is, "iteration"), while Lemma 2 follows immediately from the integration by parts. See [32].

Furthermore by this method, [32] gave the following characterization of $M^1_{p,h}$,

$_{H,a}$ by the "G - equation", in case of $1 \leqq N < \infty$. Namely, assume $N<\infty$ and

(2.10) $$(a,\phi(\cdot,\lambda_{n_\ell}))_{L^2(0,1)} = 0 \qquad (1 \leqq \ell \leqq N)$$

for $0 \leqq n_1 < n_2 < \cdots < n_N < \infty$. By definition we have $(a,\phi(\cdot,\lambda_n))_{L^2(0,1)} \neq 0$ for $n \neq n_\ell$ ($1 \leqq \ell \leqq N$). Putting

$$\Phi = \begin{pmatrix} \phi(\cdot,\lambda_{n_1}) \\ \vdots \\ \phi(\cdot,\lambda_{n_N}) \end{pmatrix} \quad \text{and} \quad \Lambda = \begin{pmatrix} \lambda_{n_1} & & 0 \\ & \ddots & \\ 0 & & \lambda_{n_N} \end{pmatrix},$$

we consider the following non-linear N-simultaneous ordinary differential equation (G - equation)

(2.11) $$\frac{d^2}{dx^2}G = [(2\frac{d}{dx}(G\cdot\Phi) + p)I - \Lambda]G,$$

and set

$$\mathcal{G} = \{ G \in C^2([0,1] \to \mathbb{R}^N) \mid G \text{ satisfies } (2.11)\},$$

where $\cdot$ and I denotes the inner product and the unit matrix in $\mathbb{R}^N$, respectively. We set, furthermore,

$$\hat{M}^1_{p,h,H,a} = \{(q,j,J) \in C^1[0,1] \times \mathbb{R} \times \mathbb{R} \mid \text{there exists } b \in L^2(0,1) \text{ such that } (q,j,J,b) \in M^1_{p,h,H,a}\}.$$

Then,

<u>Theorem 2</u> (Suzuki [32]). Under these assumptions, for each $(q,j,J) \in \hat{M}^1_{p,h,H,a}$, $b \in L^2(0,1)$ with $(q,j,J,b) \in M^1_{p,h,H,a}$ is unique. $\hat{M}^1_{p,h,H,a}$ is homeomorphic to $\mathcal{G}$:

$$\hat{M}^1_{p,h,H,a} \ni (q,j,J) \leftrightarrow G \in \mathcal{G}$$

through the relations

(2.12) $$q = p + 2\frac{d}{dx}(G\cdot\Phi), \quad j = h + (G\cdot\Phi)(0), \quad J = H - (G\cdot\Phi)(1). \quad \square$$

In particular, $M^1_{p,h,H,a}$ has 2N - degrees of freedom. G - equation (2.11) is obtained by eliminating (q,j,J) in the relations (2.7) and in some equalities on K derived from $(q,j,J)\in\hat{M}^1_{p,h,H,a}$. See [32], for the relation between b and G, and for applications of Theorem 2.

Now it is natural that we nextly consider the mapping

$$F^1 = F^1_{T_1,T_2,x_0}: \quad (p,h,H,a) \mapsto \{\langle u(t,0),\ u(t,x_0)\rangle \mid T_1 \leq t \leq T_2\}$$

for $x_0\in(0,1)$, $u=u(t,x)$ being the solution of $(E_{p,h,H,a})$. Similar M can be considered:

$$M^1_{p,h,H,a,x_0} \equiv F^1_{T_1,T_2,x_0}{}^{-1}(F^1_{T_1,T_2,x_0}(p,h,H,a))$$
$$= \{(q,j,J,b)\in C^1[0,1]\times R\times R\times L^2(0,1) \mid \text{the solution } v=v(t,x)$$
$$\text{of the equation } (E_{q,j,J,b}) \text{ satisfies}$$

$$(2.13) \qquad v(t,\xi) = u(t,\xi) \qquad (T_1 \leq t \leq T_2;\ \xi=0,x_0)\}.$$

M^1_{p,h,H,a,x_0} coincides with $M^1_{p,h,H,a}$ in case $x_0=1$. However, we unfortunately have

<u>Theorem 3</u> (Suzuki [36]). In case of $x_0\neq 1$, we always have

$$(2.14) \qquad M^1_{p,h,H,a,x_0} \supsetneq \{(p,h,H,a)\}. \quad \Box$$

Namely, non-uniqueness holds even if N=0, unless $x_0=1$.

In view of this, we nextly consider the mapping

$$F^2 = F^2_{T_1,T_2,x_0}: \quad (p,h,H,a) \mapsto \{\langle u(t,0),\ u(t,x_0),\ u_x(t,x_0)\rangle \mid T_1 \leq t \leq T_2\}$$

and obtain the following theorem, where

$$M^2_{p,h,H,a,x_0} \equiv F^2_{T_1,T_2,x_0}{}^{-1}(F^2_{T_1,T_2,x_0}(p,h,H,a))$$
$$= \{(q,j,J,b)\in C^1[0,1]\times R\times R\times L^2(0,1) \mid \text{the solution } v=v(t,x)$$
$$\text{of the equation } (E_{q,j,J,b}) \text{ satisfies}$$

$$(2.15) \qquad v(t,0) = u(t,0), \quad v(t,x_0) = u(t,x_0), \quad v_x(t,x_0) = u_x(t,x_0)$$

$(T_1 \leq t \leq T_2)\}$:

Theorem 4 (Suzuki [33,36]). Let N be the degenerate number of a with respect to $A_{p,h,H}$. Then,

(i) In the case of $x_0=1$,

$$M^2_{p,h,H,a,x_0} = \{(p,h,H,a)\} \tag{2.16}$$

holds if and only if N=0.

(ii) In the case of $\frac{1}{2}<x_0<1$, (2.16) holds whenever $N<\infty$.

(iii) In the case of $x_0=\frac{1}{2}$, (2.16) holds if and only if $N \leq 1$.

(iv) In the case of $0 \leq x_0 < \frac{1}{2}$, we always have $M^2_{p,h,H,a,x_0} \supsetneq \{(p,h,H,a)\}$. □

Thus, the position x_0 plays an important role as well as the number N. Theorems 3 and 4 are also proved by first deformation formula. Since the equation (2.7.1) is of hyperbolic type, having the properties of the domain of dependence and so on, the point $x_0=\frac{1}{2}$ comes to be important in Theorem 4.

Before concluding this section, we briefly mention the connection between (i) of Theorem 4 and Theorem 1. In this case ($x_0=1$), (2.15) is equivalent to (2.4) with J=H unless $a \neq 0$, so that M^2_{p,h,H,a,x_0} is nothing but $M^1_{p,h,H,a}$ restricted to J=H. By Theorem 2, $M^1_{p,h,H,a}$ has 2N - degrees of freedom, hence those of M^2_{p,h,H,a,x_0} will be 2N-1 in case $x_0=1$. In particular, 1 - degree of freedom remains even if N=1, which explains why (i) of Theorem 4 holds.

§3. Inverse problems for evolution equations (II).

Let us now consider the mapping

$$F^3 = F^3_{T_1,T_2,x_0} : \ (p,h,H,a) \mapsto \{\langle u(t,x_0),\ u_x(t,x_0)\rangle \mid T_1 \leq t \leq T_2\}.$$

with similar M:

$$M^3_{p,h,H,a,x_0} \equiv F^{3}_{T_1,T_2,x_0}{}^{-1}(F^3_{T_1,T_2,x_0}(p,h,H,a))$$
$$= \{(q,j,J,b)\varepsilon C^1[0,1]\times R\times R\times L^2(0,1) \mid \text{the solution } v=v(t,x)$$
$$\text{of the equation } (E_{q,j,J,b}) \text{ satisfies}$$

(3.1) $$v(t,x_0) = u(t,x_0), \quad v_x(t,x_0) = u_x(t,x_0) \qquad (T_1\leqq t\leqq T_2)\}.$$

In view of Theorem 4, we see that

(3.2) $$M^3_{p,h,H,a,x_0} = \{(p,h,H,a)\}$$

holds only if $x_0=\frac{1}{2}$ and $N\leqq 1$. In fact, if $o<x_0<\frac{1}{2}$, uniqueness $(q,j,J,b)=(p,h,H,a)$ doesn't hold even if we assume $v(t,0)=u(t,0)$ $(T_1\leqq t\leqq T_2)$ besides (3.1), by (iv) of Theorem 4. Similarly, if $\frac{1}{2}<x_0<1$, uniqueness doesn't hold even if we assume $v(t,1)=u(t,1)$ $(T_1\leqq t\leqq T_2)$ besides (3.1). Therefore, (3.2) implies $x_0=\frac{1}{2}$, hence also implies $N\leqq 1$ by (iii) of Theorem 4.

Fortunately we have

<u>Theorem 5</u> (Suzuki [34,37]). If $x_0=\frac{1}{2}$ and N=0, (3.2) holds. □

For the case $x_0=\frac{1}{2}$ and N=1, which is delicate, see [37]. Furthermore, we have

<u>Theorem 6</u> ([34,37]). Let $x_0\neq\frac{1}{2}$ and assume $\frac{1}{2}<x_0<1$ without loss of generality. Then (3.1) implies

(3.3) $$q(x) = p(x) \quad (x_0\leqq x\leqq 1), \quad J = H,$$

whenever $N<\infty$. □

<u>Theorem 7</u> ([34,37]). Under the same situation $\frac{1}{2}<x_0<1$,

(3.4) $$M^3_{p,h,H,a,x_0} \cap \{(q,j,J,b)\varepsilon C^1[0,1]\times R\times R\times L^2(0,1) \mid q(x)=p(x) \ (\tfrac{1}{2}\leqq x\leqq x_0)\}$$
$$= \{(p,h,H,a)\}$$

if and only if $N\leqq 1$. □

Similar theorems also hold for the case of $0<x_0<\frac{1}{2}$. In view of Theorems 6 and 7, we call $(x_0,1)$ the "domain of uniqueness" in the case of $\frac{1}{2}<x_0<1$, which comes to be $(0,x_0)$ in the case of $0<x_0<\frac{1}{2}$.

These theorems cannot be proved by first deformation formula. In fact, in Lemma 2 $\phi=\phi(\cdot,\lambda)$ is requested to satisfy the boundary conditions $\phi(0,\lambda)=1$ and $\phi'(0,\lambda)=h$, which are independent of λ. Without observing boundary values $u(t,0)$ or $u(t,1)$, we cannot apply this formula. Another deformation formula is needed: Set $D_1=\{(x,y) \mid 1-x<y<x, \frac{1}{2}<x<1\}$.

Lemma 3. For each $p\in C^1[0,1]$ and $q\in C^1[\frac{1}{2},1]$, there exists a unique $K=K(x,y)\in C^2(\overline{D}_1)$ such that

$$K_{xx} - K_{yy} + p(y)K = q(x)K \qquad (\text{on } \overline{D}_1) \tag{3.5.1}$$

$$K(x,x) = \frac{1}{2}\int_{\frac{1}{2}}^{x} (q(s)-p(s))ds \qquad (\tfrac{1}{2}\leqq x\leqq 1) \tag{3.5.2}$$

$$K(x,1-x) = 0 \qquad (\tfrac{1}{2}\leqq x\leqq 1). \ \square \tag{3.5.3}$$

Lemma 4 (second deformation formula, Suzuki [34]). If $\phi=\phi(x)\in C^2[0,1]$ satisfies

$$\left(-\frac{d^2}{dx^2} + p(x)\right)\phi = \lambda\phi \qquad (0\leqq x\leqq 1) \tag{3.6}$$

for $\lambda\in\mathbb{R}$, then $\psi=\psi(x)\in C^2[\frac{1}{2},1]$ defined by

$$\psi(x) = \phi(x) + \int_{1-x}^{x} K(x,y)\phi(y)dy \qquad (\tfrac{1}{2}\leqq x\leqq 1) \tag{3.7}$$

satisfies

$$\left(-\frac{d^2}{dx^2} + q(x)\right)\psi = \lambda\psi \quad (\tfrac{1}{2}\leqq x\leqq 1), \quad \psi(\tfrac{1}{2}) = \phi(\tfrac{1}{2}), \quad \psi'(\tfrac{1}{2}) = \phi'(\tfrac{1}{2}). \ \square \tag{3.8}$$

The point is that no boundary condition on ϕ is assumed in (3.6) and that instead (3.8) holds only on $[\frac{1}{2},1]$ in spite of (3.6) on $[0,1]$. To get a similar relation to (3.8) on $[0,\frac{1}{2}]$, another K has to be constructed on $\overline{D}_2$, where $D_2=\{(x,y) \mid x<y<1-x, 0<x<\frac{1}{2}\}$.

In virtue of second deformation formula, we can also study inverse problems for evolution equations on circles. Henceforth S^1 denotes the compact interval $[0,1]$ with end points identified. For $p \in C^1(S^1)$ (i.e., $p \in C^1(R)$, $p(x+1)=p(x)$) and $a \in L^2(S^1)$, we consider the following parabolic equation $(E^S_{p,a})$ on S^1:

$$\frac{\partial u}{\partial t} + (p(x) - \frac{\partial^2}{\partial x^2})u = 0 \qquad (0<t<\infty,\ x \in S^1) \tag{3.9.1}$$

$$u|_{t=0} = a(x) \qquad (x \in S^1). \tag{3.9.2}$$

Notation. A^S_p denotes the realization in $L^2(S^1)$ of the differential operator $p(x)-\frac{\partial^2}{\partial x^2}$. The eigenvalues of A^S_p are denoted by $\{\lambda_n \mid n=0,1,2,\cdots\}$ $(-\infty<\lambda_0<\lambda_1<\cdots\to\infty)$ and the multiplicitiy of λ_n is denoted by $\alpha(n)$. Note that $\alpha(n)=1$ or 2. $\{\phi_{n\ell} \mid 1\leqq\ell\leqq\alpha(n)\}$ denotes the eigenfunctions of A^S_p, corresponding to λ_n and being normalized by $||\phi_{n\ell}||_{L^2(S^1)}=1$. □

Inverse problems for $(E^S_{p,a})$ are more difficult than those of $(E_{p,h,H,a})$, partly because of the existence of double eigenvalues of A^S_p. In order to overcome this difficulty, we consider several solutions of $(E^S_{p,a})$ according to the idea of Nakagiri [24]. We thus extend the notion of the "degenerate number" as

Definition. Put $\alpha \equiv \max_n \alpha(n)$ $(=1$ or $2)$. For the set of initial values $\{a^j \mid 1\leqq j\leqq\alpha\}$, let us consider the matrix

$$A_n = ((a^j,\phi_{n\ell})_{L^2(S^1)})_{1\leqq j\leqq\alpha,\ 1\leqq\ell\leqq\alpha(n)}$$

for each $n=0,1,2,\cdots$. Then we call

$$N = \#\{\, A_n \mid \text{rank } A_n < \alpha(n)\}$$

the "degenerate number" of $\{a^j \mid 1\leqq j\leqq\alpha\}$ with respect to A^S_p. □

Let $u^j=u^j(t,x)$ be the solution of (E^S_{p,a^j}) $(1\leqq j\leqq\alpha)$. For simplicity, we henceforth assume $N=0$, where N is the degenerate number of $\{a^j \mid 1\leqq j\leqq\alpha\}$ with respect to A^S_p. Let $v=v^j(t,x)$ be another solution of (E^S_{q,b^j}) for some $q \in C^1(S^1)$ and $b^j \in L^2(S^1)$. Then,

Theorem 8 (Suzuki [35,37]). Let $x_1 \varepsilon S^1$ and $x_2 \varepsilon S^1$ satisfy the central symmetry, say $x_1=\frac{1}{2}$ and $x_2=1(=0)$. Then the equalities

$$v^j(t,x_1) = u^j(t,x_1), \quad v_x^j(t,x_1) = u_x^j(t,x_1), \quad v^j(t,x_2) = u^j(t,x_2) \quad (T_1 \leqq t \leqq T_2;\ 1 \leqq j \leqq \alpha) \tag{3.10}$$

imply

$$(q,b^j) = (p,a^j) \quad (1 \leqq j \leqq \alpha). \quad \square \tag{3.11}$$

Theorem 9 ([35,37]). Suppose that $x_1 \varepsilon S^1$ and $x_2 \varepsilon S^1$ don't satisfy the central symmetry, and let $x_1' \varepsilon S^1$ and $x_2' \varepsilon S^1$ be the symmetric points of x_1 and x_2, respectively, say $x_1=\frac{1}{2}$, $\frac{1}{2}<x_2<1$, $x_1'=1(=0)$ and $x_2'=x_2-\frac{1}{2}$. Let A, A', B and B' be the arcs $\widehat{x_1 x_2}$, $\widehat{x_1' x_2'}$, $\widehat{x_2 x_1'}$ and $\widehat{x_2' x_1}$, respectively as in Fig. 1. Then (3.10) implies

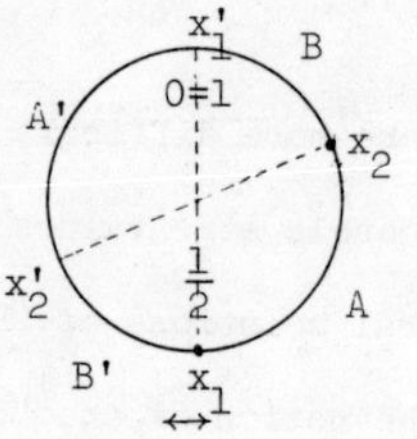

Fig. 1

$$q(x) = p(x) \quad (x \varepsilon A \cup A'). \quad \square \tag{3.12}$$

Theorem 10 ([35,37]). Under the same circumstances as those of Theorem 9, the equalities (3.10) combined with either $q(x)=p(x)$ $(x \varepsilon B)$ or $q(x)=p(x)$ $(x \varepsilon B')$ imply (3.11). □

In virtue of Theorems 8-10, we can call A∪A' the domain of uniqueness for this problem. Finally we have

Theorem 11 ([35,37]). In case of $p(x+\frac{1}{2})=p(x)$ and $q(x+\frac{1}{2})=q(x)$ $(x \varepsilon R)$, the equalities

$$v^j(t,x_1) = u^j(t,x_1), \quad v_x^j(t,x_1) = u_x^j(t,x_1) \quad (T_1 \leqq t \leqq T_2;\ 1 \leqq j \leqq \alpha) \tag{3.13}$$

imply (3.11), where $x_1 \varepsilon S^1$. □

In this case, one point $x_1 \varepsilon S^1$ is enough for the uniqueness.

§4. Inverse spectral problems for Sturm-Liouville's operators.

In this section, we consider the so-called inverse Sturm-Liouville problem. The first deformation formula will give crucial answers to them. Recall that for $p \varepsilon C^1[0,1]$, $h \varepsilon R$ and $H \varepsilon R$, $A_{p,h,H}$ denotes the realization in $L^2(0,1)$ of the differential operator $-\frac{d^2}{dx^2}+p(x)$ with the boundary condition $(\frac{d}{dx}-h)\cdot|_{x=0}=(\frac{d}{dx}+H)\cdot|_{x=1}=0$. Let $\{\lambda_n \mid n=0,1,2,\cdots\}$ $(-\infty<\lambda_0<\lambda_1<\cdots\to\infty)$ be $\sigma(A_{p,h,H})$, the eigenvalues of $A_{p,h,H}$.

Firstly,

Theorem 12 (Suzuki [36]). Suppose $q(x)=p(x)$ $(0 \leqq x \leqq \frac{1}{2})$, $j=h$ and

$$\lambda_n \ \varepsilon \ \sigma(A_{q,j,J}) \qquad (n \neq n_1) \tag{4.1}$$

for $q \varepsilon C^1[0,1]$, $j \varepsilon R$, $J \varepsilon R$ and $n_1 \varepsilon N \equiv \{0,1,2,\cdots\}$. Then

$$A_{q,j,J} = A_{p,h,H} \tag{4.2}$$

follows. □

Theorem 12 is an improvement of Hochstadt-Lieberman [11]. Actually they derived (4.2), assuming $J=H$ and $\lambda_n \varepsilon \sigma(A_{q,j,J})$ $(n=0,1,2,\cdots)$, besides $q(x)=p(x)$ $(0 \leqq x \leqq \frac{1}{2})$ and $j=h$. It is important that the converse of Theorem 12 holds:

Theorem 13 ([36]).

(i) For each $A_{p,h,H}$, $n_1 \varepsilon R$ and x_0 in $0 \leqq x_0 < \frac{1}{2}$, there exist $q \neq p$, j and J such that $q(x)=p(x)$ $(0 \leqq x \leqq x_0)$, $j=h$ and (4.1).

(ii) For each $A_{p,h,H}$ and $n_1, n_2 \varepsilon N$ with $n_1 \neq n_2$, there exist $q \neq p$, j and J such that $q(x)=p(x)$ $(0 \leqq x \leqq \frac{1}{2})$, $j=h$ and

$$\lambda_n \ \varepsilon \ \sigma(A_{q,j,J}) \qquad (n \neq n_1,\ n_2). \quad \square \tag{4.3}$$

Nextly, let us consider another Sturm-Liouville operator A_{p,h,H^*} with $\sigma(A_{p,h,H^*})=\{\lambda_n^*\}_{n=0}^{\infty}$ for $H^* \neq H$, along with $A_{p,h,H}$ and its eigenvalues $\{\lambda_n\}_{n=0}^{\infty}$.

Theorem 14 (Suzuki [36]).

(i) Suppose

(4.4) $\lambda_n \in \sigma(A_{q,j,J}),\ \lambda_n^* \in \sigma(A_{q,j,J^*}) \quad (n=0,1,2,\cdots)$

for $q \in C^1[0,1]$, $j \in R$, $J \in R$ and $J^* \in R$. Then,

(4.5) $(q,j,J,J^*) = (p,h,H,H^*)$

follows.

(ii) Suppose

(4.6) $\lambda_n \in \sigma(A_{q,j,J}) \quad (n \neq n_1), \qquad \lambda_n^* \in \sigma(A_{q,j,J^*}) \quad (n \in N)$

and either $J=H$ or $J^*=H^*$ for $q \in C^1[0,1]$, $j \in R$, $J \in R$, $J^* \in R$ and $n_1 \in N$. Then (4.5) follows. □

(ii) of Theorem 14 is an improvement of Borg [2], Levinson [17] and Hochstadt [10]. In fact, they derived (4.5) assuming $j=h$, $J=H$, $J^*=H^*$ and

(4.6') $\lambda_n = \mu_n \quad (n=1,2,\cdots), \qquad \lambda_n^* \in \sigma(A_{q,j,J^*}) \quad (n=0,1,2,\cdots),$

where $\{\mu_n\}_{n=0}^{\infty} = \sigma(A_{q,j,J})$ $(-\infty<\mu_0<\mu_1<\cdots\to\infty)$. A proof of (i) of theorem 14 can be found in Levin [16]. Levitan-Gasymov [18] reconstructed (p,h,H,H^*) from $\{\lambda_n, \lambda_n^* \mid n=0,1,2,\cdots\}$ under suitable conditions.

The converse of Theorem 14 is obtained:

Theorem 15 ([36]).

(i) For each (p,h,H,H^*) and $n_1 \in N$, there exist $(q,j) \neq (p,h)$, $J \neq H$ and $J^* \neq H^*$ such that (4.6).

(ii) For each (p,h,H,H^*) and $n_1, n_2 \in N$ with $n_1 \neq n_2$, there exist $(q,j) \neq (p,h)$, J and J^* such that $J=H$, $J^*=H^*$ and

$$(4.7) \qquad \lambda_n \in \sigma(A_{q,j,J}) \quad (n \neq n_1, n_2), \qquad \lambda_n^* \in \sigma(A_{q,j,J^*}) \quad (n \in N).$$

(iii) For each (p,h,H,H^*) and $n_1, n_2 \in N$, there exist $(q,j) \neq (p,h)$, J and J^* such that $J=H$, $J^*=H^*$ and

$$(4.8) \qquad \lambda_n \in \sigma(A_{q,j,J}) \quad (n \neq n_1), \quad \lambda_n^* \in \sigma(A_{q,j,J^*}) \quad (n \neq n_2). \quad \square$$

(i) and (ii) of Theorem 15 are generalized as follows by the G - equation. Recall the set G defined before Theorem 2:

Theorem 16 ([36]). Let (p,h,H,H^*) $(H \neq H^*)$ be given with $\{\lambda_n\}_{n=0}^{\infty} = \sigma(A_{p,h,H})$ and $\{\lambda_n^*\}_{n=0}^{\infty} = \sigma(A_{p,h,H^*})$. Let, furthermore, N be finite and $0 \leq n_1 < n_2 < \cdots < n_N < \infty$ be integers. Then, (q,j,J,J^*) satisfies

$$(4.9) \qquad \lambda_n \in \sigma(A_{q,j,J}) \quad (n \neq n_\ell;\ 1 \leq \ell \leq N), \quad \lambda_n^* \in \sigma(A_{q,j,J^*}) \quad (n \in N)$$

if and only if there exists $G \in G$ with

$$(4.10) \qquad G'(1) + (H^* - (G \cdot \Phi)(1))G(1) = 0$$

such that

$$(4.11) \qquad q = p + 2\frac{d}{dx}(G \cdot \Phi), \quad j = h + (G \cdot \Phi)(0), \quad J = H - (G \cdot \Phi)(1),$$
$$J^* = H^* - (G \cdot \Phi)(1).$$

Furthermore, the correspondence between (q,j,J,J^*) and G is homeomorphic. $\square$

Hochstadt [10] showed more weakly that if $j=h$, $J=H$, $J^*=H^*$ and

$$(4.9') \qquad \lambda_n = \mu_n \quad (n \neq n_\ell;\ 1 \leq \ell \leq N), \quad \lambda_n^* \in \sigma(A_{q,j,J^*}) \quad (n \in N),$$

then q satisfies the first equality of (4.11) for some $G \in G$.

Finally, set

$$C_s^1[0,1] = \{\, p \in C^1[0,1] \mid p(1-x) = p(x) \quad (0 \leq x \leq 1) \}.$$

We say that $A_{p,h,H}$ is (spatially) symmetric iff $p \in C_s^1[0,1]$ and $h=H$. Suppose that

$A_{p,h,h}$ and $A_{q,j,j}$ are symmetric. Put $\sigma(A_{p,h,h})=\{\lambda_n\}_{n=0}^{\infty}$ $(-\infty<\lambda_0<\lambda_1<\cdots\to\infty)$ and $\sigma(A_{q,j,j})=\{\mu_n\}_{n=0}^{\infty}$ $(-\infty<\mu_0<\mu_1<\cdots\to\infty)$. Then

Theorem 17 (Suzuki [38]).

$$\mu_n = \lambda_n \quad (n=0,1,2,\cdots) \tag{4.12}$$

implies

$$A_{q,j,j} = A_{p,h,h}. \quad \square \tag{4.13}$$

The converse of Theorem 17 holds. More precisely,

Theorem 18 ([38]). Let a symmetric operator $A_{p,h,h}$ be given with $\sigma(A_{p,h,h})=\{\lambda_n\}_{n=0}^{\infty}$. Let $n_1 \in N$. Then a symmetric operator $A_{q,j,j}$ with $\sigma(A_{q,j,j})=\{\mu_n\}_{n=0}^{\infty}$ satisfies

$$\mu_n = \lambda_n \quad (n \neq n_1) \tag{4.14}$$

if and only if there exists $g \in C^2[0,1]$ satisfying (G - equation)

$$\frac{d^2}{dx^2}g = \left(2\frac{d}{dx}(g\phi(\cdot,\lambda_{n_1})) + p - \lambda_{n_1}\right)g \quad (0 \leqq x \leqq 1) \tag{4.15}$$

with

$$g(1-x) = (-1)^{n_1+1}g(x) \quad (0 \leqq x \leqq 1), \tag{4.16}$$

such that

$$q = p + 2\frac{d}{dx}(g\phi(\cdot,\lambda_{n_1})), \qquad j = h + g(0). \tag{4.17}$$

Furthermore, the correspondence between (q,j) and g is homeomorphic. $\square$

Here the notation (2.6) of $\phi(\cdot,\lambda)$ is adopted. Since such $g \not\equiv 0$ as (4.15)-(4.16) actually exists, for each symmetric operator $A_{p,h,h}$ and $n_1 \in N$ we have a symmetric operator $A_{q,j,j}$ satisfying (4.14) in spite of $A_{q,j,j} \neq A_{p,h,h}$.

In this way, $\{\lambda_n\}_{n=0}^{\infty}$ characterizes a symmetric operator $A_{p,h,h}$. Naturally, we wonder if (4.14) combined with $j=h$ implies $q=p$. Actually, the study of (4.15) gives

Theorem 19 ([38]). In the following cases, (4.14) with j=h implies (4.13) for symmetric operators $A_{p,h,h}$ and $A_{q,j,j}$:

(i) $n_1 = 0$

(ii) $\lambda_{n_1} \in \sigma(A_p^o)$.

Henceforth A_p^o denotes the realization in $L^2(0,1)$ of $-\frac{d^2}{dx^2}+p(x)$ with Dirichlet boundary condition: $\cdot|_{x=0}=\cdot|_{x=1}=0$. □

Borg [2], Levinson [17] and Hochstadt [10] proved Theorem 19 for the case of (i).

Recently, we have succeeded in proving the converse of Theorem 19. Put $\sigma(A_p^o) =\{\lambda_n^o\}_{n=1}^{\infty}$ $(-\infty<\lambda_1^o<\lambda_2^o<\cdots\to\infty)$. It holds that $\lambda_{n_1} \in\sigma(A_p^o)$ implies $n_1\geq 1$ and $\lambda_{n_1}=\lambda_{n_1}^o$.

Theorem 20 (Suzuki [39]). Let a symmetric operator $A_{p,h,h}$ and an integer $n_1\geq 1$ be given with $\sigma(A_{p,h,h})=\{\lambda_n\}_{n=0}^{\infty}$. Suppose $\lambda_{n_1}\neq\lambda_{n_1}^o$, where $\{\lambda_n^o\}_{n=1}^{\infty}=\sigma(A_p^o)$. Then, there exists a unique symmetric operator $A_{q,j,j}$ with $\sigma(A_{q,j,j})=\{\mu_n\}_{n=0}^{\infty}$ such that $q\neq p$, (4.14) and j=h. Furthermore, such $q\in C_s^1[0,1]$ is given by

$$q = p - 2\left(\frac{L'}{L}\right)', \tag{4.17}$$

where

$$L = L(x) = \phi^{*\prime}(x,\lambda_{n_1}^o)\phi(x,\lambda_{n_1}) - \phi^*(x,\lambda_{n_1}^o)\phi'(x,\lambda_{n_1}) \quad (>0). \tag{4.18}$$

Henceforth $\phi^*=\phi^*(\cdot,\lambda)$ is the solution of

$$\left(-\frac{d^2}{dx^2} + p(x)\right)\phi^* = \lambda\phi^*, \quad \phi^*(0,\lambda) = 0, \quad \phi^{*\prime}(0,\lambda) = 1. \quad \square \tag{4.19}$$

Namely, for each $p\in C_s^1[0,1]$, $h\in\mathbb{R}$ and $n_1\in\mathbb{N}\equiv\{0,1,2,\cdots\}$, the set

$$\mathcal{Q} \equiv \{q\in C_s^1[0,1] \mid \sigma(A_{q,h,h})=\{\mu_n\}_{n=0}^{\infty} \text{ satisfies } (4.14)\}$$

coincides with {p} if and only if either $n_1=0$ or $\lambda_{n_1}=\lambda_{n_1}^o$ $(n_1\geq 1)$, and otherwise coincides with $\{p,\ p-2(L'/L)'\}$. For given $p\in C_s^1[0,1]$ and $n_1\geq 1$, Fig. 2 describes the set of (q,h) which satisfies (4.14) for $\{\mu_n\}_{n=0}^{\infty}=\sigma(A_{q,h,h})$ and $\{\lambda_n\}_{n=0}^{\infty}=\sigma(A_{p,h,h})$. A bifurcation structure can be seen. Here $h_{n_1}=s'(0)/s(0)$ with $s\neq 0$, $(-\frac{d^2}{dx^2}+p(x))s=\lambda^{o}_{n_1}s$ and $s(1-x)=(-1)^{n_1}s(x)$. It holds that $\lambda_{n_1}=\lambda^{o}_{n_1}$ if and only if $h=h_{n_1}$.

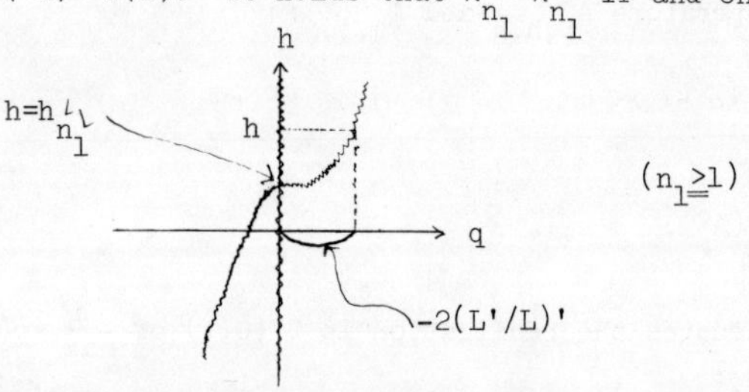

Fig. 2

The key to the proof of Theorem 20 is the following theorem, which states a relation between $\sigma(A_{p,h,h})$ and $\sigma(A^{o}_{p})$ and by itself is interesting:

<u>Theorem 21</u> ([38]). For $p,q\in C^1_s[0,1]$, let $\sigma(A^{o}_{p})=\{\lambda^{o}_n\}_{n=1}^{\infty}$, $\sigma(A^{o}_{q})=\{\mu^{o}_n\}_{n=1}^{\infty}$. Let $n_1\in N^*\equiv\{1,2,\cdots\}$. Then

(4.20) $$\mu^{o}_n=\lambda^{o}_n \quad (n\neq n_1)$$

if and only if there exists $h\in R$ such that

(4.21) $$\mu_n=\lambda_n \quad (n\neq n_1),$$

for $\{\mu_n\}_{n=0}^{\infty}=(A_{q,h,h})$, $\{\lambda_n\}_{n=0}^{\infty}=\sigma(A_{p,h,h})$. Furthermore, in case of $q\neq p$, we have

(4.22) $$\mu^{o}_{n_1}=\lambda_{n_1}, \quad \mu_{n_1}=\lambda^{o}_{n_1}. \quad \square$$

We conclude this section by the following

<u>Theorem 22</u> ([39]). Let $A_{p,h,h}$ and $A_{q,j,j}$ be symmetric operators with $\{\lambda_n\}_{n=0}^{\infty}=\sigma(A_{p,h,h})$ and $\{\mu_n\}_{n=0}^{\infty}=\sigma(A_{q,j,j})$. Assume (4.14) for $n_1\in N$ and also

(4.23) $$\frac{1}{2}(q(0)-p(0))=(j+h)(j-h).$$

Then, (4.13) follows. $\square$

Thus, (4.23) is more powerful than $j=h$ to get uniqueness.

Similar results to Theorems 19-22 are expected for the problems considered in Theorems 12-13 and Theorems 14-16. Furthermore, it would be interesting to consider

(4.14') $\mu_n = \lambda_n \qquad (n \neq n_\ell;\ 1 \leqq \ell \leqq N)$

for (1.14) and to study similar problems to those in Theorems 19, 21 and 22. We shall discuss them in a forthcoming paper.

§5. Inverse spectral problems for Hill's equations.

In this section, we refer to our results on Hill's equations obtained by second deformation formula. Recall that S^1 denotes the compact interval [0,1] with end points identified.

For $p \in C^1(S^1)$, let us consider Hill's equation

(5.1) $\left(-\frac{d^2}{dx^2} + p(x)\right)\phi = \lambda\phi \qquad (-\infty<x<\infty)$

for $\lambda \in R$. The discriminant, the trace of the monodromy matrix $M(\lambda)$ (see Magnus-Winkler [19], for example), is denoted by $\Delta(\lambda)$. The solutions of $\Delta(\lambda)=\pm 2$ are denoted by $\{\hat{\lambda}_n\}_{n=0}^{\infty}$ with $-\infty<\hat{\lambda}_0<\hat{\lambda}_1\leqq\hat{\lambda}_2<\hat{\lambda}_3\leqq\hat{\lambda}_4<\cdots\to\infty$. Let $\hat{A}_p^S$ be the realization in $L^2(\hat{S}^1)$ of $-\frac{d^2}{dx^2}+p(x)$, where $\hat{S}^1$ is the compact interval [0,2] with end points identified. It is known that $\{\hat{\lambda}_n\}_{n=0}^{\infty}$ coincides with $\sigma(\hat{A}_p^S)$, the eigenvalues of $\hat{A}_p^S$ with multiplicities counted in. Let $\hat{\phi}_n$ be the eigenvalues of $\hat{A}_p^S$ corresponding to $\hat{\lambda}_n$, normalized by $||\hat{\phi}_n||_{L^2(\hat{S}^1)}=1$. Then, it is also known

$$\hat{\phi}_n(x+1) = \hat{\phi}_n(x) \ (n\equiv 0,3 \ (\mathrm{mod}\ 4)),\quad \hat{\phi}_n(x+1) = -\hat{\phi}_n(x) \ (n\equiv 1,2 \ (\mathrm{mod}\ 4)).$$

See [19], for these facts. Set $I_n=(\hat{\lambda}_{2n-1},\hat{\lambda}_{2n})$ $(n=1,2,\cdots)$. Then, p is said to be of N - "finite band" iff $I_n=\phi$ $(n\neq n_\ell;\ 1\leqq\ell\leqq N)$ for some $0<n_1<n_2<\cdots<n_N<\infty$. Hochsatdt [9] proved that if $p\in C^1(S^1)$ is of finite band, then $p\in C^\infty(S^1)$.

Let $p\in C^\infty(S^1)$ be of N - finite band with $(\hat{A}_p^S)=\{\hat{\lambda}_n\}_{n=0}^{\infty}$ $(-\infty<\hat{\lambda}_0<\hat{\lambda}_1\leqq\hat{\lambda}_2<\hat{\lambda}_3\leqq\hat{\lambda}_4<\cdots\to\infty)$. Set $I_n=(\hat{\lambda}_{2n-1},\hat{\lambda}_{2n})$ and $[I_n]=[\hat{\lambda}_{2n-1},\hat{\lambda}_{2n}]$ for n=1,2,$\cdots$, and assume

(5.2) $I_n = \phi \qquad (n\neq n_\ell;\ 1\leqq\ell\leqq N)$

for $0<n_1<n_2<\cdots<n_N<\infty$. Take another $q\in C^\infty(S^1)$ with $\sigma(\hat{A}_q^S)=\{\hat{\mu}_n\}_{n=0}^\infty$ $(-\infty<\hat{\mu}_0<\hat{\mu}_1\leq\hat{\mu}_2<\hat{\mu}_3\leq\hat{\mu}_4$ $<\cdots\to\infty)$, and set $J_n=(\hat{\mu}_{2n-1},\hat{\mu}_{2n})$ and $[J_n]=[\hat{\mu}_{2n-1},\hat{\mu}_{2n}]$. Then,

Theorem 23 (Suzuki [40]). Suppose that for each $n\neq n_\ell$ $(1\leq\ell\leq N)$ there exists some $m(n)\in N^*$ such that

$$[J_n]=[I_{m(n)}] \qquad (n\neq n_\ell;\ 1\leq\ell\leq N). \tag{5.3}$$

Then,

$$\hat{\mu}_n=\hat{\lambda}_n \qquad (n=0,1,2,\cdots) \tag{5.4}$$

holds. □

Hochstadt [8] showed that conversely

$$[J_{n(\ell)}]=[I_{n(\ell)}] \qquad (1\leq\ell\leq N) \tag{5.5}$$

implies (5.4), assuming $q\in C^\infty(S^1)$ also to be of N - finite band.

We conclude the present article by characterizing the set

$$\hat{Q}\equiv\{q\in C^\infty(S^1)\ |\ (5.4)\text{ holds for }\{\hat{\mu}_n\}_{n=0}^\infty=(\hat{A}_q^S)\}$$

through G - equation. Set $\Phi={}^t(\hat{\phi}_0,\hat{\phi}_{n(1)},\hat{\phi}_{n(1)+1},\cdots,\hat{\phi}_{n(N)},\hat{\phi}_{n(N)+1})\in C^\infty(\hat{S}^1\to R^{2N+1})$,

$$\hat{\Lambda}=\begin{pmatrix}\hat{\lambda}_0 & & & & & 0\\ & \hat{\lambda}_{n(1)} & & & & \\ & & \hat{\lambda}_{n(1)+1} & & & \\ & & & \ddots & & \\ & & & & \hat{\lambda}_{n(N)} & \\ 0 & & & & & \hat{\lambda}_{n(N)+1}\end{pmatrix}$$

and

$$\hat{G}=\{G\in C^\infty(\hat{S}^1\to R^{2N+1})\ |\ \hat{G}={}^t(g_0,g_1,\tilde{g}_1,\cdots,g_N,\tilde{g}_N)\text{ satisfies}$$

$$\frac{d^2}{dx^2}\hat{G}=[(2\frac{d}{dx}(\hat{G}\cdot\hat{\Phi})+p)\hat{I}-\hat{\Lambda}]\hat{G} \qquad (x\in R) \tag{5.6}$$

$$g_\ell(x+1)=(-1)^{n(\ell)}g_\ell(x)\ (0\leq\ell\leq N),\quad \tilde{g}_\ell(x+1)=(-1)^{n(\ell)}\tilde{g}_\ell(x)\ (1\leq\ell\leq N) \tag{5.7}$$

(5.8) $G(x)\cdot\Phi(1-x) = 0 \qquad (x\varepsilon\hat{S}^1)\}.$

In (5.7), the notation n(0)=0 is adopted. Then,

<u>Theorem 24</u> ([40]). $q\varepsilon\mathcal{Q}$ if and only if there exists $G\varepsilon\mathcal{G}$ such that

(5.9) $$q = p + 2\frac{d}{dx}(G\cdot\Phi).$$

Furthermore, the correspondence between q and G is homeomorphic. □

Mckean-Moerbeke [21] showed that $\mathcal{G}$ is homeomorphic to $\mathcal{T}^N$, the N - dimensional torus. We conclude that so is $\mathcal{G}$ by Theorem 24. However, the direct proof of $\mathcal{G}\simeq\mathcal{T}^N$ has not been obtained yet.

Footnote

* This work was supported partly by the Fûju-kai.

References

[1] Ambarzumian, V., Uber eine Frage der Eigenwerttheorie, Z. Phys., 53 (1929) 690-695.

[2] Borg, G., Eine Umkehrung der Sturm-Liouvilleshen Eigenwertaufgabe, Acta Math. , 78 (1946) 1-96.

[3] Dobrivin, B.A., Mateev, V.B., Novikov, S.P., Nonlinear equations of Korteweg-de Vrie type, finite band operators and Abelian varieties, Russian Math. Surveys, 31 (1976) 59-146.

[4] Flashka, H., On the inverse problem for Hill's operator, Arch. Rat. Mech. Anal., 59 (1975) 293-304.

[5] Gel'fand, I.M., Levitan, B.M., On the determination of a differential equation by its spectral function (English translation), A.M.S. Transl. ser. 2., 1 (1955) 253-304.

[6] Goldberg, W., Hill's equation for finite number of instability intervals, J. Math. Anal. Appl., 51 (1975) 705-723.

[7] ——, Necessary and sufficient condition for determing a Hill's equation from its spectrum, J. Math. Anal. Appl., 55 (1976) 549-554.

[8] Hochstadt, H., Function theoretic properties of the discriminant of Hill's equation, Math. Z., 82 (1963) 237-242.

[9] ——, On the characterization of Hill's equation via its spectrum, Arch. Rat. Mech. Anal., 19 (1965) 353-362.

[10] ——, The inverse Sturm-Liouville problem, Comm. Pure Appl. Math., 26 (1973) 715-729.

[11] ——, Lieberman, B., An inverse Strum-Liouville problem with mixed given data, SIAM J. Appl. Math., 34 (1978) 676-680.

[12] Kitamura, S., Nakagiri, S., Identifiability of spatially-varying and constant parameters in distributed systems of parabolic type, SIAM J. Control & Optimization.

[13] Levrentiev, M.M., Some improperly posed problems of mathematical physics (English translation, Springer, Berlin-Heidelberg-New York, 1967).

[14] ——, Integral geometry and the problems in mathematical physics, to appear in: Glowinski, R. and Lions, J.L. (eds.) Computing Methods in Applied Sciences and Engineering, V. (Noth-Holland, Amsterdam, 1982).

[15] Lax, P.D., Periodic solutions of the KdV equations, in: Lectures of Applied Math., 15, pp.85-96, (A.M.S. Providence, Rhode Island, 1974).

[16] Levin, B.Y., Distribution of zeros of entire functions (English translation, A.M.S. Providence, Rhode Island, 1964).

[17] Levinson, N., The inverse Sturm-Liouville problem, Mat. Tidsskr, B., (1949) 25-30.

[18] Levitan, B.M., Gasymov, M.G., Determination of a differential equation by two of its spectra, Russian Math. Surveys, 19 (1964) 1-63.

[19] Magnus, W., Winkler, S., Hill's equation, (Dover, New York, 1979).

[20] Marchenko, V.A., Some questions of the theory of one-dimensional differential operators of the second order, I. (in Russian), Trudy Moskov. Mat. Obšč., 1 (1952) 327-420.

[21] Mckean, H.P., van Morebeke, P., The spectrum of Hill's equation, Inventiones Math., 30 (1975) 217-274.

[22] ——, Trubowitz, E., Hill's operator and hyperbolic function theory in the presense of infinitely many branch points. Comm. Pure Appl. Math., 29 (1976) 143 -226.

[23] Murayama, R., The Gel'fand-Levitan theory and certain inverse problems for the parabolic equation, J. Fac. Sci. Univ. Tokyo 28 (1981) 317-330.

[24] Nakagiri, S., Identifiability of linear systems in Hilbert spaces, to appear in SIAM J. Control & Optimization.

[25] ——, On the identifiability of parameters in distributed systems, preprint, 1982.

[26] Picard, E., Lecons sur quelque types simples d'equation aux dérivées partielles, (Paris-Imprimerie Gauthier-Villars, Paris, 1950).

[27] Pierce, A., Unique identification of eigenvalues and coefficients in a parabolic problem, SIAM J. Control & Optimization, 17 (1979) 494-499.

[28] Seidman, T.I., Ill-posed problems arising in boundary control and observation for diffusion equations, in: Anger, G. (ed.), Inverse and Improperly Posed

Problems in Differential Equations, pp.233-247, (Akademie, Berlin, 1979).

[29] Suzuki, T., Inverse problems for the heat equation (in Japanese), Newsletter from C&A seminar, 16 (1981) 13-20, 17 (1981) 10-11.

[30] ——, Inverse problems for the heat equaiton (in Japanese), Sûgaku, 34 (1982) 55-64.

[31] ——, Uniqueness and nonuniqueness in an inverse problem for parabolic equations, in: Glowinski, R. and Lions, J.L. (eds.), Computing Methods in Applied Sciences and Engineering, V., pp.659-668 (North-Holland, Amsterdam, 1982).

[32] ——, Uniqueness and nonuniqueness in an inverse problem for the parabolic equation, to appear in J. Differential Equations.

[33] ——, Remarks on the uniqueness in an inverse problem for the heat equation, I., Proc. Japan Acad. ser. A., 58 (1982) 93-96.

[34] ——, ditto, II., ibid, 58 (1982) 175-177.

[35] ——, On a certain inverse problem for the heat equation on the circle, ibid , 58 (1982) 243-245.

[36] ——, Inverse problems for heat equations on compact invervals and on circles, I., submitted to J. Math. Soc. Japan.

[37] ——, ditto, II., in preparation.

[38] ——, On the inverse Strum-Liouville problem for spatially symmetric operators, submitted to J. Differential Equations.

[39] ——, ditto, TT., in preparation.

[40] ——, in preparation.

[41] ——, Murayama, R., A uniqueness theorem in an identification problem for coefficients of parabolic euqations, Proc. Japan Acad., 56 (1980) 259-263.

[42] Kohn, R., Vogelius, M., Determing conductivity by boundary measurements, preprint, 1982.

Lecture Notes in Num. Appl. Anal., **5**, 313–331 (1982)
Nonlinear PDE in Applied Science. U.S.-Japan Seminar, Tokyo, 1982

STATIONARY SOLUTIONS OF THE BOLTZMANN EQUATION

Seiji Ukai* and Kiyoshi Asano**

* Department of Applied Physics, Osaka City University.

** Institute of Mathematics, Yoshida College, Kyoto University.

1. Introduction

We consider a gas flow having a prescribed constant velocity $c \in \mathbb{R}^n$ at infinity and passing by an obstacle $\mathcal{O} \subset \mathbb{R}^n$. Such a flow has been discussed by setting an exterior problem for the Euler or Navier-Stokes equation. There are many literatures on the existence and stability of stationary solutions for the incompressible case, but few for the compressible case, which gives a better description of gas flows. We mention the works [2], [8] in which the compressible Euler equation is solved for small c on the existence of two-dimensional isentropic, irrotational stationary flows, whose stability, however, is still open. We should also mention [7] which solves the compressible Navier-Stokes equation in the large in time for $c = 0$.

The aim here is to discuss the exterior problem for the Boltzmann equation describing our gas flow, and specifically to show for $n \geq 3$ that if c is small, then stationary solutions exist and are asymptotically stable in time. The special case $c = 0$ has been solved in [1], [9] in the large in time, for which stationary solutions are trivially given by Maxwellians. When $c \neq 0$, nontrivial stationary solutions appear.

Put $\Omega = \mathbb{R}^n \backslash \overline{\mathcal{O}}$. We assume that $\mathcal{O}$ is a bounded convex domain in $\mathbb{R}^n$ with piecewise smooth boundary $\partial\mathcal{O} = \partial\Omega$. Let $f = f(t,x,\xi)$ denote the (probability) density of gas molecules having the position $x \in \overline{\Omega}$ and velocity $\xi \in \mathbb{R}^n$ at time $t \in \overline{\mathbb{R}}_+$. Then we shall study the following nonlinear initial boundary value problem on f.

(1.1a) $\frac{\partial f}{\partial t} = -\xi\cdot\nabla_x f + Q[f,f]$, $(t,x,\xi) \in \mathbb{R}_+ \times \Omega \times \mathbb{R}^n$,

(1.1b) $f \to g_c(\xi) \equiv \exp(-|\xi-c|^2/2)$, $(|x| \to \infty)$, $(t,\xi) \in \mathbb{R}_+ \times \mathbb{R}^n$,

(1.1c) $\gamma^- f = M\gamma^+ f$, $(t,x,\xi) \in \mathbb{R}_+ \times S^-$,

(1.1d) $f|_{t=0} = f_0$, $(x,\xi) \in \Omega \times \mathbb{R}^n$.

The equation (1.1a) is the Boltzmann equation, where $\cdot$ denotes the inner product of $\mathbb{R}^n$ while Q, called the collision operator, is a quadratic integral operator in the velocity variable ξ whose kernel is the collision cross section specific to the intermolecular potential. We assume the cutoff hard potential in the sense of Grad, see [5].

In the boundary condition (1.1b) at infinity, $g_c(\xi)$ is a Maxwellian which describes an equilibrium state of a gas moving with the mean velocity c.

The boundary condition on $\partial\Omega = \partial\mathcal{O}$ is (1.1c), which expresses reflections of gas molecules by the wall $\partial\mathcal{O}$. Let $n(x)$ denote the unit outward normal to $\partial\Omega$ at $x \in \partial\Omega$; and put

$$S^{\pm} = \{(x,\xi) \in \partial\Omega \times \mathbb{R}^n |\ n(x)\cdot\xi \gtrless 0\} \qquad \text{(same signs)}.$$

Then $\gamma^{\pm}$ are the trace operators $\gamma^{\pm} f = f|_{S^{\pm}}$, and M is an operator which maps functions on S^+ into those on S^-. If the reflection at $x \in \partial\Omega$ causes a deterministic change of molecular velocity from $m(x,\xi)$ to ξ, then we are given a map $S^- \ni (x,\xi) \to (x,m(x,\xi)) \in S^+$ and

(1.2) $M\gamma^+ f = f(t,x,m(x,\xi))$.

For example, $m(x,\xi) = \xi-2(\xi\cdot n(x))n(x)$ for the specular reflection and $m(x,\xi) = -\xi$ for the reverse reflection. We employ the regular reflection law of [6] for the function $m(x,\xi)$. Non-deterministic reflection laws are also possible physically, an example of which is the diffuse reflection

$$(1.3)\qquad M\gamma^{+}f = \int_{n(x)\cdot\xi'>0} m(x,\xi,\xi')f(t,x,\xi')d\xi'.$$

This was discussed in [4], and we impose on the kernel $m(x,\xi,\xi')$ conditions analogous to those given there, so as to include a wider class of M than that of (1.3). Furthermore, M may be any convex linear combination of (1.2) and (generalizations of) (3).

Besides (1.1), we shall study the corresponding stationary problem:

$$(1.4)\qquad \begin{aligned} &-\xi\cdot\nabla_x f + Q[f,f] = 0, && (x,\xi) \in \Omega \times \mathbb{R}^n,\\ &f \to g_c(\xi), \quad (|x| \to \xi), && \xi \in \mathbb{R}^n,\\ &\gamma^{-}f = M\gamma^{+}f, && (x,\xi) \in S^{-}. \end{aligned}$$

That $Q[g_c,g_c] = 0$ for all c is known [5], while in general, $\gamma^{-}g_c = M\gamma^{+}g_c$ for $c = 0$ but not for $c \neq 0$, as is the case with the specular and reverse reflections. Therefore g_0 solves (1.4) for $c = 0$, and it is natural to expect solutions to (1.4) for small $c \neq 0$ which slightly differ from g_c and tend to g_0 as $c \to 0$.

The existence of such stationary solutions has been shown in [10], using a classical implicit function theorem for $n \geq 4$, but a theorem of Nash-Moser-Nirenberg type for $n = 3$. In §2, we will give a simplified proof showing that a classical one is also useful for $n = 3$. This became possible by getting non-uniform estimates in the parameter c which diverges as $c \to 0$ but can be compensated by the closeness of solutions to g_c.

Further, it will be proven in §3 that whenever the initial f_0 of (1.1d) is sufficiently close to a stationary solution, (1.1) has a unique solution in the large in time, which approaches the stationary solution as $t \to \infty$ in the order of $(1 + t)^{-\alpha}$, $\alpha > 1/2$. The proof is carried out with the aid of nice decay estimates in t of the semigroup for a linearized equation of (1.1). The proofs will be only outlined. See [11] for details.

2. Existence of Stationary Solutions

Putting $f = g_c + g_0^{1/2}u$, we rewrite (1.4) as

$$(2.1a) \qquad -\xi\cdot\nabla_x u + L_c u + \Gamma[u,u] = 0, \qquad (x,\xi) \in \Omega \times \mathbb{R}^n,$$

$$(2.1b) \qquad u \to 0 \quad (|x| \to \infty), \qquad \xi \in \mathbb{R}^n,$$

$$(2.1c) \qquad \tilde{M}_o u = h_c. \qquad (x,\xi) \in S^-.$$

where we have defined

$$L_c u = 2g_0^{-1/2}Q[g_c, g_0^{1/2}u],$$

$$\Gamma[u, v] = g_0^{1/2}Q[g_0^{1/2}u, g_0^{1/2}v],$$

$$M_o\gamma^+ u = (\gamma^- g_0^{1/2})M\gamma^+(g_0^{1/2}u), \quad \tilde{M}_o = \gamma^- - M_o\gamma^+,$$

$$h_c = (\gamma^- g_0^{-1/2})(M\gamma^+ g_c - \gamma^- g_c).$$

The operator L_c have been investigated extensively in [5] for the case $c = 0$, most results of which remain valid for $c \neq 0$. For example, L_c has the decomposition

$$L_c = -\nu_c(\xi) \times + K_c,$$

where $\nu_c(\xi) \in L^\infty_{loc}(\mathbb{R}^n_\xi)$ and

$$(2.2) \qquad 0 < \nu_1 \leq (1 + |\xi|)^{-\gamma}\nu_c(\xi) \leq \nu_2$$

with some constants $\nu_1, \nu_2 > 0$ and $0 \le \gamma \le 1$, while K_c is a compact operator on $L^p(\mathbb{R}^n_\xi)$, $1 \le p \le \infty$.

In order to state the precise definition of the solution to (2.1), we need the spaces $L^p = L^p(\Omega \times \mathbb{R}^n_\xi)$ and

$$W^p = \{u \in L^p \mid (\xi\cdot\nabla_x + \nu_c(\xi))u \in L^p\},$$

$$Y^{p,\pm} = L^p(S^\pm;\ |n(x)\cdot\xi| d\sigma_x d\xi).$$

Using (2.2), we can show the unique existence of the trace operators $\gamma^\pm$ such that

$$\gamma^+ \in \mathbb{B}(W^p, Y^{p,+}), \qquad \gamma^- \in \mathbb{B}(W^p, Y^{p,-}_{loc}),$$

$$\gamma^\pm u = u|_{S^\pm} \quad \text{if} \quad u \in C_0^\infty(\Omega\times\mathbb{R}^n_\xi).$$

Here and hereafter $\mathbb{B}(X,Y)$ will denote the set of linear bounded operators from a Banach space X into another Banach space Y. By our assumption on M,

$$M_0 \in B(Y^{p,+}, Y^{p,-}), \quad h_c \in Y^{p,-} \quad \text{with} \quad \|h_c\| \to 0(|c| \to 0),$$

for all $p \in [1,\infty]$, so the following definition makes sense.

<u>Definition.</u> Let $p \in [1,\infty)$. u is said to be an L^p-solution to (2.1) if

i) $u \in W^p$, $\gamma^- u \in Y^{p,-}$,

ii) $\Gamma[u,u] \in L^p$,

iii) u satisfies (2.1a) and (2.1c) in the spaces L^p and $Y^{p,-}$ respectively.

Note that if $p < \infty$, $u \in L^p$ satisfies (2.1b) in a generalized sense.

Define the linearized Boltzmann operator B_c by

$$(2.3)\quad \begin{aligned} &D(B_c) = \{u \in W^p \mid \gamma^- u \in Y^{p,-},\ \tilde{M}_o u = 0\}, \\ &B_c = -\xi\cdot\nabla_x + L_c. \end{aligned}$$

Suppose B_c possess an inverse B_c^{-1}. Suppose further the linear inhomogeneous boundary value problem

$$(2.4)\quad \begin{aligned} &-\xi\cdot\nabla_x\phi + L_c\phi = 0, && \text{in } \Omega\times\mathbb{R}^n_\xi, \\ &\phi \to 0 \quad (|x|\to\infty), && \text{in } \mathbb{R}^n_\xi, \\ &\tilde{M}_o\phi = h_c, && \text{on } S^-, \end{aligned}$$

possess a solution $\phi = \phi_c$. Then (2.1) can be rewritten as

$$(2.5)\quad u + B_c^{-1}\Gamma[u,u] - \phi_c = 0.$$

It is this equation to which the implicit function theorem is to be applied.

The existence of B_c^{-1} shall be established with the aid of the perturbation technique and the limiting absorption principle. The unperturbed operator which we employ is B_c for the case $\Omega = \mathbb{R}^n$, denoted as B_c^∞, namely,

$$(2.6)\quad \begin{aligned} &D(B_c^\infty) = W^p(\mathbb{R}^n_x\times\mathbb{R}^n_\xi), \\ &B_c^\infty = -\xi\cdot\nabla_x + L_c. \end{aligned}$$

For later purposes we prepare notations. We denote by $\rho(B_c^\infty)$ and $\sigma(B_c^\infty)$ the resolvent set and spectrum of B_c^∞ respectively, and for $a_o, c_o, \sigma_o > 0$, we put

$$\overline{\mathbb{C}}_+(-\sigma_o) = \{\lambda \in \mathbb{C} \mid \mathrm{Re}\lambda \ge -\sigma_o\},$$

$$\Sigma(a_o,\sigma_o) = \{\lambda \in \overline{\mathbb{C}}_+(-\sigma_o) \mid -\mathrm{Re}\lambda \le a_o|\mathrm{Im}\lambda|^2\},$$

$$B[c_o] = \{c \in \mathbb{R}^n \mid |c| \le c_o\}.$$

Further, if E is a metric space and X is a Banach space, $B^0(E;X)$ will denote the Banach space of X-valued, bounded and continuous functions on E. We need also the space

$$L^p_\beta(\mathbb{R}^n_\xi) = \{u \mid (1+|\xi|)^\beta u \in L^p(\mathbb{R}^n_\xi)\}.$$

A spectral analysis of B^∞_c is found in [3] for the special case $c = 0$ and can be carried out similarly for $c \neq 0$, yielding the

Theorem 2.1. *Let* $p = 2$. *For any* $c_o > 0$, *there are positive numbers* a_o, κ_o *and* σ_o *such that the followings hold for all* $c \in B[c_o]$.

i) $\rho(B^\infty_c) \supset \Sigma(a_o, \sigma_o) \setminus \{0\}, \quad 0 \in \sigma(B^\infty_c)$.

ii) B^∞_c *has the orthogonal decomposition such that*

$$(\lambda - B^\infty_c)^{-1} = \sum_{j=0}^{n+2} U_j(\lambda, c).$$

Here

$$U_0(\lambda, c) \in B^0(\overline{\mathbb{C}}_+(-\sigma_o) \times B[c_o] \mid \mathbb{B}(L^2)),$$

and for $1 \le j \le n+2$,

$$U_j(\lambda, c) = \mathcal{F}_x^{-1} \chi(|k| \le \kappa_o)(\lambda - \lambda_j(k,c))^{-1} P_j(k,c) \mathcal{F}_x,$$

where $\mathcal{F}_x$ *means the Fourier transformation with respect to* x, $k \in \mathbb{R}^n$ *a dual variable to* x, $\chi(|k| \le \kappa_o)$ *the characteristic function for the domain* $|k| \le \kappa_o$, *and*

$$\lambda_j(k,c) = \mu_j(|k|) + ik\cdot c \qquad i = \sqrt{-1}$$

$$\mu_j(\kappa) = i\alpha_j\kappa - \beta_j\kappa^2 + O(|\kappa|^3) \qquad (\kappa \to 0),$$

with coefficents $\alpha_j \in \mathbb{R}$, $\beta_j > 0$, *while* P_j's *are orthogonal projections on* $L^2(\mathbb{R}^n_\xi)$, $P_jP_\ell = 0$ $(j \neq \ell)$ *with*

$$P_j(k,c) \in B^0(B[\kappa_o]\times B[c_o]; \mathbb{B}(L^2(\mathbb{R}^n_\xi), L^p_\beta(\mathbb{R}^n_\xi))$$

for any $p \geq 2$ *and* $\beta \in \mathbb{R}$.

In view of (i) of the above B^∞_c is not invertible on $\mathbb{B}(L^2)$ and it is seen from (ii) that this is because $U_j \to \infty$ $(\lambda \to 0)$ in $\mathbb{B}(L^2)$ for $1 \leq j \leq n+2$. However we can establish the limiting absorption principle in the sense that U_j has a well-defined limit as $\lambda \to 0$ if different spaces are chosen. More precisely we need the space

$$L^{p,r}_\beta = L^r_\beta(\mathbb{R}^n_\xi; L^p(\mathbb{R}^n_x)).$$

<u>Theorem 2.2.</u> *Let* $1 \leq q \leq 2 \leq p \leq \infty$, $\theta \in [0,1)$, $\ell = 0, 1$. *Put* $\gamma = \frac{1}{q} - \frac{1}{p}$ *and suppose*

$$(2.7) \qquad \gamma > \frac{2-\ell}{n+\theta}\ .$$

Let a_o, c_o, σ_o *be as in Theorem 2.1. Then for* $1 \leq j \leq n+2$,

$$|c|^{\theta\gamma}U_j(\lambda,c)(I-P_j(0,c))^\ell \in B^0(\Sigma(a_o, \sigma_o) \times B[c_o]; \mathbb{B}(L^{q,2}_0, L^{p,\infty}_\beta)).$$

Proof. Write $\hat{u} = \mathcal{F}_x u$. It is not hard to see that

$$\|u\|_{L^{p,r}_{\beta}} \le C\|\hat{u}\|_{L^{p',r}_{\beta}}$$

holds if $p \ge 2$ and $\frac{1}{p} + \frac{1}{p'} = 1$. Therefore, by virtue of Theorem 2.1,

$$\begin{aligned}
\|U_j(\lambda,c)u\|_{L^{p,\infty}_{\beta}} &\le C\|\mathcal{F}_x U_j(\lambda,c)u\|_{L^{p',\infty}_{\beta}} \\
&\le C\Big(\int_{|k|\le\kappa_o} |\lambda-\lambda_j(k,c)|^{-p'}|\kappa|^{p'}\|\hat{u}(k,\cdot)\|^{p'}_{L^2_{\xi}}dk\Big)^{1/p'} \\
&\le C\, I_{\gamma}\|\hat{u}\|_{L_0^{q',2}} \qquad \Big(\frac{1}{q'} + \frac{1}{q} = 1, \text{ Hölder}\Big) \\
&\le C\, I_{\gamma}\|u\|_{L^{q,2}} ,
\end{aligned}$$

where

$$\begin{aligned}
I_{\gamma} &= \Big(\int_{|k|\le\kappa_o} |\lambda-\lambda_j(k,c)|^{-1/\gamma}|k|^{1/\gamma}dk\Big)^{\gamma} \\
&\le C\Big(\int_{|k|\le\kappa_o} |\lambda-(i\alpha_j|k|-\beta_j|k|^2+ik\cdot c)|^{-1/\gamma}|k|^{1/\gamma}dk\Big)^{\gamma} \\
&\le C|c|^{-\theta\gamma}
\end{aligned}$$

The last inequality is valid only if (2.7) is fulfilled. The continuity in λ and c can be proved similarly.

In order to link B_c^{∞} to B_c, we need the solution operator $R_c(\lambda)$ to the inhomogeneous boundary value problem

$$\begin{cases}
(\lambda + \xi\cdot\nabla_x + \nu_c(\xi))u = 0, & \text{in } \Omega\times\mathbb{R}^n, \\
\gamma^- u = h, & \text{on } S^- .
\end{cases}$$

It can be shown that a unique L^p-solution $u = R_c(\lambda)h$ exists and $R_c(\lambda) \in \mathbb{B}(Y^{p,-}, W^p)$. Let e, r be the extension and restriction operators;

$$eu = u \quad \text{in} \quad \Omega, \; = 0 \quad \text{in} \quad \mathbb{R}^n \setminus \Omega, \; u \in L^p(\Omega),$$

$$ru = u|_{\Omega}, \quad u \in L^p(\mathbb{R}^n_x).$$

Define the operator

$$T_c(\lambda) = \tilde{M}_o r(\lambda - B_c^\infty)^{-1} eK_c R_c(\lambda),$$

which can be found to be in $B^0(\Sigma(a_o, \sigma_o) \times B[c_o]; \mathbb{B}(Y^{p,-}))$ for all $p \in [2,\infty]$ where a_o, c_o, σ_o are those of Theorem 2.2. Furthermore, we can see the

Proposition 2.3. *Let* $n \geq 3$. *There exist positive numbers* $\bar{a}_o$, $\bar{c}_o$, $\bar{\sigma}_o$, *such that* $1 \in \rho(T_c(\lambda))$ *for all* $p \in [2,\infty]$, $\lambda \in \Sigma(\bar{a}_o, \bar{\sigma}_o)$ *and* $c \in B[\bar{c}_o]$. *Accordingly there exists*

$$(I - T_c(\lambda))^{-1} \in B^0(\Sigma(\bar{a}_o, \bar{\sigma}_o) \times B[\bar{c}_o]; \mathbb{B}(Y^{p,-})).$$

This fact is essential for the proof of the existence of $(\lambda - B_c)^{-1}$. Indeed, solve the equation $(\lambda - B_c)u = f$ by letting $u = r(\lambda - B_c^\infty)^{-1} ef + u_1$. After some manipulations, we get

$$(2.8) \quad (\lambda - B_c)^{-1} = r(\lambda - B_c^\infty)^{-1} e + (\gamma^- r(\bar{\lambda} - B_c^{\infty *})^{-1} e)^* (I - T_c(\lambda))^{-1} \tilde{M}_o r(\lambda - B_c^\infty)^{-1} e$$

where $*$ denotes adjoints. This is a substitute for the second resolvent equation in the perturbation theory of linear operators. Combined with Theorems 2.1, 2.2 and Proposition 2.3, it provides necessary informations of B_c. First, we see that in L^2,

$$(2.9) \quad \rho(B_c) \supset \Sigma(\bar{a}_o, \sigma_o) \setminus \{0\}$$

for $c \in B[\bar{c}_o]$. Moreover, applying to U_0 of Theorem 2.1 the iterative scheme of Grad [5] which makes possible to deduce L^∞-estimates from L^2-ones with the aid of nice properties of the operator K_c, we can establish the limiting absorption principle for B_c at $\lambda = 0$ and obtains the inverse B_c^{-1}, regardless of (2.9). Put

$$X_\beta^p = L_\beta^{\infty,\infty} \cap L^{p,\infty}_{\beta-\frac{1}{p}} ,$$

$$Z^q = L^2 \cap L_0^{q,2} ,$$

and define $\Lambda_c = \nu_c(\xi)\times$, $P_c = \sum_{j=1}^{n+2} P_j(0,c)$.

Theorem 2.4. *Let* $n \geq 3$ *and* $\bar{a}_o, \bar{c}_o, \sigma_o > 0$ *be as before. Suppose* $1 \leq q \leq 2 \leq p \leq \infty$, $\beta > \frac{n}{2}$, $\theta \in [0,1)$, $\ell = 0, 1$ *satisfy*

$$\gamma = \frac{1}{q} - \frac{1}{p} > \frac{2-\ell}{n+\theta} , \quad \frac{1}{p} < 1 - \frac{2}{n+\theta} .$$

Put for $\alpha \in [0,1]$

$$v(\lambda,c) = (\lambda - B_c)^{-1}(I-P_c)^\ell \Lambda_c^\alpha u(c).$$

(i) If $u(c) \in L^\infty(B[\bar{c}_o]; X_\beta^p)$ *and* $\Lambda_c^\alpha u(c) \in L^\infty(B[\bar{c}_o]; Z^q)$, *then*

$$\| v(\lambda,c)\|_{X_\beta^p} \leq C|c|^{-\theta\gamma}(\| u(c)\|_{X_\beta^p} + \|\Lambda_c^\alpha u(c)\|_{Z^q}),$$

for all $\lambda \in \Sigma(\bar{a}_o, \bar{\sigma}_o)$, $c \in B[\bar{c}_o]$.

(ii) If in addition $u(c) \in B^0(B[\bar{c}_o]; X_{\beta-\varepsilon}^p)$, $\Lambda_c u(c) \in B^0(B[\bar{c}_o]; Z^q)$ *with* $\varepsilon > 0$, *then*

$$|c|^{\theta\gamma} v(\lambda,c) \in B^0(\Sigma(\bar{a}_o, \bar{\sigma}_o) B[\bar{c}_o]; X_{\beta-\varepsilon}^p).$$

(iii) Under the condition of (i) (ii) with $\alpha = \theta = \ell = 0$, $v(0,c) \in W^p$, $\gamma^- v(0,c) \in Y^{p,-}$, $\tilde{M}_0 v(0,c) = 0$ *and* $-B_c v(0,c) = u(c)$ *holds in* L^p.

Combining this with the estimate in [5] for Γ and noting that the nullspace of P_c is invariant in c, we have the

Proposition 2.5. *Let* $n \geq 3$ *and* $\bar{c}_o$ *be as before. Let*

$$(2.10)\quad \theta \in [0,1),\ p \in [2,4] \cap \left(\frac{n+\theta}{n+\theta-2}, n+\theta\right), \qquad \beta > \frac{n}{2} + 1,$$

Then there is a constant $C \geq 0$ *and for any* $c \in B[\bar{c}_o]$,

$$\| B_c^{-1}\Gamma[u,v]\|_{X^p_\beta} \leq C|c|^{-\alpha_1}\|u\|_{X^p_\beta}\|v\|_{X^p_\beta}.$$

with $\alpha_1 = \theta(1 + \frac{2}{p})$.

The inverse B_c^{-1} obtained in Theorem 2.4 is also useful to solve (2.4). If $\phi_c = R_c(0)h_c + \tilde{\phi}$ is substituted, it is reduced to $B_c\tilde{\phi} = -K_cR_c(0)h_c$. Therefore

$$(2.11)\quad \phi_c = R_c(0)h_c - B_c^{-1}K_cR_c(0)h_c.$$

Proposition 2.6. *Let* $n \geq 3$ *and* $\bar{c}_o$ *be as before and let*

$$\theta \in [0,1),\quad p \in [2,\infty] \cap \left(\frac{n+\theta}{n+\theta-2}, \infty\right],\quad \beta \geq 0.$$

(i) ϕ_c *of* (2,11) *is a unique* L^p*-solution to* (2.4) *for* $c \in B[\bar{c}_o]$.

(ii) $\phi_c \in B^0(B[\bar{c}_o]; X^p_\beta)$ *with*

$$\|\phi_c\|_{X^p_\beta} \leq C|c|^{\alpha_2},\quad \alpha_2 = 1 - \theta(2 - \frac{1}{p}).$$

These two propositions enable us to apply the contraction mapping principle to (2.5). It should be noted, however, that if $\theta > 0$, (2.11) becomes meaningless when $c \to 0$, and that if $n = 3$, $\theta = 0$ is excluded since then (2.10) becomes vacuous for p. But this difficulty can be removed as follows. If $\theta \in [0, 2/7)$, then $\alpha_1 < \alpha_2$ for $p \geq 2$, and we can choose an α such that $\alpha_1 < \alpha < \alpha_2$. Put $u = |c|^{\alpha} v$ in (2.5) to write

$$v = G(v,c) \equiv -|c|^{\alpha} B_c^{-1} \Gamma[v,v] + |c|^{-\alpha} \phi_c$$

for $c \neq 0$. Put $G(v,0) = 0$. For functions $v = v(c)$, define the nonlinear map $\tilde{G}[v]$ by $\tilde{G}[v](c) = G(v(c),c)$. What is to be proved is that $\tilde{G}$ is a contraction on a ball of the space

$$V^p_{\beta,\varepsilon} = L^{\infty}(B[\bar{\bar{c}}_o];\ X^p_{\beta}) \cap B^0(B[\bar{\bar{c}}_o];\ X^p_{\beta-\varepsilon}),$$

if $\varepsilon > 0$ and if $\bar{\bar{c}}_o > 0$ is sufficiently small. By virtue of Theorem 2.4 (i) (ii) and Proposition 2.6 (ii), $\tilde{G}$ maps $V^p_{\beta,\varepsilon}$ into itself, and by the aid of Proposition 2.5 and writing the norm of $V^p_{\beta,\varepsilon}$ as $|||\ |||$,

$$|||\tilde{G}[v]||| \leq C_1 |\bar{\bar{c}}_o|^{\sigma} |||v|||^2 + C_2 |\bar{\bar{c}}_o|^{\tau},$$

$$|||\tilde{G}[v] - \tilde{G}[v']||| \leq C_1 |\bar{\bar{c}}_o|^{\sigma} (|||v||| + |||v'|||) |||v - v'|||,$$

where $\sigma = \alpha - \alpha_1 > 0$, $\tau = \alpha_2 - \alpha > 0$, whence the desired conclusion readily follows. Now $\tilde{G}$ has a unique fixed point $v = v(c) \in V^p_{\beta,\varepsilon}$ and $u_c = |c|^{\alpha} v(c)$ solves (2.5) uniquely. Theorem 2.4(iii) and Proposition 2.6 (i) then complete the proof of the

<u>Theorem 2.7.</u> *Let* $n \geq 3$ *and let*

$$\theta \in [0,\tfrac{2}{7}), \quad p \in [2,4] \cap (\tfrac{n+\theta}{n+\theta-2}, n+\theta),$$

$$\alpha \in [0, 1-\theta(2-\tfrac{1}{p})), \quad \beta > \tfrac{n}{2} + 1.$$

Then there exists a positive number $\bar{\bar{c}}_o$ *and a constant* $C \geq 0$ *such that for each* $c \in B[\bar{\bar{c}}_o]$, (2.1) *has a unique* L^p*-solution* u_c. *Moreover* $u_c \in V^p_{\beta,\varepsilon}$ *for any* $\varepsilon > 0$ *and*

$$\|u_c\|_{X^p_\beta} \leq C|c|^\alpha.$$

Obviously $f_c = g_c + g_0^{1/2}u_c$ is a desired stationary solution to (1.1) and $f_c \to g_0$ $(c \to 0)$.

3. Stability of Stationary Solution.

In (1.1), put

$$f = f_c + g_0^{1/2}w = g_c + g_0^{1/2}(u_c+w).$$

Then $w = w(t,x,\xi)$ should solve

$$\begin{aligned} &\frac{\partial w}{\partial t} = -\xi\cdot\nabla_x w + L_c w + 2\Gamma[u_c,w] + \Gamma[w,w], \\ &w \to 0 \ (|x| \to \infty), \\ (3.1) \quad &\tilde{M}_o w = 0, \\ &w|_{t=0} = w_o. \end{aligned}$$

If we would have a nice decay estimate in t of the linear semigroup for the linearized equation to (3.1), then we could prove the existence in the large in time for the nonlinear problem (3.1) by the technique developed for the case

$c = 0$ (see e.g. [1,9]). However such an estimate is difficult to deduce because of the presence of the term $\Gamma[u_c,\cdot]$ which is an operator with "variable coefficient", and so we have to linearize (3.1) ignoring also this term. Thus we again meet the operator B_c of (2.3). Suppose it generates a semigroup

$$E_c(t) = e^{tB_c}.$$

Then if $w = w(t)$ is a solution to (3.1), it satisfies

$$\text{(3.2)} \qquad w(t) = E_c(t)w_o + \int_0^t E_c(t-s)\{2\Gamma[u_c,w(s)]+\Gamma[w(s),w(s)]\}ds.$$

When $u_c = 0$ as is the case with $c = 0$, the existence in the large in t for (3.2) can be shown if the decay rate $E_c(t) = 0(t^{-\gamma})$ is available with $\gamma > 1/2$, but in order to dispose of the extra linear term $\Gamma[u_c, w]$ when $u_c \neq 0$, the decay with $\gamma > 1$ is required as well as the smallness of u_c.

The desired decay shall be found starting from the semigroup

$$E_c^{\infty}(t) = e^{tB_c^{\infty}}$$

generated by B_c^{∞} of (2.6). Recall that a semigroup e^{tA} is the inverse Laplace transform $\mathcal{L}^{-1}[(\lambda-A)^{-1}]$ of the resolvent of the generator A. Apply this to B_c^{∞}. By virtue of Theorem 2.1, we have the orthogonal decomposition

$$E_c^{\infty}(t) = \sum_{j=0}^{n+2} V_j(t,c), \qquad V_j = \mathcal{L}^{-1}[U_j]$$

on L^2, and can find that

$$\text{(3.3)} \qquad \| V_0(t,c)\|_{\mathbb{B}(L^2)} \leq Ce^{-\sigma_o t},$$

while for $1 \leq j \leq n+2$,

$$\mathcal{F}_x V_j(t,c)u = \chi(|k| \le \kappa_o) e^{\lambda_j(k,c)t} P_j(k,c)\hat{u}(k,\cdot).$$

Proposition 3.1. *Suppose* $1 \le q \le 2 \le p \le \infty$ *and* $m = 0, 1$. *Put*

$$\gamma_1 = \frac{n}{2}\left(\frac{1}{q} - \frac{1}{p}\right).$$

Then there is a constant $C \ge 0$ *and for all* $c \in \mathbb{R}^n$, $t \ge 0$ *and* $1 \le j \le n+2$,

$$\|V_j(t,c)(I - P_j(0,c))^m\|_{\mathbb{B}(L_0^{q,2},\ L_\beta^{p,\infty})} \le C(1+t)^{-\gamma_1 - \frac{m}{2}}.$$

Proof. It suffices to proceed exactly in the same way as in the proof of Theorem 2.2, in this time with

$$I_{\gamma_1}(t) = \left(\int_{|k|\le_o} e^{-\mathrm{Re}\lambda_j(k,c)t/\sigma} |k|^{m/\sigma} dk\right)^{\delta}$$

$$\le \left(\int_{|k|\le\kappa_o} e^{-\beta_j|k|^2 t/\delta} |k|^{m/\delta} dk\right)^{\delta} \le C(1+t)^{-\gamma_1},$$

where $\delta = \frac{1}{q} - \frac{1}{p}$.

Take the inverse Laplace transform of (2.8) to obtain

$$E_c(t) = rE_c^\infty(t)e + (\gamma^- rE_c^\infty(t)^* e)^* * D_c(t) * \tilde{M}_o rE_c^\infty(t)e \tag{3.4}$$

where, writing $T_c^1(\lambda) = T_c(\lambda)(I - T_c(\lambda))^{-1}$,

$$D_c(t) = I + D_c^1(t), \quad D_c^1(t) = \mathcal{L}^{-1}[T_c^1(\lambda)],$$

and $*$ means the convolution in t;

$$f(t)*g(t) = \int_0^t f(t-s)g(s)ds.$$

No confusions arise with the adjoint symbol *.

Proposition 3.2. *Let* $n \geq 3$ *and* $\bar{c}_o > 0$ *be as in Proposition 2.3. For* $\theta \in [0,1)$, *there is a constant* $C \geq 0$ *and for any* $c \in B[\bar{c}_o]$,

$$\| D_c(t) \|_{\mathbb{B}(Y^\infty_\beta, -)} \leq C|c|^{-\theta}(1+t)^{-\gamma_2}$$

with $\gamma_2 = \frac{1}{2}(n-1+\theta)$ *if* n *is odd and* $= \frac{1}{2}(n-1)$ *if* n *is even.*

Note that $\gamma_2 > 1$ is possible for $n = 3$ only if $\theta > 0$. Propositions 3.1, 3.2 and (3.3), substituted into (3.4), give desired estimates of $E_c(t)$ and $E_c(t)^*$, by the aid of the scheme in [5] stated earlier. More precisely if we write the right side of (3.2) as $H[w](t)$, we get the following estimates. In Theorem 2.7, rewrite p, α as p_o, α_o and impose the additional condition $p_o < n$. Let

$$p \in [2,4] \cap ((1 - \tfrac{2}{n})^{-1}, \ (\tfrac{1}{2} - \tfrac{1}{p_o})^{-1}),$$

$$q \in [1,2] \cap [1, \ (\tfrac{1}{p} + \tfrac{1}{n})^{-1}),$$

$$\theta \in (0,\alpha_o), \quad \beta > \tfrac{n}{2} + 1$$

$$\gamma = \min\,(\tfrac{n}{2}(\tfrac{1}{q} - \tfrac{1}{p}), \ \tfrac{1}{2}(\tfrac{n}{p_o} + 1), \ \tfrac{1}{2}(\tfrac{n}{p} + 1)).$$

Then $\gamma > 1/2$ and

$$\| H[w](t) \|_{X^p_\beta} \leq C(1+t)^{-\gamma}\{\| w_o \|_{X^p_\beta \cap Z^q} + (|c|^{-\theta}a + |||w|||)|||w|||\},$$

$$\| H[w](t) - H[w'](t) \|_{X^p_\beta} \leq C(1+t)^{-\gamma}(|c|^{-\theta}a + |||w||| + |||w'|||)|||w-w'|||$$

where $a = \|u_c\|_{X^p_{\beta o}}$ and $|||w||| = \sup(1+t)^{\gamma}\|w(t)\|_{X^p_\beta}$. By Theorem 2.7, $|c|^{-\theta}a$ $\le C|c|^{\alpha_o-\theta} \to 0$ $(c \to 0)$.

It then follows that if w_o is small in $X^p_\beta \cap Z^q$, and c is small, H is a contraction map on a ball of the Banach space of functions $w(t) \in B^0([0,\infty); X^p_\beta)$ such that $|||w||| < \infty$. Its unique fixed point $w = w(t)$ is the desired solution to (3.2). In this way we prove the

Theorem 3.3. *Let* $n \ge 3$ *and* p, q, β, γ *be as above. Then there is positive numbers* a_o, a_1, c_o *such that for each* $c \in B[c_o]$ *and if* $\|w_o\|_{X^p_\beta \cap Z^q} \le a_o$, *then* (3.2) *has a unique global solution* $w = w(t) \in B^0([0, \infty); X^p_\beta)$ *with*

$$\| w(t)\|_{X^p_\beta} \le a_1(1+t)^{-\gamma}.$$

Obviously $f(t) = f_c + g_o^{1/2}\, w(t)$ solves (1.1) and $f(t) \to f_c$ as $t \to \infty$, showing the asymptotic stability of the stationary solution f_c.

References

[1] Asano,K.; On the initial boundary value problem of the nonlinear Boltzmann equation in an exterior domain, to appear.

[2] Brezis,H. and Stampacchia,G; The hodograph method in fluid dynamics in the light of variational inequalities, Arch. Rat. Mech. Anal. 61 (1975) 125-156.

[3] Ellis,R.S. and Pinsky,M.A.; The first and second fluid approximations to the linearized Boltzmann equation, J. Math. Pures et Appl. 54 (1975) 125-156.

[4] Giraud,J.P.; An H theorem for a gas of rigid spheres, Theories Cinetiques Classiques et Rilativistes, C. N. R. S., Paris (1975).

[5] Grad,H.; Asymptotic theory of the Boltzmann equation, in Laurmann, J.A.(ed.), Rarefied Gas Dynamics, I (Academic Press, New York, 1963).

[6] Kaniel,S. and Shinbrot,M.; The Boltzmann equation I, Comm. Math. Phys. 57 (1978) 1-20.

[7] Matsumura,A. and Nishida,T.; Global solutions to the initial boundary value problem for the equation of compressible, viscous and heat conductive fluids, to appear.

[8] Morawetz,C.;Mixed equations and transonic flows, Rend. Math. 25 (1966) 482-509.

[9] Ukai,S. and Asano,K; On the initial boundary value problem of the linearized Boltzmann equation in an exterior domain, Proc. Japan Acad. 56 (1980) 12-17.

[10] _______ and _______; On the existence and stability of stationary solutions of the Boltzmann equation for a gas flow past an obstacle. Research notes in Math. 60 (Pitman 1982) 350-364.

[11] _______ and _______; Stationary solutions of the Boltzmann equation for a gas flow past an obstacle, I Existence and II Stability, preprints.

Lecture Notes in Num. Appl. Anal., **5**, 333–344 (1982)
Nonlinear PDE in Applied Science. U.S.-Japan Seminar, Tokyo, 1982

On the Linear Stability Analysis of Magnetohydrodynamic System *)

Teruo Ushijima

Department of Information Mathematics
The University of Electro-Communications
1-5-1 Chofugaoka, Chofu-shi, Tokyo, 182
JAPAN

A mathematical description of linear stability analysis of magnetohydrodynamic (MHD, in short) system of equations is presented. A way of linearization of ideal MHD system in a vicinity of an equilibrium was physically established. We show that the obtained linearized MHD system of equations can be treated as the 2nd order linear evolution equation in a Hilbert space under appropriate conditions on the equilibrium. A justification of energy principle is also established.

§1 THE INITIAL BOUNDARY VALUE PROBLEM OF THE LINEARIZED MAGNETOHYDRODYNAMIC SYSTEM OF EQUATIONS.

The governing system of equations is the following ideal magnetohydrodynamic system.

In the plasma region Ω_p,

(1) $\frac{D\rho}{Dt} = -\rho \operatorname{div} v$,

(2) $\rho \frac{Dv}{Dt} = -\nabla P + J\times B$,

(3) $\frac{D}{Dt}(P\rho^{-\gamma}) = 0$,

(4) $\frac{\partial B}{\partial t} = -\operatorname{rot} E$,

(5) $\operatorname{div} B = 0$,

(6) $\operatorname{rot} B = \mu J$,

(7) $E + v\times B = 0$.

---

*) This work was partly supported by Grant-in Aid for Special Project Research on Energy (Nuclear Fusion) of the Ministry of Education, Japan.

In this model the plasma is considered to be a fluid spreading over the bounded domain Ω_p in the space R^3 with the density ρ and the velocity vector v. The scalar function P is the pressure, and the 3-vector functions B, E and J are the magnetic flux density, the electric field and the current density, respectively. The positive constant μ, and γ are the magnetic permeability of the vacuum, and the specific heat ratio. Here we used the material derivative: $\frac{D}{Dt} = \frac{\partial}{\partial t} + (v, \nabla)$ with the gradient operator ∇ and the 3-dimensional inner product: (,). The symbol $\times$ denotes the exterior product of 3-vectors. The equation (1) is the conservation of mass, (2) is the equation of motion, (3) is the adiabatic condition coming from the conservation of energy. And (4), (5) and (6) are derived from the Maxwell equation under the assumption of the superiority of the magnetic field. The constitutive law (7) is Ohm's law with infinite electric conductivity.

Now we consider the closure of plasma region Ω_p is completely contained in a bounded region Ω, whose boundary Γ is considered to be a completely electrically conducting wall. The domain $\Omega - \overline{\Omega}_p$ is the vacuum region Ω_v. Let Γ_p be the boundary of Ω_p. Then the boundary $\partial\Omega_v$ of Ω_v is the union of Γ and Γ_p (see Fig. 1). Hereafter Ω_v is assumed to be connected.

In the vacuum region Ω_v, the following conditions are assumed to hold:

(8) $\rho = P = 0, \; v = J = 0,$

$(4)_v$ $\frac{\partial B}{\partial t} = -\mathrm{rot}\, E,$

$(5)_v$ $\mathrm{div}\, B = 0,$

$(6)_v$ $\mathrm{rot}\, B = 0.$

On the boundary Γ, the following boundary condition is assumed:

(9) $n \times E = 0$

where n is the unit normal

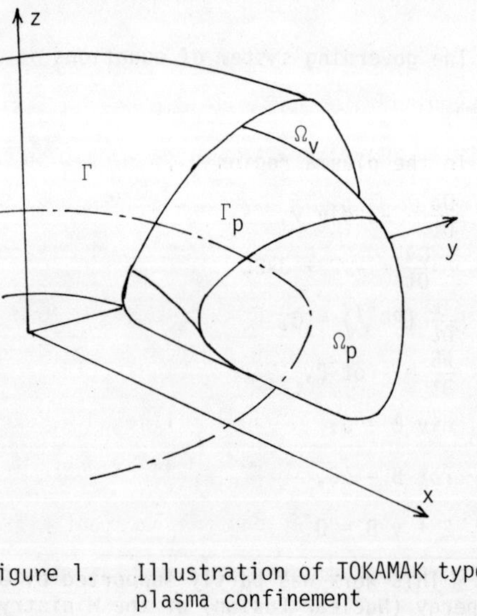

Figure 1. Illustration of TOKAMAK type plasma confinement

vector of Γ. This represents the effect of complete conductivity of the wall Γ. It is noted that the boundary condition

(10) $(B,n) = 0$

is compatible with (9) since (9) implies (rot $E,n) = 0$ which may imply $\frac{\partial}{\partial t}(B,n) = 0$ by $(4)_v$.

On the boundary Γ_p the following 3 connection conditions are imposed:

$(10)_{p,v}$ $(B_p,n) = (B_v,n) = 0,$

(11) $(E_p + v_p \times B_p) \times n = (E_v + v_p \times B_v) \times n,$

(12) $P_p + \frac{1}{2\mu} B_p{}^2 = \frac{1}{2\mu} B_v{}^2$.

Here the subscripts p, and v, represent the limiting values of the subscripted quantities from the interior of the plasma region Ω_p, and the vacuum region Ω_v, respectively. $(10)_{p,v}$ corresponds to the requirement that Γ_p should be a magnetic surface. (11) represents the continuity of the tangential part of the electric field in the coordinate system attached to the plasma. (12) is called the pressure balance condition. The sum $P + \frac{1}{2\mu} B^2$ is said to be the total pressure, where $B^2 = (B,B)$.

Thus we have an evolution problem: (1)∿(12) for the ideal MHD system.

A time independent solution of this system with zero velocity and zero electric field is said to be an equilibrium:

(13) $\{\rho, v=0, P, J, B, E=0\}$.

Then an equilibrium triple $\{P,J,B\}$ satisfies

$$
(14) \quad \begin{cases} \nabla P = J \times B,\ \text{div}\, B = 0,\ \text{rot}\, B = \mu J & \text{in } \Omega_p, \\ P = 0, \quad \text{div}\, B = 0,\ \text{rot}\, B = 0 & \text{in } \Omega_v, \\ (B,n) = 0, & \text{on } \Gamma_p \cup \Gamma, \\ P + \frac{1}{2\mu} B_p{}^2 = \frac{1}{2\mu} B_v{}^2 & \text{on } \Gamma_p. \end{cases}
$$

In the work of Bernstein et al. [1], a way of linearization of ideal MHD system in a vicinity of an equilibrium was formulated. The obtained linearized magnetohydrodynamic (LMHD, in short) system of equations describes the time evolution of the Lagrangean displacement $\xi(t,r)$ of a point r of the plasma region Ω_p at time t=0. We encounter the following initial boundary value problem:

PROBLEM 1 Find a pair $\{\xi,\alpha\}$ of vector valued functions

$$\xi = \xi(t,r): [0,\infty)\times\overline{\Omega}_p \to \mathbb{R}^3 \quad \text{and } \alpha = \alpha(t,r): (0,\infty)\times\overline{\Omega}_v \to \mathbb{R}^3$$

satisfying the following conditions:

$$(15)\quad \begin{cases} \rho \dfrac{\partial^2\xi}{\partial t^2} = -K\xi & \text{in } (0,\infty)\times\Omega_p, \\ \xi(0,r) = 0 \ , \ \dfrac{\partial}{\partial t}\xi(0,r) = v(r) & \text{in } \Omega_p, \\ L(\xi,\alpha) = 0 \ , -(\xi,n)B_v = v\times\alpha & \text{on } \Gamma_p, \\ \text{rot rot } \alpha = 0 & \text{in } \Omega_v, \\ n\times\alpha = 0 & \text{on } \Gamma. \end{cases}$$

In the above problem, we used the notations:

$$(16)\quad K\xi = -\nabla\{(\xi,\nabla P)+\gamma P \operatorname{div} \xi\} - \frac{1}{\mu}\{\text{rot } B\times \text{rot}(\xi\times B)+[\text{rot rot}(\xi\times B)]\times B\} ,$$

$$(17)\quad L(\xi,\alpha) = -\gamma P \operatorname{div} \xi + \frac{1}{\mu}(B_p, \text{rot}(\xi\times B_p)+(\xi,\nabla)B_p) - \frac{1}{\mu}(B_v,\text{rot } \alpha + (\xi,\nabla)B_v)$$

with specified equilibrium quantites P and B. Operators K, and L, are formal operators defined in Ω_p, and on Γ_p, respectively. The boundary condition: $L(\xi,\alpha)=0$ on Γ_p, is derived from the pressure balance condition (12). And the boundary condition : $-(\xi,n)B_v=n\times\alpha$, comes from the continuity condition (11). The vector rot α represents the variation of magnetic field B_v in the vacuum region. The requirement: rot rot α = 0 corresponds to $(6)_v$. Finally the boundary condition: $n\times\alpha=0$ corresponds to (9).

Hereafter we fix an equilibrium $\{P,J,B\}$ satisfying the following Assumption 2.

ASSUMPTION 2.

(A.1) Ω,Ω_p and Ω_v are bounded connected domains with C^∞ class boundaries.

(A.2) $B|_{\Omega_p}$, and $B|_{\Omega_v}$, have extensions belonging to $C^2(\overline{\Omega}_p)$, and $C^2(\overline{\Omega}_v)$, respectively.

(A.3) $P|_{\Omega_p}$ has the extension belonging to $C^2(\overline{\Omega}_p)$, being positive on Ω_p, whose critical point set $C_P=\{r\in\Omega_p:(\nabla P)(r)=0\}$ is composed of the finite number of connected components, each of which is either a C^3-class simple curve, or a C^3-surface at every point of which B is contained in the tangential plane.

(A.4) Either B_p never vanishes on Γ_p, or P is bounded below by a positive constant in Ω_p.

(A.5) For the exterior normal derivative $\frac{\partial}{\partial n}$ on Γ_p along the unit normal vector n from Ω_p to Ω_v, the gap: $G= \frac{\partial}{\partial n}\frac{B^2}{2\mu}\Big|_v - \frac{\partial}{\partial n}(\frac{B^2}{2\mu}+P)\Big|_p$ is a nonnegative C^1 class function on Γ_p, where the first, and the second, terms are the one sided derivatives from Ω_v, and from Ω_p, respectively.

§2 THE PLASMA ENERGY BILINEAR FORM AND THE LINEARIZED MAGNETOHYDRODYNAMIC OPERATOR.

Let $H^m(\Omega)$ be the usual m-th order Sobolev space of the real valued scalar functions on the domain Ω in $\mathbb{R}^3$. The totality of 3-vector functions whose components belong to $H^m(\Omega)$ is denoted by $\mathbb{H}^m(\Omega)$. Now we introduce the spaces of pairs $\{\xi,\alpha\}$ of 3-vector functions, $\hat{W}$, $\hat{V}$ and $\hat{D}$ as follows.

(1) $\hat{W} = \{\hat{\xi}=\{\xi,\alpha\}\in \mathbb{H}^1(\Omega_p)\times\mathbb{H}^1(\Omega_v)$: $n\times\alpha = -(\xi,n)B_v$ on Γ_p, $n\times\alpha = 0$ on $\Gamma\}$.

(2) $\hat{V} = \{\hat{\xi}=\{\xi,\alpha\}\in\hat{W}$: rot rot $\alpha = 0$ in $\Omega_v\}$.

(3) $\hat{D} = \{\hat{\xi}=\{\xi,\alpha\}\in\hat{V}$: $\xi\in\mathbb{H}^2(\Omega_p)$, $L(\xi,\alpha) = 0$ on $\Gamma_p\}$.

Define the bilinear form $\hat{a}(\hat{\xi},\hat{\eta})$ for $\hat{\xi} = \{\xi,\alpha\}$, $\hat{\eta} = \{\eta,\beta\}\in \mathbb{H}^1(\Omega_p)\times\mathbb{H}^1(\Omega_v)$ as follows:

$$
(4)\quad\begin{cases}
\hat{a}(\hat{\xi},\hat{\eta}) = a_p(\xi,\eta)+a_s(\xi,\eta)+\hat{a}_v(\alpha,\beta),\\
a_p(\xi,\eta) = a_1(\xi,\eta)+a_2(\xi,\eta),\\
a_1(\xi,\eta) = \int_{\Omega_p}\{\gamma P \operatorname{div}\xi \operatorname{div}\eta + \frac{1}{\mu}(\operatorname{rot}(\xi\times B),\operatorname{rot}(\eta\times B))\}dr\\
a_2(\xi,\eta) = \int_{\Omega_p}\{(\xi,\nabla P)\operatorname{div}\eta - \frac{1}{\mu}(\operatorname{rot}(\xi\times B),\eta\times\operatorname{rot} B)\}dr,\\
a_s(\xi,\eta) = \int_{\Gamma_p}\{\frac{\partial}{\partial n}\frac{B^2}{2\mu}\Big|_v - \frac{\partial}{\partial n}(\frac{B^2}{2\mu}+P)\Big|_p\}(\xi,n)(\eta,n)d\Gamma,\\
\hat{a}_v(\alpha,\beta) = \int_{\Omega_v}\frac{1}{\mu}(\operatorname{rot}\alpha,\operatorname{rot}\beta)dr,
\end{cases}
$$

where n in the expression of $a_s(\xi,\eta)$ is the unit outward normal vector on Γ_p from Ω_p to Ω_v. It is easily checked that $\hat{a}(\hat{\xi},\hat{\eta})$ is well-defined for $\hat{\xi},\hat{\eta}\in \mathbb{H}^1(\Omega_p)\times\mathbb{H}^1(\Omega_v)$.

Using the property of the equilibrium in (1.14), and the integration by part, it can be proved that

$$\int_{\Omega_p}(K\xi,\eta)dr = \hat{a}(\hat{\xi},\hat{\eta}) + \int_{\Gamma_p} L(\xi,\alpha)(\eta,n)d\Gamma$$

for any $\hat{\xi}\in\hat{V}$ with $\xi\in \mathbb{H}^2(\Omega_p)$ and $\hat{\eta}\in\hat{V}$. From this formula we can conclude the following Proposition 1.

<u>Proposition 1</u> The following identities hold.

(5) $\hat{a}(\hat{\xi},\hat{\eta}) = \hat{a}(\hat{\eta},\hat{\xi})$ for $\hat{\xi},\hat{\eta}\in \mathbb{H}^1(\Omega_p)\times \mathbb{H}^1(\Omega_v)$.

(6) $\int_{\Omega_p}(K\xi,\eta)dr = \hat{a}(\hat{\xi},\hat{\eta})$ for $\hat{\xi}\in\hat{D}$, $\hat{\eta}\in\hat{V}$.

(7) $\int_{\Omega_p}(K\xi,\eta)dr = \int_{\Omega_p}(\xi,K\eta)dr$ for $\hat{\xi}\in\hat{D},\hat{\eta}\in\hat{D}$.

Now we introduce function spaces V_0 and D_0 as follows.

(8) $V_0 = \{\xi\in \mathbb{H}^1(\Omega_p)$: there is an $\alpha\in \mathbb{H}^1(\Omega_v)$ such that $\hat{\xi} = \{\xi,\alpha\}\in\hat{V}\}$.

(9) $D_0 = \{\xi\in \mathbb{H}^2(\Omega_p)$: there is an $\alpha\in \mathbb{H}^1(\Omega_v)$ such that $\hat{\xi} = \{\xi,\alpha\}\in\hat{D}\}$.

From the theory of harmonic vector fields written in the book of Morrey [3], we have the following fact.

<u>Proposition 2</u> $V_0 = \mathbb{H}^1(\Omega_p)$.

For $\xi\in V_0$, rot α is uniquely determined. Hence we can define a symmetric form $a(\xi,\eta)$ defined on $V_0\times V_0$ by the relation:

(10) $a(\xi,\eta) = \hat{a}(\hat{\xi},\hat{\eta})$, $\xi,\eta\in V_0$.

For $\mathbb{r}\in\Omega_p-C_p$, define 3-vector functions $E=(\text{rot } B)\times e$, $F=2(B,\nabla)e+B\times\text{rot } e$ where $e=\nabla P/|\nabla P|$. For $\xi,\eta\in \mathbb{H}^1(\Omega_p)$ and $\mathbb{r}\in\Omega_p-C_p$, let

(11) $$A_p(\xi,\eta) = \frac{1}{\mu}(\text{rot}(\xi\times B) + (\xi,e)E, \text{rot}(\eta\times B) + (\eta,e)E) + \gamma P \text{ div } \xi \text{ div } \eta - \frac{1}{\mu}(\xi,e)(\eta,e)(E,F).$$

Then we can show

(12) $a_p(\xi,\eta) = \int_{\Omega_p} A_p(\xi,\eta)d\mathbb{r}$ for $\xi,\eta\in \mathbb{H}^1(\Omega_p)$.

From this representation of Plasma energy, the following Proposition 3 can be deduced.

<u>Proposition 3</u> Let

$$(13)\begin{cases} \underline{M} = \inf\limits_{r\in\Omega_p - C_p} (E,F)\ ,\ \overline{M} = \sup\limits_{r\in\Omega_p - C_p} (E,F)\ ,\ N = \sup\limits_{r\in\Omega_p - C_p} |E|^2\ , \\ \kappa_0 = \frac{1}{\mu}\max(|\frac{N}{2} + \underline{M}|, |\frac{3N}{2} + \overline{M}|). \end{cases}$$

Then we have

$$(14)\quad \frac{1}{2} a_1(\xi,\xi)+(\kappa-\kappa_0)(\xi,\xi)_X \le a_p(\xi,\xi)+\kappa(\xi,\xi)_X \le \frac{3}{2} a_1(\xi,\xi)+(\kappa+\kappa_0)(\xi,\xi)_X,$$
$$\text{for } \xi\in H^1(\Omega_p),$$

where the notation $(\xi,\eta)_X = \int_{\Omega_p}(\xi,\eta)dr$ is used.

COROLLARY 4 Assume that

$$(15)\quad G = \frac{\partial}{\partial n}\frac{B^2}{2\mu}\Big|_v - \frac{\partial}{\partial n}(\frac{B^2}{2\mu} + P)\Big|_p \ge 0 \quad \text{on } \Gamma_p.$$

Then the symmetric bilinear form $a_\kappa(\xi,\eta)$ defined on V_0:

$$(16)\quad a_\kappa(\xi,\eta) = a(\xi,\eta) + \kappa(\xi,\eta)_X,\quad \xi,\eta\in V_0,$$

is positive definite if

$$(17)\quad \kappa>\kappa_0.$$

The following Proposition 5 is fundamental to our discussion, the proof of which was given in [4].

Proposition 5 Let X be the Hilbert space $\{L^2(\Omega_p)\}^3$. Under Assumption 1.1, a_κ is closable in X if $\kappa>\kappa_0$. Namely, if the saquence $\{\xi_m: m=1,2,\cdots\}$ contained in V_0 satisfies

$$(18)\quad \lim_{m\to\infty} \|\xi_m\|_X = 0$$

and

$$(19)\quad \lim_{m,m'\to\infty} a_\kappa(\xi_m-\xi_{m'}) = 0,$$

then we have

$$(20)\quad \lim_{m\to\infty} a_\kappa(\xi_m) = 0,$$

where $\|\xi\|_X = (\xi,\xi)_X^{1/2}$ and $a_\kappa(\xi) = a_\kappa(\xi,\xi)^{1/2}$.

The preceding argument can be transformed in the frame work of complex Hilbert space theory. Under trivial modifications, we can consider the hermitian form $a_\kappa(\xi,\eta)$ with the domain $V_0 = H^1(\Omega_p)$ in the complex Hilbert space $X=\{L^2(\Omega_p)\}^3$.

Let $\kappa>\kappa_0$ then a_κ is closable by Proposition 5. Thererfore we have the following statements (21) and (22) for the completion V of V_0 by a_κ-norm.

(21) V is densely imbedded in X.

(22) $a_\kappa(\xi,\eta)$, which is considered to be defined on V by continuity, is a positive definite closed hermitian form.

From Theorem 2.23 of Chapter VI in Kato [2], there exists the unique positive definite selfadjoint operator A_κ satisfying the following properties (23)~(25).

(23) The domain $D(A_\kappa)$ is dense in V with a_κ-norm.

(24) $(A_\kappa\xi,\eta)_X = a_\kappa(\xi,\eta)$ for $\xi\in D(A_\kappa)$, $\eta\in V$.

(25) $D(A_\kappa{}^{1/2}) = V$, and $(A_\kappa{}^{1/2}\xi, A_\kappa{}^{1/2}\eta)_X = a_\kappa(\xi,\eta)$ for $\xi,\eta\in V$.

Now we define the LMHD operator A by

(26) $A = A_\kappa - \kappa$, $D(A) = D(A_\kappa)$.

By a standard argument, this definition (26) is independent of κ if $\kappa>\kappa_0$. It is to be noted that a_κ are mutually equivalent if $\kappa>\kappa_0$. On the other hand we can define the operator K_0 by

(27) $K_0\xi = K\xi$, $\xi\in D(K_0) = D_0$,

which is symmetric by (7) of Proposition 1.

Theorem 6 A is an extension of K_0 satisfying the following properties.

(28) If $\xi\in D_0$, then $\xi\in D(A)$ and $A\xi = K_0\xi$.

(29) If $\xi\in D(A)\cap \mathbb{H}^2(\Omega_p)$, then $\xi\in D_0$.

Namely we have

(30) $D_0 = D(A)\cap\mathbb{H}^2(\Omega_p)$.

Now Problem 1.1 can be regarded as an evolution equation in the Hilbert space $X = \{L^2(\Omega_p)\}^3$:

$$(31)\quad \begin{cases} M\dfrac{d^2\xi}{dt^2} + A\xi = 0,\ t>0, \\ \xi(0) = \xi^1,\ \dfrac{d\xi}{dt}(0) = \xi^0, \end{cases}$$

with the initial data $\{\xi^1,\xi^0\} = \{0,v\}$ under Assumption 1.2 on the equilibrium quantites. Here M is the multiplying operator with multiplier ρ and A is the LMHD operator.

§3 A REMARK ON THE ENERGY PRINCIPLE.

Assumption 1 Let $\{\rho, v=0, P, J, B, E=0\}$ is an equilibrium of ideal MHD system (1.14) with the properties that for the plasma density $\rho=\rho(r)$, being measurable in Ω_p, there are positive constants $\underline{\rho}$ and $\overline{\rho}$ such that

$$0 < \underline{\rho} \le \rho(r) \le \overline{\rho} < \infty \quad \text{on } \Omega_p,$$

and that Assumption 1.2 is satisfied for $\{P, J, B\}$.

Let us define

$$m(\xi,\eta) = \int_{\Omega_p} \rho(r)(\xi,\eta)_{C^3} dr \qquad \text{for } \xi,\eta \in \{L^2(\Omega_p)\}^3,$$

$$M\xi = \rho\xi \qquad \text{for } \xi \in \{L^2(\Omega_p)\}^3.$$

Denote the Hilbert space $\{L^2(\Omega_p)\}^3$ with the inner product $m(\xi,\eta)$ by X_ρ.

Define the operator A_ρ acting in X_ρ by

$$A_\rho\xi = \rho^{-1}A\xi \quad \text{for } \xi \in D(A_\rho) = D(A),$$

where A is the LMHD operator established in Theorem 2.6.

Proposition 2 1) The operator A_ρ is selfadjoint in the space X_ρ.

2) Let κ_0 be as in (2.13). For $\kappa_\rho > \kappa_0/\underline{\rho}$, $A_\rho + \kappa_\rho$ is positive definite, satisfying

$$D((A_\rho + \kappa_\rho)^{1/2}) = V,$$

and

$$a(\xi,\eta) + \kappa_\rho m(\xi,\eta) = m((A_\rho + \kappa_\rho)^{1/2}\xi, (A_\rho + \kappa_\rho)^{1/2}\eta) \quad \text{for } \xi,\eta \in V,$$

and

$$a(\xi,\eta) = m(A_\rho\xi,\eta) \quad \text{for } \xi \in D(A_\rho),\ \eta \in V.$$

Let $\sigma(A_\rho)$ be the spectrum set of A_ρ, which is the complement of the resolvent set $\rho(A_\rho)$ of A_ρ in the complex plane C:

$$\rho(A_\rho) = \{\lambda \in C: \text{there is a bounded inverse } (\lambda - A_\rho)^{-1} \text{ from } X_\rho \text{ into } X_\rho\}.$$

By Proposition 2 $\sigma(A_\rho)$ is a part of real line which is bounded below.

Let $\underline{\lambda} = \inf \sigma(A_\rho)$. Then we have

$$(1) \quad \underline{\lambda} = \inf_{\xi\in D(A)} \frac{a(\xi,\xi)}{m(\xi,\xi)} = \inf_{\xi\in V} \frac{a(\xi,\xi)}{m(\xi,\xi)} = \inf_{\xi\in V_0} \frac{a(\xi,\xi)}{m(\xi,\xi)}.$$

Definition 3 An equilibrium $\{\rho, v=0, P, J, B, E=0\}$ is said to be stable, marginal, and unstable if $\underline{\lambda}>0$, $\underline{\lambda}=0$, and $\underline{\lambda}<0$, respectively.

The following Theorem may be considered as a justification of the energy principle formulated by Bernstein et al. [1].

Theorem 4 An equilibrium is stable or marginal if $\hat{a}(\hat{\xi},\hat{\xi}) \geq 0$ for any $\hat{\xi} = \{\xi,\alpha\}\in\hat{W}$. And it is unstable if $\hat{a}(\hat{\xi},\hat{\xi}) < 0$ for some $\hat{\xi} = \{\xi,\alpha\}\in\hat{W}$. Here $\hat{W}$ is defined in (2.1).

To prove this Theorem we utilize the next Lemma which can be obtained from Theorem 7.8.2 of Morrey [3].

Lemma 5 For any $\xi \in \mathbb{H}^1(\Omega_p)$, there is an $\alpha_0 \in \mathbb{H}^1(\Omega_v)$ satisfying

$$(2) \begin{cases} \text{rot rot } \alpha_0 = 0 & \text{in } \Omega_v, \\ n\times\alpha_0 = -(n,\xi)B_v & \text{on } \Gamma_p, \\ n\times\alpha_0 = 0 & \text{on } \Gamma. \end{cases}$$

Proof of Theorem 4 The first statement of Theorem follows from the last equality in (1) since we have $\hat{V}\subset\hat{W}$ and $a(\xi,\xi) = \hat{a}(\hat{\xi},\hat{\xi})$ for $\xi\in V_0$. To prove the second statement, let $\hat{\xi} = \{\xi,\alpha\}\in\hat{W}$ satisfy $\hat{a}(\hat{\xi},\hat{\xi}) < 0$. By Lemma 5, there is an $\alpha_0 \in \mathbb{H}^1(\Omega_v)$ satisfying (2). Let $\hat{\xi}_0 = \{\xi, \alpha_0\}$. Then $\hat{\xi}_0 \in \hat{V}$.
Let $\alpha = \alpha_0 + \alpha_1$. Then

$$\begin{aligned} \hat{a}_v(\alpha_0,\alpha_1) &= \frac{1}{\mu}\int_{\Omega_v} (\text{rot } \alpha_0, \text{ rot } \alpha_1)d\mathbf{r} \\ &= \frac{1}{\mu}\int_{\Omega_v} (\text{rot rot } \alpha_0, \alpha_1)d\mathbf{r} + \frac{1}{\mu}\int_{\Gamma_p\cup\Gamma} (\text{rot } \alpha_0, n\times\alpha_1)d\Gamma \\ &= 0 \end{aligned}$$

since rot rot $\alpha_0 = 0$ on Ω_v and $n\times\alpha_1 = 0$ on $\Gamma_p\cup\Gamma$. Hence

$$\hat{a}_v(\alpha,\alpha) = \hat{a}_v(\alpha_0,\alpha_0) + \hat{a}_v(\alpha_1,\alpha_1).$$

Therefore

$$\begin{aligned} a(\xi,\xi) &= \hat{a}(\hat{\xi}_0,\hat{\xi}_0) \\ &= a_p(\xi,\xi) + a_s(\xi,\xi) + \hat{a}_v(\alpha_0,\alpha_0) \leq a_p(\xi,\xi) + a_s(\xi,\xi) + \hat{a}_v(\alpha,\alpha) \end{aligned}$$

$= \hat{a}(\hat{\xi},\hat{\xi}) < 0.$

The last equality in (1) again implies the assertion.

Finally we add the following Proposition 6 which may give an explanation of the energy principle.

Proposition 6 If the plasma energy $\hat{a}(\hat{\xi},\hat{\xi})$ attains the minimum $\hat{\lambda}$ on the set $\hat{B}_\rho = \{\hat{\xi}\in\hat{W} : m(\xi,\xi) = 1\}$, then we have

(3) $\underline{\hat{\lambda}} = \underline{\lambda} = \min_{\xi\in V_0} \frac{a(\xi,\xi)}{m(\xi,\xi)} = \min_{\xi\in V} \frac{a(\xi,\xi)}{m(\xi,\xi)} = \min_{\xi\in D(A)} \frac{a(\xi,\xi)}{m(\xi,\xi)}$

In this case $\underline{\lambda}$ is an eigenvalue of the operator A_ρ. The corresponding eigenfunction ϕ belongs to V_0 determined by the following generalized eigenvalue problem:

(4) Find $\phi\in V_0-\{0\}$, such that $a(\phi,\xi) = \underline{\lambda} m(\phi,\xi)$ for $\xi\in V_0$.

Proof The inequality $\underline{\hat{\lambda}} \leq \underline{\lambda}$ follows from the facts $\hat{V}\subset\hat{W}$ and $a(\xi,\xi) = \hat{a}(\hat{\xi},\hat{\xi})$ for $\xi\in V_0$.

The inequality $\underline{\hat{\lambda}} \geq \underline{\lambda}$ is shown as follows. Let $\hat{\xi} = \{\xi,\alpha\} \in \hat{B}_\rho$ give the minimum $\underline{\hat{\lambda}}$ of $\hat{a}(\hat{\xi},\hat{\xi})$ in $\hat{B}_\rho$:

(5) $\hat{a}(\hat{\xi},\hat{\xi}) \leq \hat{a}(\hat{\eta},\hat{\eta})$ for $\hat{\eta}\in\hat{B}_\rho$.

Choose $\beta \in \mathbb{H}^1(\Omega_v)$ such that $n\times\beta = n\times\alpha$ on $\Gamma \cup \Gamma_\rho$. Then $\hat{\eta} = (\xi,\beta)\in\hat{B}_\rho$. By (5) we have

$$\hat{a}_v(\alpha,\alpha) \leq \hat{a}_v(\beta,\beta).$$

For any $\phi\in\{H_0^1(\Omega_v)\}^3$ and $t \in \mathbb{R}^1$, we can choose β as $\beta=\alpha+t\phi$. For this β, it holds

$$\hat{a}_v(\alpha+t\phi,\ \alpha+t\phi) = \hat{a}_v(\phi,\phi)t^2 + 2\mathrm{Re}\ \hat{a}_v(\phi,\alpha)t + \hat{a}_v(\alpha,\alpha).$$

So we have as the stationary condition

(6) $\hat{a}_v(\phi,\alpha) = 0$ for any $\phi\in\{H_0^1(\Omega_v)\}^3$ with $\|\mathrm{rot}\ \phi\|_{L^2(\Omega_v)} \neq 0.$

On the other hand, we have the identity:

$$\hat{a}_v(\phi,\alpha) = \frac{1}{\mu}\int_{\Omega_v}(\mathrm{rot\ rot}\ \phi,\ \alpha)dr - \frac{1}{\mu}\int_{\partial\Omega_v}(\mathrm{rot}\ \phi,\ \alpha\times n)d\Gamma$$

for any $\phi\in\{\mathcal{D}(\overline{\Omega}_v)\}^3$,

where the boundary integral in the right hand side is understood to be the duality pair of $\mathrm{rot}\ \phi\in\mathbb{H}^{1/2}(\partial\Omega_v)$ and $\alpha\times n\in\mathbb{H}^{-1/2}(\partial\Omega_v)$. Hence we have

(7) $\hat{a}_v(\phi,\alpha) = \frac{1}{\mu}\int_{\Omega_v}(\mathrm{rot\ rot}\ \phi,\ \alpha)dr$ for any $\phi\in\{\mathcal{D}(\Omega_v)\}^3$.

Let $\Phi\in\mathcal{D}(\Omega_V)-\{0\}$. Then $\phi = (\Phi,0,0) \in \{\mathcal{D}(\Omega_V)\}^3$ satisfies the condition: $\|\text{rot}\,\phi\|_{L^2(\Omega_V)} \neq 0$. (In fact, unless the condition holds, we have $\frac{\partial\Phi}{\partial x_2} = \frac{\partial\Phi}{\partial x_3} = 0$ in Ω_V. Hence $\Phi = \Phi(x_1)$. And Φ vanishes on the boundary of support of Φ. Hence $\Phi=0$.) From (6) and (7)

(8) $\int_{\Omega_V}(\text{rot rot}\,\phi,\alpha)dr = 0$

for this $\phi= (\Phi,0,0)$. By the same reason, (8) holds for $\phi= (0,\Phi,0)$ and $\phi= (0,0,\Phi)$. Hence it holds for any $\phi \in \{\mathcal{D}(\Omega_V)\}^3$. Namely we have rot rot $\alpha= 0$ in $\mathcal{D}'(\Omega_V)$, which implies $\hat{\xi}\in\hat{V}$ and $\xi\in V_0$. Thus we obtain $\hat{\underline{\lambda}} \geq \underline{\lambda}$.

Since we have $\hat{\underline{\lambda}} = \underline{\lambda}$, the second equality of (3) is valid. Hence the third equality of (3) is obtained from that of (1). The fourth equality and the remaining statement of the present Proposition follows from standard arguments. It is noted that

$$\underline{\lambda} = \frac{a(\phi,\phi)}{m(\phi,\phi)}, \quad \phi\in V - \{0\}$$

implies that ϕ satisfies

(9) $a(\phi,\xi) = \underline{\lambda}m(\phi,\xi),\ \xi\in V.$

Using the density argument, we have (4) from (9).

References

[1] Bernstein, I., Frieman, E., Kruskal, M., and Kulsrud, R., An energy principle for hydromagnetic stability problems, Proc. Royal Soc. A. 244 (1958), 17-40.

[2] Kato, T., Perturbation theory for linear operators (Springer, Berlin, 1966).

[3] Morrey, C., Jr., Multiple integrals in the calculus of variations. (Springer, Berlin, 1966).

[4] Ushijima, T., On the linearized magnetohydrodynamic systems of equations for a contained plasma in a vacuum region, in: Glowinski, R. and Lions, J. L. (eds) Computing Method in Applied Sciences and Engineering, V, (North-Holland, Amsterdam, 1982).

Lecture Notes in Num. Appl. Anal., **5**, 345–359 (1982)
Nonlinear PDE in Applied Science. U.S.-Japan Seminar, Tokyo, 1982

A simple system with a continuum of stable inhomogeneous steady states

H.F. Weinberger
Institute for Mathematics and its Applications
University of Minnesota

I. Introduction

The system

$$u_t = \{(1 + \alpha v)u\}_{xx} + (R_1 - au - bv)\,u \tag{1.1}$$

$$v_t = (R_2 - bu - av)v$$
$$\{(1 + \alpha v)u\}_x = 0 \quad \text{at} \quad x = 0 \quad \text{and} \quad x = 1$$

with

$$\frac{1}{2}\left(\frac{a}{b} + \frac{b}{a}\right) < \frac{R_1}{R_2} < \frac{a}{b} \tag{1.2}$$

and

$$\alpha > \frac{a(a^2 - b^2)}{2abR_1 - (a^2 + b^2)R_2} \tag{1.3}$$

was considered by M. Mimura [2] as a model for the population densities of two competing species, one of which increases its migration rate in response to crowding by the other species. It is a special case of the model of N. Shigesada, K. Kawasaki, and E. Teramoto [3].

Numerical computation by D.G. Aronson and P.N. Brown seems to indicate that the solution converges to a steady state in which v has one or more discontinuities, and that these discoutinuities move continuously with changes in the initial conditions.

The existence of a continuum of discontinuous solutions of the system (1.1) was proved by Mimura. The purpose of this lecture is to prove that there are, indeed, whole one parameter families of discontinuous solutions which are stable in a suitable topology.

The family of piecewise continuous steady states is described in Section 2.

We shall show in Section 3 that a somewhat unusual topology is needed for this problem and prove that linearized stability implies stability in this topology.

In Section 4 we give a sufficient condition for stability and show that a continuum of the discontinuous steady states satisfies this criterion.

Section 5 discusses the evolutionary consequences of the existence of stable nonconstant steady states.

This work is a part of ongoing joint research with D.G. Aronson and A. Tesei.

I am grateful to Don Aronson for getting me interested in this problem and for a great deal of useful discussion and criticism.

2. The steady state.

M. Mimura [2] introduced the new independent variable

$$w = (1 + \alpha v)u$$

in (1.1) to obtain the system

$$v_t = G(v,w)$$

$$w_t = (1 + \alpha v)(w_{xx} + H(v,w)) \tag{2.1}$$

where

$$G(v,w) = v(R_2 - av - \frac{bw}{1 + \alpha v}) ,$$

$$H(v,w) = \frac{w}{1 + \alpha v} \{R_1 - \frac{aw}{1 + \alpha v} - bv + \frac{\alpha v}{1 + \alpha v} (R_2 - av - \frac{bw}{1 + \alpha v})\} . \tag{2.2}$$

The no-flux boundary conditions are

$$w_x = 0 \text{ at } x = 0 \text{ and } x = 1 . \tag{2.3}$$

The corresponding steady-state equations are

$$G(v,w) = 0 ,$$

$$w'' + H(v,w) = 0 , \tag{2.4}$$

$$w' = 0 \text{ at } x = 0,1 .$$

When

$$R_2/b < w < w_m \equiv (\alpha R_2 + a)^2/4\alpha ab \tag{2.5}$$

the equation $G(v,w) = 0$ has the three nonnegative branches of solutions

$$v_0 \equiv 0$$

$$v_1(w) = \frac{1}{2} (\frac{R_2}{a} - \frac{1}{\alpha}) - (\frac{b}{a\alpha})^{1/2}(w_m - w)^{1/2} , \tag{2.6}$$

$$v_2(w) = \frac{1}{2} (\frac{R_2}{a} - \frac{1}{\alpha}) + (\frac{b}{a\alpha})^{1/2}(w_m - w)^{1/2} .$$

If we substitute these in the second equation of (2.4), we find the three ordinary differential equations

$$
\begin{aligned}
&w'' + H(0,w) = 0\\
&w'' + H(v_1(w),w) = 0 \qquad (2.7)\\
&w'' + H(v_2(w),w) = 0
\end{aligned}
$$

when w lies in the interval $(R_2/b, w_m)$. It is easy to see from the first equation of (2.1) that v tends to move away from the branch $v = v_1(w)$ for $w \varepsilon (R_2/b, w_m)$ and from $v = 0$ for $w < R_2/b$. Thus only the first and last of these equations can yield stable steady states.

Easy computations show that $H(0,w) < 0$ for $w \geq R_2/b$ while $H(v_2(w),w) > 0$ for $w \varepsilon (0,w_m)$. If we integrate the first and third equations of (2.7) and use the boundary conditions $w_x = 0$, we see that neither one can have a solution with $w \varepsilon (R_2/b, w_m)$.

One can, however, obtain solutions by letting v jump from the branch v_0 to the branch v_2 and back again while keeping w and w_x continuous. This can be seen from the method of first integrals (see [2]) or from phase plane diagrams. The points of discontinuity are rather arbitrary, so that one obtains a large continuum of steady-state solutions with $w \varepsilon (R_2/b, w_m)$ and discontinuous v. In particular, by introducing sufficiently many discontinuities one can keep w arbitrarily close to any constant in $(R_2/b, w_m)$.

These solutions with discontinuous v are the only candidates for stable solutions.

3. The stability of discontinuous solutions.

We shall investigate the stability of a steady-state solution $(\tilde{v}(x), \tilde{w}(x))$ with piecewise continuous $\tilde{v}$, as discussed in the preceding Section. It is not difficult to prove the asymptotic stability of such a solution in the norm $\|v\|_{L_\infty} + \|w\|_{H^1}$ when the linearized operator is stable. (See Remark 2 after Theorem 1.)

This fact seems surprising because the points of discontinuity of v can be chosen rather freely. However, in the L_∞ norm the distance between a function with a jump at x_0 and one without a jump at x_0 is at least half the magnitude of the jump. Therefore in this norm solutions with discontinuities at different points are isolated from each other.

For the same reason a sufficiently small neighborhood of a discontinuous function contains no continuous functions. Since a solution of the system (2.1), (2.3) with smooth initial data remains smooth, it cannot converge to a discontinuous solution in the L_∞ norm. Thus the L_∞ topology on the v-component of the solution is not appropriate for this problem.

We shall, instead, use the weaker topology with the neighborhood base

$$N_\varepsilon(\tilde{v}) = \{r \in L_\infty : \text{meas}\{x : |r(x) - \tilde{v}(x)| \geq \varepsilon\} < \varepsilon^4\}$$

for the v-component. The closure of the set of continuous functions in this topology is the set of functions which are almost everywhere continuous.

It is usual to relate the stability of the steady-state solution of the system (2.1) to the spectrum of the linearized operator

$$L\begin{pmatrix}\eta\\ \zeta\end{pmatrix} \equiv \begin{pmatrix}G_v(\tilde{v},\tilde{w})\eta + G_w(\tilde{v},\tilde{w})\zeta\\ [1+\alpha\tilde{v}][\zeta'' + H_v(\tilde{v},\tilde{w})\eta + H_w(\tilde{v},\tilde{w})\zeta]\end{pmatrix} . \tag{3.1}$$

with the boundary conditions $\zeta' = 0$ at 0 and 1. We shall show that this can also be done in our topology by proving the following result.

<u>THEOREM</u> 1 Let $(\tilde{v},\tilde{w})$ be a steady-state solution of (2.1), (2.3) with $\tilde{v}$ piecewise continuous, $\tilde{w}$ continuously differentiable, $0 \leq \tilde{v} < R_2/a$, and $R_2/b < \tilde{w} < w_m$. Suppose that the spectrum of the operator L in (3.1) lies uniformly in the left half-plane

$$\text{Re}\,\lambda \leq -2k < 0 . \tag{3.2}$$

Then there exist positive constants ε_0 and A with the property that if (v,w) is a solution of (2.1), (2.3) such that $0 \leq v(x,0) \leq R_2/a$,

$$\| v(x,0) - \tilde{v}(x) \|^2_{L_\infty(S)} + \| w(x,0) - \tilde{w}(x) \|^2_{H^1} \leq \varepsilon^2 \tag{3.3}$$

where the measure $|S'|$ of the complement of the subset S of $[0,1]$ satisfies

$$|S'| \leq \varepsilon^4 , \tag{3.4}$$

and if $\varepsilon < \varepsilon_0$, the inequality

$$\| v(x,t) - \tilde{v}(x) \|^2_{L_\infty(S)} + \| w(x,t) - \tilde{w}(x) \|^2_{H^1} < A\varepsilon^2 . \tag{3.5}$$

is valid for all $t > 0$.

Proof We first observe that

$$G(0,w) = 0 \quad , G(R_2/a,w) \leq 0$$
$$H(v,0) = 0 \quad , H(v,w_M) \leq 0$$

when

$$\begin{aligned} &0 \leq v \leq R_2/a \\ &0 \leq w \leq w_M \equiv (\alpha R_1 + b)^2/4\alpha ab . \end{aligned} \tag{3.6}$$

It follows [4] that the set (3.6) is an invariant set for the system (2.1), (2.3). That is, if $(v(x,0) , w(x,0))$ satisfies these inequalities, so does $(v(x,t),w(x,t))$ for all $t > 0$.

We agree to choose ε_0 so small that for $\varepsilon \leq \varepsilon_0$ the inequalities (3.3) imply that the initial values, and hence also the solution, satisfy (3.6). (The inequality (1.3) implies that $w_M > w_m$.)

We now define

$$\eta(x,t) = v(x,t) - \tilde{v}(x) \quad , \quad \zeta(x,t) = w(x,t) - \tilde{w}(x)$$

$$\eta_0(x) = v(x,0) - \tilde{v}(x) \quad , \quad \zeta_0(x) = w(x,0) - \tilde{w}(x)$$

and write the system (2.1), (2.3) in the form

$$\frac{\partial}{\partial t}\binom{\eta}{\zeta} - L\binom{\eta}{\zeta} = \binom{\rho}{\sigma} \tag{3.7}$$

$$\zeta_x(0,t) = \zeta_x(1,t) = 0$$

$$\eta(x,0) = \eta_0(x)$$

$$\zeta(x,0) = \zeta_0(x)$$

where L is defined by (3.1) and

$$\rho = G(\tilde{v} + \eta, \tilde{w} + \zeta) - [G(\tilde{v},\tilde{w}) + G_v(\tilde{v},\tilde{w})\eta + G_w(\tilde{v},w)\zeta] ,$$

$$\sigma = \frac{\eta\zeta t}{1 + \alpha(\tilde{v} + \eta)} + (1 + \alpha\tilde{v})\{H(\tilde{v} + \eta, \tilde{w} + \zeta) - [H(\tilde{v},\tilde{w}) + H_v(\tilde{v},\tilde{w})\eta + H_w(\tilde{v},\tilde{w})\zeta]\} . \tag{3.8}$$

We first treat (3.7) as a linear system. Because the spectrum of L is in the half-plane (3.2), a standard estimate (see e.g. [1, Theo. 1.3.4]) shows that

$$\|\eta(\cdot,t)\|_{L_1} + \|\zeta(\cdot,t)\|_{L_1} \leq c\{[\|\eta_0\|_{L_1} + \|\zeta_0\|_{L_1}]e^{-kt} + \int_0^t [\|\rho(\cdot,\tau)\|_{L_1} + \|\sigma(\cdot,\tau)\|_{L_1}]e^{-k(t - \tau)}d\tau\} . \tag{3.9}$$

Here and in all that follows c stands for any constant which depends only on bounds for G and H and their partial derivatives on the set (3.6), and on k .

Because the second equation of (3.7) is parabolic, we can find a bound of the form

$$\|\zeta\|_{L_2}^2 \leq c\{\|\zeta_0\|_{L_2}^2 e^{-3kt} + \int_0^t[\|\sigma\|_{L_1} + \|\eta\|_{L_1} + \|\zeta\|_{L_1}]^2 e^{-3k(t-\tau)}d\tau\} . \quad (3.10)$$

The first equation of (3.7) can be solved for η by quadratures in terms of ρ and ζ. It is easily seen that the closure of the range of $G_v(v,w)$ lies in the spectrum of L so that the spectral bound (3.2) implies that $G_v \leq -2k$. Consequently we find that

$$|\eta(x,t)| \leq |\eta_0(x)|e^{-2kt} + c\int_0^t[|\zeta(x,\tau)| + |\rho(x,\tau)|]e^{-2k(t-\tau)}d\tau . \quad (3.11)$$

It follows that

$$\|\eta\|_{L_2}^2 \leq c\{\|\eta_0\|_{L_2}^2 e^{-3kt} + \int_0^t [\|\zeta\|_{L_2}^2 + \|\rho\|_{L_2}^2]e^{-3k(t-\tau)}d\tau\} .$$

We combine this inequality with (3.9) and (3.10) to obtain

$$\|\eta\|_{L_2}^2 + \|\zeta\|_{L_2}^2 \leq c\{[\|\eta_0\|_{L_2}^2 + \|\zeta_0\|_{L_2}^2]e^{-2kt} \quad (3.12)$$

$$+ \int_0^t[\|\rho\|_{L_2}^2 + \|\sigma\|_{L_1}^2]e^{-2k(t-\tau)}d\tau\} .$$

We see from (3.8) that

$$\|\rho\|_{L_2}^2 + \|\sigma\|_{L_1}^2 \leq c[\|\eta\|_{L_2}^2\|\zeta_t\|_{L_2}^2 + \|\eta\|_{L_4}^4 + \|\zeta\|_{L_4}^4] . \quad (3.13)$$

In order to use (3.12) and (3.13) we need bounds for an integral of $\|\zeta_t\|_{L_2}^2$ and for $\|\zeta\|_{L_4}^4$. For this purpose we write the second equation of (2.1) in the form

$$\frac{1}{1+\alpha v}\zeta_t - \zeta_{xx} + \zeta = H(v+\eta, w+\zeta) - H(v,w) + \zeta .$$

We multiply by $(\zeta_t + \frac{1}{2} k\zeta)e^{kt}$, integrate by parts, and use (3.12) and (3.13) to find that

$$\int_0^t \|\zeta_t\|^2 e^{-k(t-\tau)}d\tau + \|\zeta\|^2_{H^1} \leq c_1\{[\|\zeta_0\|^2_{H^1} + \|\eta_0\|^2_{L_2}]e^{-kt} + \int_0^t [\|\eta\|^2_{L_2}\|\zeta_t\|^2_{L_2} + \|\eta\|^4_{L_4} + \|\zeta\|^4_{L_4}]e^{-k(t-\tau)}d\tau\} . \tag{3.14}$$

Suppose that on a time interval $[0,t_1)$ we have

$$c_1\|\eta\|^2_{L_2} \leq 1 . \tag{3.15}$$

Then (3.14) shows that on this interval

$$\|\zeta\|^2_{H^1} \leq c_1\{[\|\zeta_0\|^2_{H^1} + \|\eta_0\|^2_{L_2}]e^{-kt} + \int_0^t [\|\eta\|^4_{L_4} + \|\zeta\|^4_{L_4}]e^{-k(t-\tau)}d\tau \quad . \tag{3.16}$$

We now observe that because of the bound (3.6)

$$\|\eta\|^p_{L_p} \leq \|\eta\|^p_{L_\infty(S)} + (R_2/a)|S'| . \tag{3.17}$$

We find from (3.11) and Sobolev's inequality that

$$\|\eta\|_{L_\infty(S)} \leq c\{\|\eta_0\|_{L_\infty(S)}e^{-2kt} + \int_0^t [\|\zeta\|_{H^1} + \|\zeta\|^2_{H^1} + \|\eta\|^2_{L_\infty(S)}]e^{-2k(t-\tau)}d\tau\} .$$

We combine this with (3.16) and (3.17) to obtain the inequality

$$\begin{aligned}\|\eta\|^2_{L_\infty(S)} + \|\zeta\|^2_{H^1} \leq c_2\{[&\|\eta_0\|^2_{L_\infty(S)} + \|\zeta_0\|^2_{H^1} + \\ &+ (\|\eta_0\|^2_{L_\infty(S)} + \|\zeta_0\|^2_{H^1})^2]e^{-kt} \\ &+ \int^t [(\|\eta\|^2_{L_\infty(S)} + \|\zeta\|^2_{H^1})^2 + (\|\eta\|^2_{L_\infty(S)} + \|\zeta\|^2_{H^1})^4]e^{-k(t-\tau)}d\tau \\ &+ |S'| + |S'|^2\} .\end{aligned} \tag{3.18}$$

Choose any A such that

$$A > \max(c_2, 1)$$

where c_2 is the constant in (3.18). Let ε_0 be so small that the inequalities

$$c_1 A \varepsilon_0^2 \leq 1 ,$$

$$1 + (2 + \frac{A^2}{k})\varepsilon_0^2 + (1 + \frac{A^4}{k})\varepsilon_0^6 < A/c_2$$

are valid. Then the inequality (3.3) implies (3.5) for sufficiently small t. Moreover, if (3.5) holds in an interval $[0,t_1)$, then (3.15) is valid there and (3.18) implies that this inequality is still valid at $t = t_1$. Thus the maximal interval where (3.5) holds is both open and closed. That is, (3.5) is valid for all positive t, which proves the Theorem.

<u>REMARKS</u>: 1. The first equation of (2.1) and (3.5) imply that there is a constant C such that if the inequality $v \leq v_2(w) + C\varepsilon$ is valid on the whole interval $[0,1]$ for one value of t, it is valid for all larger t, and that this inequality holds for all sufficiently large t.

2. If $|S'| = 0$, that is, if one works in the norm $\|\eta\|_{L_\infty} + \|\zeta\|_{H^1}$, the bounds

$$\|\eta\|_{L_\infty} e^{1/4kt} \leq A\varepsilon , \quad \|\zeta\|_{H^1} e^{1/4kt} \leq B\varepsilon$$

follow from (3.18), so that (v,w) is asymptotically stable in this topology.

4. A sufficient condition for stability.

We wish to derive a sufficient condition for the spectrum of the operator L defined by (3.1) to be bounded away from the right half-plane, so that the conclusion of Theorem 1 is valid.

We consider the system

$$\begin{aligned} &G_v(\tilde{v},\tilde{w})\eta + G_w\zeta - \lambda\eta = \rho \\ &(1 + \alpha\tilde{v})[\zeta'' + H_v\eta + H_w\zeta] - \lambda\zeta = \sigma \qquad (4.1) \\ &\zeta'(0) = \zeta'(1) = 0 \end{aligned}$$

whose solution gives the inverse of $L - \lambda I$ at points of its resolvent set. As we have already mentioned, the closure of the range of G_v can be shown to lie in the spectrum. Consequently, a necessary condition for (3.2) is

$$G_v(\tilde{v},\tilde{w}) \leq -2k < 0 \,. \qquad (4.2)$$

Our criterion will involve the solution of the initial value problem

$$\begin{aligned} &r'' + \{H_w(\tilde{v},\tilde{w}) + [\frac{G_w(\tilde{v},\tilde{w})H_v(\tilde{v},\tilde{w})}{-G_v(\tilde{v},\tilde{w})}]_+\}r = 0 \\ & \qquad\qquad\qquad\qquad\qquad\qquad\qquad (4.3) \\ &r(0) = 0 \\ &r'(0) = 1 \end{aligned}$$

As usual,

$$[s]_+ = \begin{cases} 0 & \text{if } s \leq 0 \\ s & \text{if } s \geq 0 \,. \end{cases}$$

<u>THEOREM</u> 2. Suppose that G_v satisfies an inequality of the form (4.2) and that the solution of (4.3) has the properties

$$\begin{aligned} &r > 0 \quad \text{on} \quad [0,1] \\ &r'(1) > 0 \end{aligned} \tag{4.4}$$

Then the steady-state solution v, w is stable in the sense of Theorem 1.

<u>PROOF</u>. If λ is outside the closure of the range of G_v, we can solve the first equation for v and substitute in the second to obtain the problem

$$\zeta'' + [H_w - \frac{\lambda}{1+\alpha\tilde{v}} + \frac{G_w H_v}{\lambda - G_v}]\zeta = \frac{\sigma}{1+\alpha\tilde{v}} - \frac{H_v \rho}{\lambda - G_v} \tag{4.5}$$

$$\zeta'(0) = \zeta'(1) = 0$$

This equation can always be solved unless there is a nontrivial solution of the equation with $\rho = \sigma = 0$. Therefore the spectrum outside the closure of the range of G_v is discrete. To locate this part of the spectrum we set $\rho = \sigma = 0$ in (4.5), multiply by the complex conjugate $\overline{\zeta}$, and integrate by parts to find an equation on whose real and imaginary parts are

$$\int\{-|\zeta'|^2 + [H_w - \frac{\mathrm{Re}(\lambda)}{1+\alpha\tilde{v}} + \frac{(\mathrm{Re}(\lambda) - G_v)G_w H_v}{|\lambda - G_v|^2}]\}|\zeta|^2 dx = 0 \tag{4.6}$$

and

$$\mathrm{Im}(\lambda)\int_0^1 [\frac{1}{1+\alpha\tilde{v}} + \frac{G_w H_v}{|\lambda - G_v|^2}]|\zeta|^2 dx = 0 \quad . \tag{4.7}$$

The second equation shows that any complex spectrum is confined to the union as x goes from 0 to 1 of the discs

$$|\lambda - G_v|^2 \leqslant -(1+\alpha v)G_w H_v \ ,$$

which is a bounded set. Therefore it is sufficient to show that there are no eigenvalues with $\mathrm{Re}\,\lambda \geqslant 0$.

If $\operatorname{Re}\lambda \geq 0$, we see from (4.2) that

$$-\frac{\operatorname{Re}\lambda}{1+\alpha\tilde{v}} + \frac{(\operatorname{Re}\lambda - G_v)G_wH_v}{|\lambda - G_v|^2} \leq \left[\frac{G_wH_v}{-G_v}\right]_+ .$$

Thus (4.6) yields the inequality

$$\int\{-|\zeta'|^2 + (H_w + \left[\frac{G_wH_v}{-G_v}\right]_+)\ |\zeta|^2\}dx \geq 0 , \tag{4.8}$$

We now define

$$q = \zeta/r ,$$

integrate (4.8) by parts, and substitute for ζ to find that

$$-|q(1)|^2 r(1)r'(1) - \int_0^1 |q|^2 r^2\ dx \geq 0 ,$$

Since the eigenfunction ζ cannot satisfy $\zeta(1) = \zeta'(1) = 0$, since $|q(1)|^2 > 0$, and since $r(1)r'(1) > 0$, this leads to a contradiction.

We conclude that if the solution of (4.3) satisfies the conditions (4.4), and if (4.2) is satisfied, then the solution (v,w) is stable, which proves our Theorem.

The conditions (4.4) are obviously satisfied when the coefficient of r in (4.3) is nonpositive. Because $H_w(0,w) < 0$ for $w > R_1/a$, $G_w(0,w) = 0$, $G_v(v_2(w),w) < 0$, and $G_w(v_2(w),w) < 0$, this is the case if $H_v > 0$ and $H_w < 0$ on the part of the range of $(\tilde{v},\tilde{w})$ where $\tilde{v} = v_2(\tilde{w})$. Computation shows that $H_v > 0$, $H_w < 0$ at $(v_2(w_m),w_m)$, so that one can construct a family of stable solutions by keeping w near a constant just below w_m.

For the limiting solutions computed by Aronson and Brown the sufficient conditions (4.4) are found to be valid in most though not all cases.

REMARK. Replacing the factor $(1 + \alpha v)$ by 1 in the second equation of (2.1), yields a semilinear system with the same steady states. The above analysis shows that this system also has a continuum of stable steady states, so that quasilinearity is not needed to produce this phenomenon.

5. A biological consequence.

The system (1.1) is a model for a pair of competitors, one of which avoids the other to such an extent that the homogeneous steady state solution

$$\bar{u} = \frac{R_1 a - R_2 b}{a^2 - b^2} \quad , \quad \bar{v} = \frac{R_2 a - R_1 b}{a^2 - b^2}$$

is rendered unstable and inhomogeneous stable steady states (u,v) occur instead.

It is reasonable to ask whether this mechanism is advantageous to the two species.

We integrate the steady state form of (1.1) to find that

$$\int_0^1 \tilde{u}(R_1 - a\tilde{u} - b\tilde{v})dx = 0 \; . \tag{5.1}$$

The second equation of (1.1) becomes $R_2 - b\tilde{u} - a\tilde{v} = 0$ when $\tilde{v} = v_2(\tilde{w})$, while $R_2 - b\tilde{u} < 0$ on the branch $\tilde{v} = 0$. Thus on both branches

$$\tilde{v} \geq \frac{R_2 - b\tilde{u}}{a} \quad , \tag{5.2}$$

so that

$$\int_0^1 \tilde{u}(R_1 - a\tilde{u} - \frac{b}{a}(R_2 - b\tilde{u}))dx \geq 0 \; .$$

Since $R_1 = a\bar{u} + b\bar{v}$ and $R_2 = b\bar{u} + a\bar{v}$, we obtain the inequality

$$(a - \frac{b^2}{a}) \int_0^1 \tilde{u}(\bar{u} - \tilde{u})dx \geq 0$$

so that

$$\bar{u} \int (\bar{u} - \tilde{u})dx \geq \int (\bar{u} - \tilde{u})^2 dx .$$

Thus if $\tilde{u}(x)$ is not constant, we find that

$$\int_0^1 \tilde{u}\, dx < \bar{u} . \tag{5.3}$$

Thus the avoidance mechanism reduces the total population of the organism that possesses it.

On the other hand, (5.2) can be written in the form

$$a(\tilde{v} - \bar{v}) \geq b(\bar{u} - \tilde{u}) .$$

Thus, (5.3) implies that

$$\int_0^1 \tilde{v}\, dx > \bar{v} ,$$

so that the second species profits from the nervousness of the first one. The second species might thus evolve a mechanism to frighten its competitors away.

BIBLIOGRAPHY

1. Henry, D., Geometric Theory of Semilinear Parabolic Equations, (Lecture Notes in Mathematics #840, Springer, New York, 1982).

2. Mimura, M., Stationary pattern of some density-dependent diffusion system with competitive dynamics, Hiroshima Math. J. 11 (1981), 621-635.

3. Shigesada, N., Kawasaki, K., and Teramoto, E., Spatial segregation of interacting species, J. Theor. Biol. 79 (1979), 83-99.

4. Weinberger, H.F., Invariant sets for weakly coupled parabolic and elliptic systems. Rendiconti di Mat. 8(VI)(1975), 295-310.

This work was partly supported by the National Science Foundation through Grant MCS 8102609 and by a Japan Society for the Promotion of Science Fellowship for Research.

Lecture Notes in Num. Appl. Anal., **5**, 361–375 (1982)
Nonlinear PDE in Applied Science. U.S.-Japan Seminar, Tokyo, 1982

Chaos arising from the discretization of O.D.E. and an age dependent population model.

Masaya YAMAGUTI
and
Masayoshi HATA

Department of Mathematics
Faculty of Science
Kyoto University
Kyoto 606
JAPAN

In the first part of our paper, we review our recent studies of some usual finite difference schemes for the autonomous system, which can produce chaotic dynamical system. In the second part, we present an age dependent discrete population model whose solution exhibits some significant chaotic behaviors.

1. Finite difference schemes.

1.1 Definition of Chaos.

First we begin with a recall for the notion of Chaos. Let us consider a most simple dynamical system which is described as follows:

$$(1.1)\quad y_{n+1} = \begin{cases} 2y_n & (0 \le y_n \le \frac{1}{2}) \\ 2(1 - y_n) & (\frac{1}{2} < y_n \le 1) \end{cases} \equiv F(y_n)$$

whose graph is shown in Figure 1.

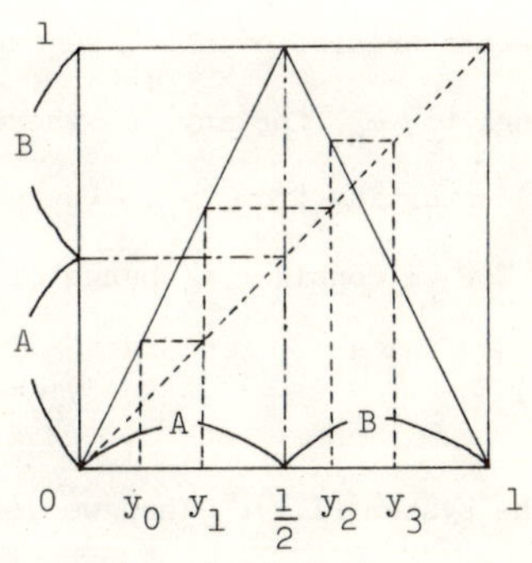

Figure 1. Graph of F.

We denote A the domain $[0,\frac{1}{2}]$ and B the domain $[\frac{1}{2},1]$. Then we remark the following property of the mapping F.

(1.2) $$F(A) \supset A \cup B, \quad F(B) \supset A \cup B.$$

This property means that for any point y which belongs to $[0,1]$, there exist always y' and y'' such that $F(y') = y$, $y' \in A$ also $F(y'') = y$, $y'' \in B$.

Now we consider an orbit $\{y_n\}$ which starts from y_0. We list up this sequence of values as follows:

(1.3) $$y_0, y_1, y_2, \ldots, y_n, \ldots$$

And we also list up the sequence of the symbols of domains to which y_n belongs.

$$A, A, B, B, B, \ldots$$

We write this sequence as follows:

(1.4) $$\omega_0, \omega_1, \omega_2, \ldots, \omega_n, \ldots$$

where ω_j = A or B corresponds to y_j.

Conversely, using the property (1.2), we can prove that for any arbitrary given sequence $\{\omega_k\}_{k=0}^{\infty}$ we get

(1.5) $$\bigcap_{k=0}^{\infty} F^{-k}(\omega_k) \neq \phi.$$

This means that if the sequence of symbols $\{\omega_k\}$ is given by some record of coin-toss trials, even so, we can decide an initial point y_0 such that y_n belongs to ω_n for any n where $\{y_n\}$ is an orbit of the dynamical system (1.1) starting from y_0. (Here we use ω_n as the name of domain.)

Now we consider a change of variables:

(1.6) $$y_n = \frac{2}{\pi} \arcsin \sqrt{x_n}$$

to the system (1.1). Then we get

(1.7) $$x_{n+1} = 4x_n(1 - x_n).$$

This mapping maps [0,1] to [0,1]. This is a special case $a = 4$ of the more general dynamical system:

(1.8) $$x_{n+1} = ax_n(1 - x_n) \equiv F_a(x_n)$$

which also maps [0,1] into [0,1] for $0 < a \leq 4$. We show the graph of this dynamical system in Figure 2.

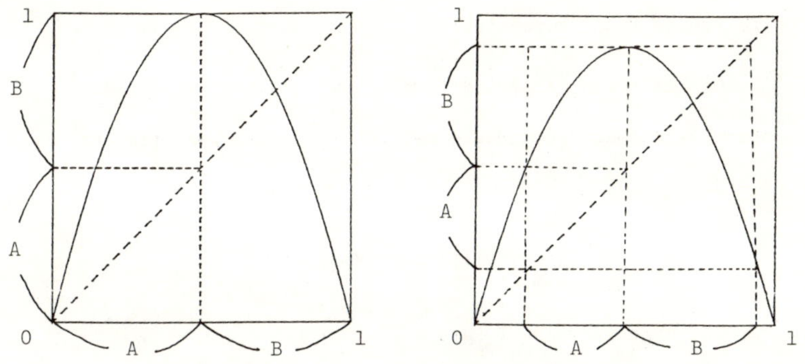

Figure 2. (1) Graph of F_4. (2) Graph of F_a where a is sufficiently near 4.

As is easily seen, the fact (1.5) holds for the system (1.7) by exactly similar reason as in the case of (1.1). But how about for (1.8)? If a is sufficiently near 4, then some weak property (1.9) follows.

(1.9) $$F_a(A) \supset B, \quad F_a(B) \supset A \cup B.$$

It produces any symbolic sequence $\{\omega_k\}$ where $\omega_j = A$, $\omega_{j+1} = A$ never arise.

Remark. This simple dynamical system (1.8) was considered by R. May who studied a discrete population model of some insects population which has non-overlapping generation. We are going to explain this fact introducing an age dependent population model in the second part of this paper.

Now we explain the notion of Chaos mathematically, that is, the definition of Chaos in the sense of Li-Yorke and Marotto. We consider a dynamical system which described by

(1.10) $$X_{n+1} = F(X_n)\ ,\quad X_n \in R^m,$$

where F is a continuous mapping from R^m to R^m.

Definition. We say F is chaotic in the sense of Li-Yorke and Marroto if F has the following four properties:

(1) (1.10) has infinite periodic orbits with distinct periods.

(2) there exists an uncountable set $S \subset R^m$ such that for any $X, Y \in S$, $X \neq Y$,

$$\overline{\lim_{n\to\infty}}\ |\ F^n(X) - F^n(Y)\ | > 0$$

(3) for the same X and Y as in (2),

$$\underline{\lim_{n\to\infty}}\ |\ F^n(X) - F^n(Y)\ | = 0$$

(4) for any $X \in S$, X is not even asymptotically periodic.

Now we can state very briefly the result of Li-Yorke[1] which is that our condition (1.9) for R^1 implies "Chaos" in the above sense. Also Marotto[2] has shown this "Chaos" in R^m (for any m) under the assumption that (1.10) has a snap-back repeller. Here we recall the definition of the snap-back repeller.

Definition. Assume that F is continuously differentiable. Then we call a fixed point Z of (1.10) a snap-back repeller if Z is expanding in some neighbourhood U of Z and there exists a point $X_0 \in U$ with $X_0 \neq Z$, $F^M(X_0) = Z$ and $|DF^M(X_0)| \neq 0$ for some positive integer M where $|DF^M(X_0)|$ is a Jacobian determinant of F^M at X_0.

1.2 Chaos arising from the discretization of ordinary differential equations.

Here we mention a review of the results of our group about the "Chaos" which are obtained by some simple discretization of ordinary differential equations.

Our first result was that of Yamaguti-Matano[3] which is stated as follows.

Theorem 1. For a given differential equation:

$$\frac{dy}{dt} = f(y) \tag{1.11}$$

where $f(y)$ is continuous function of y in R^1. If $f(y)$ has at least two zeros, one of which is globally asymptotically stable, then the Euler's difference scheme:

$$y_{n+1} = y_n + \Delta t\ f(y_n) \equiv F_{\Delta t}(y_n) \tag{1.12}$$

is chaotic if we take Δt sufficiently large. More precisely, there exist two positive values τ_1, τ_2 such that for Δt which satisfies $\tau_1 \le \Delta t \le \tau_2$, the mapping (1.12) maps a finite interval into itself and this dynamical system is chaotic in this interval.

After this result, we study several generalizations of this result. Here, we limit ourselves to list up a series of our recent results. Let us consider the system of differential equations:

$$\frac{dU}{dt} = G(U), \quad U(0) = U_0 \tag{1.13}$$

where U is unknown m-vector and G is a continuously differentiable mapping from R^m into R^m. Then Hata succeeded to prove that the Euler's difference scheme for (1.13):

$$U_{n+1} = U_n + \Delta t\ G(U_n) \equiv G_{\Delta t}(U_n) \tag{1.14}$$

is chaotic for sufficiently large Δt under the following conditions:

(1.15) there exist $\overline{U} \neq \overline{V}$ such that $G(\overline{U}) = G(\overline{V}) = 0$, $|DG(\overline{U})| \neq 0$ and $|DG(\overline{V})| \neq 0$.

For the proof of this theorem, see [4].

Nextly, S. Ushiki and Yamaguti[5] studied a central discretization of the following differential equation:

$$\frac{dx}{dt} = x(1 - x). \tag{1.16}$$

The central difference scheme of this equation is

$$\frac{x_{n+1} - x_{n-1}}{2\Delta t} = x_n(1 - x_n). \tag{1.17}$$

Putting $x_{n-1} = y_n$, we get a mapping from R^2 into R^2 as follows.

$$\begin{cases} x_{n+1} = y_n + 2\Delta t\, x_n(1 - x_n) \\ y_{n+1} = x_n \end{cases} \tag{1.18}$$

S. Ushiki[6] proved that this dynamical system shows some chaotic behavior for any mesh size Δt.

Similar result as Yamaguti-Matano's has been proved by Y. Oshime for the modified Euler scheme of (1.11). Also the above result of Hata has been generalized by himself for Runge-Kutta scheme of (1.13).

Before finishing our review of the results, we sketch the proof of the above theorem by Hata.

Lemma 1. For any $\delta > 1$, there exist $r > 0$ and $c(\delta) > 0$ such that

$$\| G_{\Delta t}(X) - G_{\Delta t}(Y) \| \geq \delta \| X - Y \|$$

for any $\Delta t > c(\delta)$ and $X, Y \in B(\overline{U}, r)$ where $B(\overline{U}, r)$ is a ball whose center is $\overline{U}$ and its radius r.

Proof. Because of our conditions (1.15), we get

$$|DF(\overline{U})^*DF(\overline{U})| \neq 0.$$

Therefore we can show easily

$$\| DF(\overline{U})X\| \geq \sqrt{\lambda_{min}} \| X \| \quad (\text{ for all } X \in R^m)$$

Here λ_{min} means the minimum eigenvalue of $DF(\overline{U})^*DF(\overline{U})$. By the continuity,

$$\| DF(X) - DF(\overline{U})\| < \frac{1}{2} \sqrt{\lambda_{min}} \quad (\text{ for all } X \in B(\overline{U},r))$$

Then we get the following series of inequalities

$$\begin{aligned} \| G_{\Delta t}(X) - G_{\Delta t}(Y)\| &\geq \Delta t \| F(X) - F(Y)\| - \| X - Y\| \\ &\geq (\frac{\Delta t}{2} \sqrt{\lambda_{min}} - 1) \| X - Y\| \\ &\geq \delta \| X - Y\| \end{aligned}$$

where

$$\Delta t \geq c(\delta) \equiv \frac{2}{\sqrt{\lambda_{min}}} (1 + \delta). \qquad \text{c.q.f.d.}$$

Lemma 2. For sufficiently small open neighbourhood W of $\overline{U}$ and any bounded set B, there exists a positive constant $c(W,B)$ such that the equation

$$G_{\Delta t}(w) = b$$

has at least one solution $w \in W$ for any $\Delta t > c(W,B)$ and for any $b \in B$.

Using these lemmas we can constract a snap-back repeller. Thus we can prove the conclusion of the theorem.

2. An age dependent population model.

We consider here an age dependent population model which is described by the following equation:

$$(2.1)\qquad U^{n+1} = F(U^n) = \begin{pmatrix} (\sum_{k=1}^{N} b(k)u_k^n)(R - \sum_{k=1}^{N} b(k)u_k^n) \\ (1 - d(1))u_1^n \\ \vdots \\ (1 - d(N-1))u_{N-1}^n \end{pmatrix}$$

where we denote

$$U^n = {}^t(u_1^n, u_2^n, \ldots, u_N^n) \in R^N,$$

u_k^n is the population of k-age animals at n-th year, $b(k)$ is the birthrate of k-age population, $d(k)$ is the deathrate of k-age population, R is a positive constant which means a saturation, and $b(i)$, $d(i)$ satisfy the following condition;

$$(2.2)\qquad \begin{cases} 0 \le b(i) < 1 & \text{for } 1 \le i \le N,\ b(N) \neq 0, \\ 0 \le d(i) < 1 & \text{for } 1 \le i \le N-1. \end{cases}$$

Now it is convenient to introduce new variables by the following formulae;

$$(2.3)\qquad \begin{cases} v_j^n = (\frac{4}{R^2} \prod_{k=1}^{j-1} (1 - d(k))^{-1})u_j^n \\ a(j) = b(j) \prod_{k=1}^{j-1} (1 - d(k)) \end{cases} \qquad \text{for } 1 \le j \le N.$$

Then we have new equation;

$$(2.4)\qquad V^{n+1} = G(V^n) = \begin{pmatrix} \frac{R^2}{4}(\sum_{j=1}^{N} a(j)v_j^n)(\frac{4}{R} - \sum_{j=1}^{N} a(j)v_j^n) \\ v_1^n \\ \vdots \\ v_{N-1}^n \end{pmatrix}$$

where we denote $\qquad V^n = {}^t(v_1^n, v_2^n, \ldots, v_N^n).$

Also we have

$$0 \le a(j) < 1 \quad \text{for} \quad 1 \le j \le N, \; a(N) \neq 0.$$

We assume that

$$AR \le 4 \quad \text{where} \quad A = \sum_{k=1}^{N} a(k).$$

Under this assumption, it is easily verified that the N-dimensional mapping G in (2.4) has the following invarient domain;

$$Q = \underbrace{[0,1]\times[0,1]\times\ldots\times[0,1]}_{N \text{ times}} \subset R^N,$$

that is, G maps Q into itself. Then we find the fixed points of G in Q as follows.

(a) For the case in which $0 < AR \le 1$, the only one fixed point in Q is the origin $0 = (0, 0, \ldots, 0)$.

(b) For the case in which $1 < AR \le 4$, the fixed points in Q are the origin and

$$\bar{V} = \left(\frac{4}{AR}\left(1 - \frac{1}{AR}\right), \ldots, \frac{4}{AR}\left(1 - \frac{1}{AR}\right)\right). \tag{2.5}$$

The local stability of these fixed point of G is easily studied by standard linearization techniques. From (2.4), we obtain

$$DG(V) = DG(v_1, v_2, \ldots, v_N) = \begin{pmatrix} \mu a(1) & \mu a(2) & \cdots & \mu a(N) \\ 1 & & & \\ & 1 & & O \\ & & \ddots & \\ O & & & 1 \end{pmatrix} \tag{2.6}$$

where

$$\mu = \mu(v_1, v_2, \ldots, v_N) = R - \frac{R^2}{2} \sum_{k=1}^{N} a(k) v_k.$$

Especially, for the fixed points of G, we have

$$\mu(0) = R \quad \text{and} \quad \mu(\overline{V}) = \frac{2}{A} - R.$$

It follows easily from (2.6) that the characteristic equation of $DG(V)$ is

$$\lambda^N - \mu(V) \sum_{j=1}^{N} a(j)\lambda^{N-j} = 0. \tag{2.7}$$

Putting

$$f(\lambda) = \lambda^N \quad \text{and} \quad g(\lambda) = \mu(V) \sum_{j=1}^{N} a(j)\lambda^{N-j}, \tag{2.8}$$

we have the following estimates on the unit circle $|\lambda| = 1$;

$$|f(\lambda)| = 1, \quad |g(\lambda)| \le |\mu(V)|A.$$

Therefore, if $|\mu(V)|A < 1$, we have

$$|f(\lambda)| > |g(\lambda)|$$

on the unit circle. By the theorem of Rouché, the all roots of the polynomial $f(\lambda) - g(\lambda)$ lie inside the unit circle. So we have the followings.

(c) For the case in which $0 < AR < 1$, the origin 0 is locally stable.

(d) For the case in which $1 < AR < 3$, the non-trivial fixed point $\overline{V}$ is locally stable.

Similarly, we can study when all roots of the polynomial $f(\lambda) - g(\lambda)$ lie outside the unit circle. Put

$$\xi = \frac{1}{\lambda}.$$

Then from (2.7) we have the following new polynomial of ξ;

$$a(N)\xi^N + \sum_{k=1}^{N-1} a(k)\xi^k - \frac{1}{\mu(V)} = 0. \tag{2.9}$$

Putting

$$f_1(\xi) = a(N)\xi^N \quad \text{and} \quad g_1(\xi) = \sum_{k=1}^{N-1} a(k)\xi^k - \frac{1}{\mu(V)}, \tag{2.10}$$

we have the following estimates on the unit circle $|\xi| = 1$;

$$|f_1(\xi)| = a(N), \quad |g_1(\xi)| \le \sum_{k=1}^{N-1} a(k) + \frac{1}{|\mu(V)|}.$$

By the theorem of Rouché, we have the followings.

(e) If $a(N) > \frac{5}{3} \sum_{j=1}^{N-1} a(j)$ and $\frac{A}{2a(N) - A} < AR \le 4$, then all eigenvalues of $DG(0)$ exceed 1 in magnitude.

(f) If $a(N) > 3 \sum_{j=1}^{N-1} a(j)$ and $\frac{4a(N) - A}{2a(N) - A} < AR \le 4$, then all eigenvalues of $DG(\overline{V})$ exceed 1 in magnitude.

The above conditions (e) and (f) about the distribution of $a(j)$ include some insects population which has non-overlapping generation as a special case. Actually, we can prove that G is chaotic for sufficiently small $a(1), a(2), \ldots a(N-1)$ and some $a(N)$ which satisfy our assumptions by showing that the non-trivial fixed point $\overline{V}$ of G is a snap-back repeller.

First, we choose a special parameters as follows;

$$a(1) = a(2) = \ldots = a(N-1) = 0, \quad \alpha \equiv a(N)R \in (0,4].$$

And G_α denotes the corresponding mapping defined in (2.4) with a parameter α, that is,

$$G_\alpha(v_1, v_2, \ldots, v_N) = \begin{pmatrix} \alpha v_N - \frac{\alpha^2}{4} v_N^2 \\ v_1 \\ \cdot \\ \cdot \\ \cdot \\ v_{N-1} \end{pmatrix}. \tag{2.11}$$

Putting $h_\alpha(x) = \alpha x - \frac{\alpha^2}{4} x^2$, we get

$$H_\alpha(v_1,v_2,\dots,v_N) \equiv G_\alpha^N(v_1,v_2,\dots,v_N) = \begin{pmatrix} h_\alpha(v_1) \\ h_\alpha(v_2) \\ \vdots \\ h_\alpha(v_N) \end{pmatrix} \tag{2.12}$$

where G_α^N denotes the N-fold iteration by G_α. Also we obtain from (2.12),

$$DH_\alpha(v_1,v_2,\dots,v_N) = \begin{pmatrix} h_\alpha'(v_1) & & & O \\ & h_\alpha'(v_2) & & \\ & & \ddots & \\ O & & & h_\alpha'(v_N) \end{pmatrix}. \tag{2.13}$$

Therefore

$$|DH_\alpha(v_1,v_2,\dots,v_N)| = h_\alpha'(v_1)h_\alpha'(v_2)\dots h_\alpha'(v_N). \tag{2.14}$$

The condition of a snap-back repeller clealy includes the existence of a homoclinic orbit $\{z_{-k}\}_{k\geq 0}$ such that

z_0 is a fixed point of F, $z_{-1} \neq z_0$,

$F(z_{-k}) = z_{-k+1}$ for $k \geq 1$, and $F(z_{-k}) \to z_0$ as $k \to \infty$.

Even for one-dimensional continuous mapping, it is sometimes simpler to find a homoclinic orbit than to verify (1.9) or some odd period conditions. In this case, for one-dimensional continuous mapping h_α, it is easily shown that there exists a positive constant $\alpha_0 < 4$ such that, for any fixed $\alpha \in [\alpha_0, 4]$, h_α has a transversal homoclinic orbit $\{p_{-n}\}_{n\geq 0}$ such that

$$h_\alpha'(p_{-n}) \neq 0 \quad \text{for any} \quad n \geq 0.$$

The dotted lines in Figure 3 represent a homoclinic orbit of h_α which is found by starting in a fixed point and iterating backward.

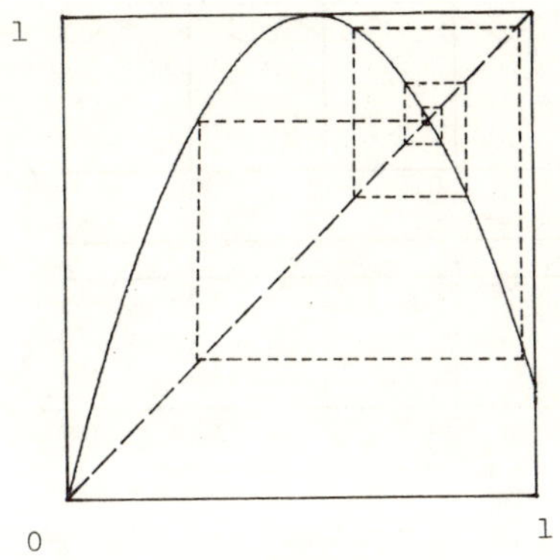

Figure 3.

Homoclinic orbit of h_α.

Using a transversal homoclinic orbit $\{p_{-n}\}_{n\geq 0}$ of h_α, we construct a transversal homoclinic orbit $\{P_{-n}\}_{n\geq 0}$ of H_α as follows;

$$P_{-n} = {}^t(\, p_{-n}, p_{-n}, \ldots, p_{-n}\,) \in R^N \quad \text{for} \quad n \geq 0,$$

since from (2.14),

$$|DH_\alpha(P_{-n})| = (h_\alpha'(p_{-n}))^N \neq 0 \quad \text{for} \quad n \geq 0.$$

Since the existence of a snap-back repeller is a stable property under small C^1-perturbations and the orbit $\{P_{-n}\}_{n\geq 0}$ of H_α is contained in the interior of Q, we have the following.

Theorem. There exists a constant $0 < \varepsilon < 1$ such that for any

$$(a(1), \ldots, a(N-1), a(N)R) \in [0,\varepsilon]^{N-1} \times [\alpha_0, 4],$$

which satisfies $AR \leq 4$, the corresponding mapping G in (2.4) is chaotic in Q in the sense of Li-Yorke.

Finally, by computer simulations (see Figure 4), we conjecture that G is sometimes chaotic even for the case of over-lapping generations.

(4 - a)

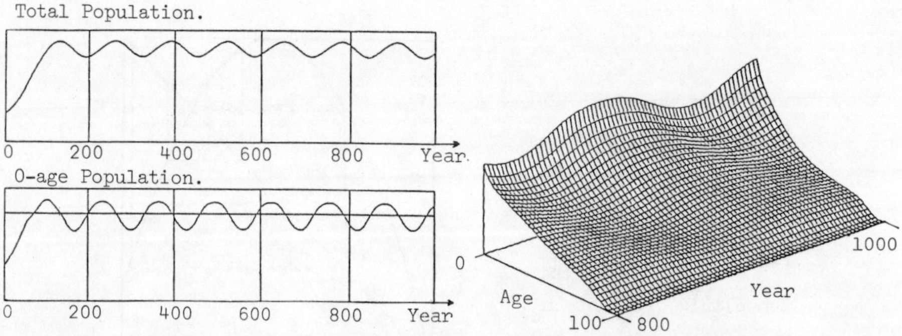

(4 - b)

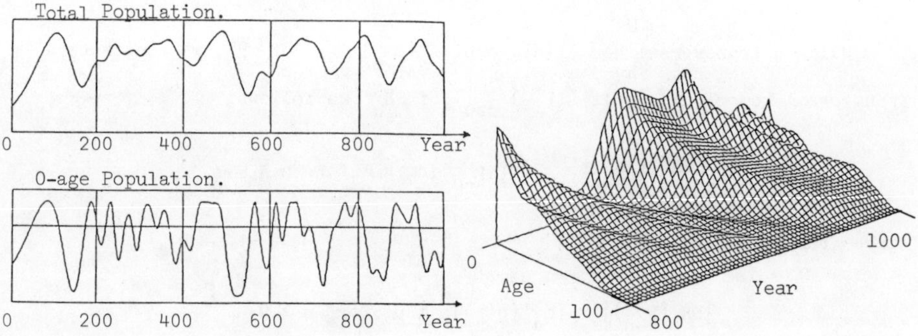

Figure 4. Numerical computations of our model (2.1) where we put $N = 100$ and birthrate and deathrate are shown in (4 - c).

(4 - c)

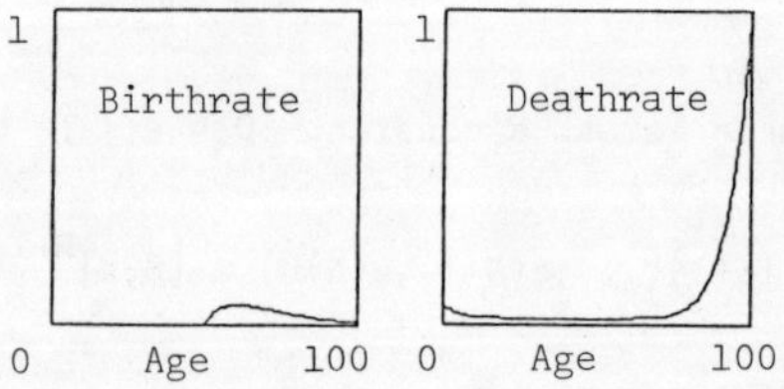

First, we choose a saturation value $R = 9.0$ in (4 - a) and presumably the periodic structure is caused by a Hopf bifurcation from a non-trivial fixed point. Secondly, we choose a saturation value $R = 11.3$ in (4 - b) and the chaotic phenomenon will be appear.

References

[1] T.-Y. Li and J. A. Yorke, Period three implies chaos, Amer. Math. Monthly 82 (1975), 985-992.

[2] F. R. Marotto, Snap-back repellers imply chaos in R^n, J. Math. Anal. Appl. 63 (1978), 199-223.

[3] M. Yamaguti and H. Matano, Euler's finite difference scheme and chaos, Proc. Japan Acad. 55A (1979), 78-80.

[4] M. Hata, Euler's finite difference scheme and chaos in R^n, Proc. Japan Acad. 58A (1982), 178-181.

[5] M. Yamaguti and S. Ushiki, Chaos in numerical analysis of ordinary differential equations, Physica 3D, 3 (1981), 618-626.

[6] S. Ushiki, Central difference scheme and chaos, Physica 4D (1982), 407-424.

Lecture Notes in Num. Appl. Anal., **5**, 377–397 (1982)
Nonlinear PDE in Applied Science. U.S.-Japan Seminar, Tokyo, 1982

STABILITY, REGULARITY AND NUMERICAL ANALYSIS OF THE NONSTATIONARY NAVIER-STOKES PROBLEM

John G. Heywood

University of British Columbia

Vancouver, B.C., Canada

In this paper I will describe some results relating the stability and regularity of solutions of the Navier-Stokes equations with the long term error and stability of numerical approximations. These results were obtained jointly with Rolf Rannacher and are presented in full detail, along with related results, in Part II of our work on finite element approximation of the nonstationary Navier-Stokes problem [1]. They are of two general types, both utilizing stability assumptions to extend results which were known locally in time to ones which are global in time. First, that if the solution of the initial boundary value problem is stable, then the error in its discrete approximation remains small uniformly in time, as $t \to \infty$. Second, that from the stability of a discrete solution, for a single sufficiently small choice of the mesh size, one can infer the global existence of a closely neighboring smooth solution. The concepts of stability which are dealt with are formulated to describe the stability of such phenomena as Taylor cells and von-Kármán vortex shedding, and also the partial stability observed in some flows exhibiting slight or incipient turbulence. I will include an account of some of the stability theory developed in [1], particularly as exemplified by a new proof of the principle of linearized stability appropriate to nonstationary solutions.

1. The Continuous Problem

We consider the nonstationary Navier-Stokes problem

$$
\begin{aligned}
&u_t - \Delta u + u\cdot\nabla u + \nabla p = f\ , \\
(1)\qquad &\nabla\cdot u = 0\ ,\ \text{for}\ \ (x,t)\in\Omega\times(0,\infty)\ , \\
&u_{|t=0} = a\ ,\ u_{|\partial\Omega} = 0\ ,
\end{aligned}
$$

in a bounded two or three-dimensional domain Ω. Here u represents the velocity of a viscous imcompressible fluid, p the pressure, f the prescribed external force, and a the prescribed initial velocity. The boundary values are zero. The fluid's density and viscosity have been normalized, as is always possible, by changing the scales of space and time.

As usual, $L^p(\Omega)$, or simply L^p, denotes the space of functions defined and p^{th}-power summable in Ω, and $\|\cdot\|_{L^p}$ its norm. We denote the inner product in L^2 by $(\cdot,\cdot)$ and let $\|\cdot\| = \|\cdot\|_{L^2}$. C^∞ is the space of functions continuously differentiable any number of times in Ω, and C_o^∞ consists of those members of C^∞ with compact support in Ω. The Sobolev space H^m is obtained by the completion in the norm

$$\|u\|_m = \{\sum_{0\le|\alpha|\le m} \|D^\alpha u\|^2\}^{1/2}\ ,$$

expressed in multi-index notation, of those members of C^∞ for which the norm is finite. H_o^1 is the closure of C_o^∞ in H^1. Spaces of R^n-valued functions will be denoted with boldface type. We use

$$(\nabla u,\nabla v) = \sum_{1\le i,j\le n} (\partial_j u_i, \partial_j v_i)\ ,\quad \|\nabla u\| = (\nabla u,\nabla u)^{1/2}\ ,$$

as inner product and norm for $\mathbf{H}_o^1$. Finally, we need the spaces

$$\mathbf{J} = \{\phi\in\mathbf{L}^2 : \nabla\cdot\phi = 0\ \text{ in }\ \Omega\ \text{ and }\ \phi\cdot n_{|\partial\Omega} = 0\ ,\ \text{weakly}\}\ ,$$

$$\mathbf{J}_1 = \{\phi\in\mathbf{H}_o^1 : \nabla\cdot\phi = 0\}$$

of solenoidal functions.

Denoting the orthogonal projection of $\mathbf{L}^2$ onto $\mathbf{J}$ by P, we introduce the "Stokes operator" $\tilde{\Delta} = P\Delta$. Assuming the boundary $\partial\Omega$ is sufficiently regular, the mapping $\tilde{\Delta} : \mathbf{J}_1 \cap \mathbf{H}^2 \to \mathbf{J}$ is one-to-one and onto, and

$$\|v\|_2 \le c\|\tilde{\Delta}v\| \tag{A1}$$

holds for all $v \in \mathbf{J}_1 \cap \mathbf{H}^2$. We assume this as well as some regularity of the prescribed data, namely that

$$\begin{aligned} &a \in \mathbf{J}_1 \cap \mathbf{H}^2 , \\ &f, f_t \in L^\infty(0,\infty;\mathbf{L}^2) . \end{aligned} \tag{A2}$$

For the sake of simplicity in our presentation, we have assumed the boundary values vanish. All of our results remain valid in the case of inhomogeneous boundary conditions if one assumes an appropriate degree of smoothness of the boundary values, as well as the same conditions of spatial and temporal invariance as may be required of f.

Finally, we assume that the strong solution u, p of problem (1) exists globally and satisfies

$$\sup_{[0,\infty)} \|\nabla u\| < \infty . \tag{A3}$$

Once this much regularity is known or assumed of a solution, its full regularity can be proved so far as is permitted by the data. In particular, the following is proven in Theorem 2.3 of [2].

Proposition 1. Given Ω *satisfying* (A1), *there exists a continuous increasing function* F *of three variables, such that every solution* u *of* (1) *satisfies*

$$\sup_{t\le t^*} \|\tilde{\Delta}u\| \le F(\|\tilde{\Delta}a\| , \sup_{t\le t^*} \|f\| + \|f_t\| , \sup_{t\le t^*} \|\nabla u\|) , \tag{2}$$

provided $\sup_{t\le t^*} \|\nabla u\| < \infty$. *The function* F *is independent of* t^*.

We mention that to bound higher order derivatives of u, uniformly as $t \to 0$, requires nonlocal compatibility conditions of the prescribed data, conditions usually unverifiable in practice. For instance, $\|\nabla u_t\|$ and $\|u\|_3$ tend to infinity as $t \to 0$ unless there exists a solution p_o of the overdetermined Neumann problem

$$\Delta p_o = \nabla\cdot(f(\cdot,o) - a\cdot\nabla a) \quad \text{in } \Omega ,$$

$$\nabla p_o = \Delta a + f(\cdot,0) - a\cdot\nabla a \quad \text{on } \partial\Omega .$$

The loss of regularity as $t \to 0$ complicates the proof of error estimates of higher than second order for numerical approximations. While such estimates are proven in Part III of this work, independently of any nonlocal compatibility conditions, the present discussion will be restricted to second order error estimates.

2. The Discrete Problem

We suppose that $\mathbf{H}_h$ and L_h are finite dimensional subspaces of $\mathbf{L}^2$ and L^2, respectively, corresponding to a sequence of values, tending to zero, of a discretization parameter h, $0 < h \le 1$. The space $\mathbf{H}_h$ is considered as a discrete analogue of $\mathbf{H}_o^1$. In order to include the consideration of nonconforming finite elements, it is not required that $\mathbf{H}_h \subset \mathbf{H}_o^1$, but merely that the gradient operator ∇ has an extension ∇_h to the algebraic sum $\mathbf{H}_o^1 + \mathbf{H}_h$, such that $\nabla_h \equiv \nabla$ on $\mathbf{H}_o^1$, and such that $\|\nabla_h\cdot\|$ is a norm on $\mathbf{H}_h$. Another frequently used norm is

$$|||\cdot|||_h \equiv \|\cdot\| + h\|\nabla_h\cdot\| .$$

A discrete analogue of the space $\mathbf{J}_1$ is introduced by setting

$$\mathbf{J}_h \equiv \{v_h \in \mathbf{H}_h : (\chi_h, \nabla_h\cdot v_h) = 0 \text{ , for all } \chi_h \in L_h\} .$$

We also set

$$N_h \equiv \{\chi_h \in L_h : (\chi_h, \nabla_h\cdot v_h) = 0 \text{ , for all } v_h \in \mathbf{H}_h\} .$$

A discrete analogue $\Delta_h : \mathbf{H}_h \to \mathbf{H}_h$ of the Laplacian operator is defined by requiring

$$(\Delta_h v_h, \phi_h) = -(\nabla_h v_h, \nabla_h \phi_h) \ , \text{ for } v_h, \phi_h \in \mathbf{H}_h \ .$$

Letting $P_h : L^2 \to \mathbf{J}_h$ denote the L^2-projection onto $\mathbf{J}_h$, we introduce a discrete analogue, $\tilde{\Delta}_h \equiv P_h \Delta_h$, of the Stokes operator $\tilde{\Delta} = P\Delta$. The restriction of $\tilde{\Delta}_h$ to $\mathbf{J}_h$ is automatically invertible, with inverse denoted by $\tilde{\Delta}_h^{-1}$. The invertibility of $\tilde{\Delta}_h$ and $\tilde{\Delta}$ permits us to introduce operators

$$R_h \equiv \tilde{\Delta}_h^{-1} P_h \tilde{\Delta} : \mathbf{J}_1 \cap \mathbf{H}^2 \to \mathbf{J}_h \ ,$$
$$R^h \equiv \tilde{\Delta}^{-1} P \tilde{\Delta}_h : \mathbf{J}_h \to \mathbf{J}_1 \cap \mathbf{H}^2 \ ,$$

associating discrete solenoidal functions with smooth ones and vice versa. We assume there are constants c, independent of h, such that for $v \in \mathbf{J}_1 \cap \mathbf{H}^2$ and $v_h \in \mathbf{J}_h$, there holds

$$\begin{aligned} &|||v - R_h v|||_h \le c\, h^2 ||\tilde{\Delta} v|| \ , \\ (3) \qquad & \\ &|||v_h - R^h v_h|||_h \le c\, h^2 ||\tilde{\Delta}_h v_h|| \ . \end{aligned}$$

These inequalities were proven in Corollary 4.3 of [2] under the detailed assumptions of [2].

We suppose we have a discrete analogue of the Navier-Stokes equations determining, for any given $a_h \in \mathbf{J}_h$ and $t_o \ge 0$, unique functions $u_h(\cdot,t) \in \mathbf{J}_h$ and $p_h(\cdot,t) \in L_h/N_h$, defined for all $t \ge t_o$, such that $u_h(\cdot,t_o) = a_h$. Our notation here is that for semidiscrete approximation, with the time variable remaining continuous and a "discrete" analogue of the Navier-Stokes equations consisting of a system of ordinary differential equations. However, in [1] we adapted our notation and argument to apply to a full discretization of the equations, at least in a simple case of backward Euler time discretization of the ordinary differential equations.

We assume a "local" error estimate is already known, "local" meaning with error constants that grow (exponentially) with time:

Proposition 2. If u,p *and* u_h,p_h *are continuous and discrete solutions defined on some time interval* $[t_o,t_*)$ *, then*

$$(4)\qquad \begin{aligned} |||(u-u_h)(t)|||_h &\le h^2 K e^{K(t-t_o)} , \\ \|(p-p_h)(t)\|_{L^2/N_h} &\le hK(t-t_o)^{-1/2} e^{K(t-t_o)} \end{aligned}$$

hold for $t \in [t_o,t_*)$, *with constants* K *dependent on* $\|\tilde{\Delta} u(t_o)\|$,

$\sup_{[t_o,t_*)} \|f\| + \|f_t\|$, $\sup_{[t_o,t_*)} \|\nabla u\|$, *and* $h^{-2}|||u(t_o) - u_h(t_o)|||_h + \|\tilde{\Delta}_h u_h(t_o)\|$.

We proved Proposition 2 for a class of semidiscrete finite element approximations in [2]. The exponential growth of the error constants is unavoidable if the solution u,p under consideration is unstable.

Our proof of "global" error estimates, exemplified by Theorem 3 below, requires an estimate for the regularity of discrete solutions analogous to Proposition 1. The following was proven for a class of semidiscrete finite element approximations in Lemma 5.5 of [2].

Proposition 3. There exists a continuous increasing function $\tilde{F}$ *of three variables, such that every solution* u_h *of the discrete Navier-Stokes equations, defined on any time interval* $[t_o,t_*]$, *satisfies*

$$(5)\qquad \sup_{[t_o,t_*]} \|\tilde{\Delta}_h u_h\| \le \tilde{F}(\|\tilde{\Delta}_h u_h(t_o)\| , \sup_{[t_o,t_*]} \|f\| + \|f_t\| , \sup_{[t_o,t_*]} \|\nabla_h u_h\|) .$$

The function F *is independent of* h , *as well as of* t_o *and* t_* .

3. Exponential Stability and Global Error Estimates

Questions about the stability of u concern the behaviour of "perturbed solutions", by which we mean any solution v of the Navier-Stokes problem

$$(6)\qquad \begin{gathered} v_t - \Delta v + v\cdot\nabla v + \nabla q = f , \\ \nabla\cdot v = 0 \ \text{ for } \ (x,t) \in \Omega \times (t_o,\infty) , \\ v|_{t_o} = v_o , \ v|_{\partial\Omega} = 0 , \end{gathered}$$

starting at an initial time $t_o \geq 0$, with an initial value v_o near $u(t_o)$. We refer to $w = v - u$ as a "perturbation" of u, and to t_o and $w_o = v_o - u(t_o)$ as the "initial time" and "initial value" of the perturbation w. To avoid any doubt about the global existence of v, and hence of w, it is necessary to define it first as a weak solution in the sense of Hopf. We will not belabour this point, as a proof of the regularity of any small perturbation of a stable strong solution is implicit in Theorem 1 below.

The ordinary, simplest, notion of stability is the following.

Definition 1. The solution u *of problem (1) is said to be stable if, for every* $\varepsilon > 0$, *there exists a number* $\delta > 0$ *such that every perturbation* w, *with* $w_o \in \mathbf{J}$ *and* $\|w_o\| < \delta$, *satisfies* $\sup_{[t_o,\infty)} \|w\| < \varepsilon$.

Here, in speaking of "every perturbation", it should be understood that we are referring to every perturbation, starting at every initial time $t_o \geq 0$. A stronger notion of stability is required upon which to base error estimates which are uniform in time.

Definition 2. The solution u *of problem (1) is said to be exponentially stable if there exist numbers* $\delta, T > 0$ *such that every perturbation* w, *with* $w_o \in \mathbf{J}$ *and* $\|w_o\| < \delta$, *satisfies* $\|w(t_o+T)\| \leq \frac{1}{2}\|w_o\|$.

If u satisfies the conditions of either of these definitions with $\delta = \infty$, we say u is unconditionally stable.

An example of an exponentially stable flow is provided by simple axially symmetric Taylor cells occurring in flow between rotating coaxial cylinders. The situation is one in which, if the data are steady, there exist multiple steady solutions. If the difference between two such solutions is considered as a perturbation, it certainly will not decay. Thus Taylor cells are not unconditionally stable. Further, there generally exist even small perturbations whose decay in the L^2-norm is not monotonic. However, the cells are certainly stable in some sense, and intuitive considerations of linearization suggest that the decay of small perturbations is exponential.

Our development of a stability theory is based on several lemmas asserting the continuous dependence of solutions on their initial values. Below, c is a generic constant depending only on Ω , and

$$M = \sup_{t\geq 0} \|\nabla u\| .$$

<u>Lemma 1</u>. *For every perturbation* w *of* u *there holds*

$$\|w(t)\|^2 + \int_{t_o}^{t} \|\nabla w\|^2 d\tau \leq \|w(t_o)\|^2 e^{cM^4(t-t_o)} ,$$

for $t \geq t_o$.

<u>Lemma 2</u>. *For every* $T > 0$, *there exists a number* $\delta > 0$ *such that every perturbation* w *of* u , *with* $w_o \in J_1$ *and* $\|\nabla w_o\| < \delta$, *satisfies*

$$\|\nabla w(t)\|^2 + \int_{t_o}^{t} \|\tilde{\Delta} w\|^2 d\tau \leq \|\nabla w(t_o)\|^2 e^{c(1+M^4)(t-t_o)} ,$$

for $t_o \leq t \leq t_o + T$.

<u>Lemma 3</u>. *For every* $T > 0$, *there exists numbers* $\rho, B > 0$ *such that every perturbation* w *of* u , *with* $\|w(t_o)\| < \rho$, *satisfies*

$$\|\nabla w(t_o+T)\| \leq B\|w(t_o)\| .$$

To prove these lemmas, one begins by writing the perturbation equation

$$w_t - \Delta w + w\cdot\nabla w + u\cdot\nabla w + w\cdot\nabla u = -\nabla q , \tag{7}$$

for the difference $w = v - u$ of the solutions of (6) and (1). Multiplying (7) by w and integrating leads to Lemma 1. Multiplying (7) by $\tilde{\Delta} w$ and integrating leads to Lemma 2. In both cases, the constants c depend on Sobolev's inequality. Lemma 3 is obtained by combining Lemmas 1 and 2. All three lemmas need somewhat more precise statements if v is understood only as a "weak solution."

Using the preceding lemmas, we can establish the equivalence of various definitions of stability. We prove the following simple theorem to indicate the nature of argument.

Theorem 1. The stability condition of Definition 1 is equivalent to the following: For every $\varepsilon > 0$ *, there exists a number* $\delta > 0$ *such that every perturbation* w *, with* $w_o \in \mathbf{J}_1$ *and* $\|\nabla w_o\| < \delta$ *, satisfies* $\sup_{[t_o,\infty)} \|\nabla w\| < \varepsilon$.

Proof. First we check that the condition of Definition 1 implies that of Theorem 1. According to Lemma 2, one may guarantee that $\|\nabla w(t)\|$ is small, for $t_o \le t \le t_o + 1$, by taking $\|\nabla w_o\|$ small. Mindful of Poincare's inequality $\|w_o\| \le c\|\nabla w_o\|$, we see that if $\|\nabla w_o\|$ is taken small, then the condition of Definition 1 ensures that $\|w(t)\|$ is small for all $t \ge t_o$, and hence Lemma 3 ensures that $\|\nabla w(t)\|$ is small for all $t \ge t_o + 1$. Thus the condition of Theorem 1 is satisfied.

Next we check that the condition of Theorem 1 implies that of Definition 1. According to Lemma 1, one may guarantee that $\|w(t)\|$ is small, for $t_o \le t \le t_o + 1$, by taking $\|w_o\|$ small. But then, $\|\nabla w(t_o+1)\|$ is also small, according to Lemma 3. Hence the condition of Theorem 1, considered with starting time $t_o + 1$, implies $\|\nabla w(t)\|$ is small for $t \ge t_o + 1$. Thus, remembering Poincare's inequality, $\|w(t)\|$ is small for $t \ge t_o + 1$. This completes the proof.

The next theorem is more complicated, but proved by a similar type of argument.

Theorem 2. The stability condition of Definition 2 is equivalent to any one of the following conditions

(i) *There exist numbers* $\delta, T > 0$ *such that every perturbation* w *, with* $w_o \in \mathbf{J}_1$ *and* $\|\nabla w_o\| < \delta$ *, satisfies*

$$\|w(t_o+T)\| \le \frac{1}{2}\|w_o\| .$$

(ii) *There exist numbers* $\delta, \alpha, A > 0$ *such that every perturbation* w , *with* $w_o \in \mathbf{J}$ *and* $\|w_o\| < \delta$, *satisfies*

$$\|w(t)\| \le A\, e^{-\alpha(t-t_o)} \|w_o\| , \quad \textit{for all} \quad t \ge t_o .$$

(iii) There exist numbers $\delta, \alpha, A > 0$ *such that every perturbation* w , *with* $w_o \in \mathbf{J}_1$ *and* $\|\nabla w_o\| < \delta$, *satisfies*

$$\|\nabla w(t)\| \le A\, e^{-\alpha(t-t_o)} \|\nabla w_o\| , \quad \textit{for all} \quad t \ge t_o .$$

One of our principal results about the numerical analysis of problem (1) is that the error constants of Proposition 2 remain bounded as $t \to \infty$, if the solution u being approximated is exponentially stable.

Theorem 3. If u, p *and* u_h, p_h *are continuous and discrete solutions defined for* $t \ge 0$, *and if* u *is exponentially stable, then there exist constants* $K, h_o > 0$ *such that*

$$\text{(8)} \qquad \begin{aligned} |||(u-u_h)(t)|||_h &\le h^2 K , \\ \|(p-p_h)(t)\|_{L^2/N_h} &\le h\, K \max(1, t^{-1/2}) \end{aligned}$$

hold for all $t > 0$ *and* $0 < h \le h_o$. *The constants* K *and* h_o *depend on* $\|\tilde{\Delta} a\|$, $\sup_{t \ge 0} \|f\| + \|f_t\|$, $\sup_{t \ge 0} \|\nabla u\|$, *a bound for* $h^{-2} |||a - a_h|||_h + \|\tilde{\Delta}_h a_h\|$, *and the stability parameters of Definition* 2.

Proof. Rather than actually choosing δ, T as in Definition 2, it will be more convenient to choose them in accordance with Theorem 2, so that for any solution v of (6) satisfying $\|\nabla(v-u)(t_o)\| < \delta$, there holds

$$\text{(9)} \qquad |||(v-u)(t_o+T)|||_h \le \frac{1}{4} |||(v-u)(t_o)|||_h ,$$

$$\text{(10)} \qquad \sup_{t \ge t_o} \|\nabla(v-u)\| \le 1 .$$

The main point to be established is an induction step for the velocity error estimate. We claim there exist constants K and h_o such that, for any choice of $h < h_o$ and $t_o \ge 0$, if

(11) $$\sup_{t \le t_o} |||u - u_h|||_h \le h^2 K ,$$

then

(12) $$|||(u - u_h)(t_o + T)|||_h \le h^2 K .$$

Since $\sup_{t \ge 0} ||\nabla u|| < \infty$, it is clear that (11) implies

(13) $$\sup_{t \le t_o} ||\nabla_h u_h|| \le C_1 ,$$

where C_1 depends on h and K only through their product hK . Using Proposition 3, one sees that (13) implies

(14) $$||\tilde{\Delta}_h u_h(t_o)|| \le C_2 ,$$

with C_2 also depending on h and K only through their product hK . Clearly

(15) $$||\tilde{\Delta} R^h u_h(t_o)|| = ||P\tilde{\Delta}_h u_h(t_o)|| \le C_2 .$$

Further, using (3), it is seen that (14) implies

(16) $$|||(R^h u_h - u_h)(t_o)|||_h \le h^2 C_3 ,$$

with C_3 again depending on h and K only through their product hK . Finally, taking (11) and (16) together, it is evident that

(17) $$||\nabla(u - R^h u_h)(t_o)|| \le hK + hC_3 < \delta ,$$

provided hK and h are small enough.

Let v be the solution of (6) satisfying $v(t_o) = R^h u_h(t_o)$. Then v satisfies (9) and (10), provided hK and h are small enough to ensure (17).

In view of (15), (10), (16) and (14), we can apply Proposition 2 to obtain an error estimate

$$\text{(18)} \qquad |||(v-u_h)(t)|||_h \le h^2 \hat{K} e^{\hat{K}(t-t_o)} , \quad \text{for} \quad t \ge t_o ,$$

between v and u_h, with constants $\hat{K}$ depending on h and K only through C_2 and C_3, i.e., only through their product hK. Thus, for K sufficiently large and all h sufficiently small, there will hold

$$\hat{K} e^{\hat{K}T} \le \frac{1}{2} K , \ C_3 \le K ,$$

while at the same time both h and hK will be small enough to ensure (17). Now (16) and (11) imply $|||(v-u)(t_o)|||_h \le 2h^2 K$, so that together (9) and (18) imply (12). This completes the proof of the velocity error estimate (8). The pressure error estimate (8) is a relatively easy consequence of it.

Much of the existing theory of hydrodynamic stability rests upon the "principle of linearized stability". This is a general assertion that in determining the stability of a solution u it suffices to consider the linearized perturbation equation

$$\text{(19)} \qquad \bar{w}_t - \Delta\bar{w} + u\cdot\nabla\bar{w} + \bar{w}\cdot\nabla u = -\nabla q ,$$

in place of the full nonlinear perturbation equation (7). In the following theorem we give a precise statement of the principle of linearized stability appropriate in the general context of the nonstationary problem. The proof is a direct and simple one, entirely bypassing spectral methods, as indeed one must in the nonstatinary case.

Theorem 4. The solution u *of problem* (1) *is exponetially stable if and only if there exist numbers* $\alpha, A > 0$, *such that every solution* $\bar{w}$ *of the linearized perturbation equation* (19) *satisfies*

$$\text{(20)} \qquad \|\bar{w}(t)\| \le A e^{-\alpha(t-t_o)} \|\bar{w}_o\| , \quad \textit{for} \quad t \ge t_o .$$

Proof. Let $\psi = \bar{w} - w$, where $\bar{w}$ and w are solutions of (19) and (7), respectively, satisfying $\bar{w}(t_o) = w(t_o) = w_o$. Subtracting (7) from (19) gives

$$\psi_t - \Delta\psi + u\cdot\nabla\psi + \psi\cdot\nabla u - w\cdot\nabla w = -\nabla q \ ,$$

for some scalar function q . Multiplying by ψ and integrating, this leads to

$$\frac{d}{dt}\|\psi\|^2 + \|\nabla\psi\|^2 \le c\|\nabla u\|^4\|\psi\|^2 + c\|\nabla w\|^4 \ .$$

Using Gronwall's inequality now yields

$$\|\psi(t_o+T)\|^2 \le c\,e^{cM^4T}\int_{t_o}^{t_o+T}\|\nabla w\|^4 d\tau$$

$$\le c\,e^{cM^4T}\sup_{[t_o,t_o+T]}\|\nabla w\|^2\int_{t_o}^{t_o+T}\|\nabla w\|^2 d\tau \ ,$$

for any fixed $T > 0$. Thus, if $\|\nabla w_o\|$ is sufficiently small, depending on T , Lemmas 2 and 1 imply

(21) $$\|\psi(t_o+T)\|^2 \le c\,e^{c(M_3^4+1)T}\|\nabla w_o\|^2\|w_o\|^2 \ .$$

Now suppose the condition of Theorem 4 holds. Choose T above such that (20) implies

$$\|\bar{w}(t_o+T)\| \le \frac{1}{4}\|w_o\| \ .$$

Then, also, provided $\|\nabla w_o\|$ is sufficiently small, (21) implies

$$\|\psi(t_o+T)\| \le \frac{1}{4}\|w_o\| \ .$$

Combining these gives

$$\|w(t_o+T)\| \le \|\bar{w}(t_o+T)\| + \|\psi(t_o+T)\| \le \frac{1}{2}\|w_o\| \ ,$$

showing that condition (i) of Theorem 2 is satisfied, implying the exponential stability of u .

To show that exponential stability implies linearized stability, we argue similarly, starting again with (21). This completes the proof.

In [1], we applied Theorem 4 to show that the set of initial values for u , which give rise to solutions that are exponentially stable and have bounded

Dirichlet norms, is open with respect to the Dirichlet norm. All solutions starting within a common connectivity component of this set converge together as $t \to \infty$. We also showed that an exponentially stable solution necessarily tends to a steady or time periodic motion, if the forces and boundary conditions are steady or time periodic. These results combined with Theorem 3 were shown to provide a justification of time stepping as a means of calculating steady or time periodic solutions.

4. Quasi-Exponential Stability

Below, ϕ will represent the angular variable about an axis of symmetry common to both Ω and f, if there is one. For simplicity, we will write $u = u(\phi,t)$, suppressing in our notation the usually nontrivial dependence of u on the other spatial variables. The symbol ω will also denote an angle about the axis of symmetry, thought of as a rotation. If f and Ω do not possess a common axis of symmetry, it will be understood that $\omega = 0$. Further, for any Ω, if f is time independent we will consider time shifts denoted by s. If f is not time independent, it will be understood that $s = 0$.

Definition 3. We say u *is quasi-exponentially stable if there are numbers* $\delta, T, B > 0$ *such that for every perturbation* w, *with* $w_o \in J$ *and* $\|w_o\| < \delta$ *there exists a time shift* s *and a spatial rotation* ω *satisfying*

$$|s| + |\omega| \le B\|w_o\| , \tag{22}$$

$$\|(v-\tilde{u})(t_o+T)\| \le \frac{1}{2}\|w_o\| , \tag{23}$$

where v *is the solution of the perturbed problem* (6) *corresponding to the perturbation* w, *and* $\tilde{u}(x,t) = u(\phi+\omega, t+s)$.

A simple example of quasi-exponential stability occurs in the Taylor experiment. At certain rotational speeds of the cylinders, the convection cells loose rotational symmetry, taking on a wavy appearance in the angular variable. Clearly, if the boundary values and forces are rotationally symmetric, a small angular shift in the pattern of waves will constitute an admissible perturbation with no tendency to decay. However, the same reasoning that leads one to believe simple Taylor cells are exponentially stable leads to the conclusion that wavy Taylor cells are quasi-exponentially stable "modulo spatial rotations", meaning that there is a fixed length of time T during which the difference between a slightly disturbed flow v and a slightly rotated image $\tilde{u} = u(\phi+\omega,t)$ of the original undisturbed flow will decay to half the size of the initial perturbation $w_o = v(t_o) - u(t_o)$, and further that the required rotation ω should be less than a fixed constant B times the size of the initial perturbation. In this case the time shift s in Definition 3 is taken to be zero. Alternatively, if the waves are precessing about the axis of symmetry, and if the forces and boundary values are time independent, the flow can be considered as quasi-exponentially stable "modulo time shifts", meaning that there exists a time shift s such that the difference between v and $\tilde{u} = u(\phi,t+s)$ decays to half the size of w_o in time T . An important example of a flow which is quasi-exponentially stable modulo time shifts, but not modulo rotations, is provided by von-Kármán vortex shedding behind a cylinder. Small perturbations decay modulo slight shifts in the time phase.

Definition 3 permits consideration of quasi-exponential stability modulo both time shifts and spatial rotations simultaneously. An example occurs in the Taylor experiment, when at certain rotational speeds of the cyliners wavy cells are observed to undergo a further time periodic oscillation, odd and even numbered cells alternately expanding and contracting. Though these cells are sometimes referred to as doubly time periodic, it is clear that the second time periodicity is possible only because the first one is equivalent to a spatial periodicity.

In [1] we proved a result concerning the discrete approximation of quasi-exponentially stable solutions, analogous to Theorem 3. Its conclusion differs from that of Theorem 3 in that it provides error estimates modulo rotations and time shifts. More precisely, it asserts the existence of time dependent rotations $\omega_h(t)$ and time shifts $s_h(t)$, in addition to the constants K and h_o, such that for $0 < h \le h_o$ and $t > 0$, there holds

$$|||(\tilde{u} - u_h)(t)|||_h \le h^2 K ,$$

$$\|(\tilde{p} - p_h)(t)\|_{L^2/N_h} \le h K \max(1, t^{-1/2}) ,$$

where $\tilde{u}(\phi,t) = u(\phi+\omega_h(t), t+s_h(t))$ and $\tilde{p}(\phi,t) = p(\phi+\omega_h(t), t+s_h(t))$. Moreover, $\omega_h(o) = 0$, $s_h(o) = 0$, and their time derivatives satisfy

$$|\omega_h'(t)| + |s_h'(t)| \le h^2 K .$$

Thus the rates of angular precession and of drift in the time phase, of the discrete solution relative to the continuous solution, are of order h^2.

The theory of quasi-exponential stability has been developed in [1] similarly to that exponential stability, with similar consequences for discrete approximations. We will only state here the corresponding principle of linearized stability. To understand the modification needed in Theorem 4, note that if f is independent of time, and/or Ω and f possess a common axis of rotational symmetry with the corresponding angular variable ϕ, then the derivatives u_t and/or u_ϕ are necessarily solutions of the linearized perturbation equation (19).

Theorem 5. The solution u *of problem* (1) *is quasi-exponentially stable if and only if there exist numbers* $\alpha, A, B > 0$, *such that every solution* $\bar{w}(t)$ *of the linearized perturbation equation* (19) *satisfies*

$$\|\bar{w}(t) - \sigma u_t(t) - \rho u_\phi(t)\| \le A e^{-\alpha(t-t_o)} \|\bar{w}(t_o)\| \tag{24}$$

for $t \ge t_o + 1$, *where* σ *and* ρ *are scalar multipliers satisfying*

$$|\sigma| + |\rho| \le B\|\bar{w}_o(t)\| . \tag{25}$$

Nonzero multipliers σ *and* ρ *are required in* (24) *if and only if nonzero time shifts* s *and nontrivial rotations* ω , *respectively, are required in* (23).

5. Contractive Stability to a Tolerance and Long Term A Posteriori Error Estimates

We turn now to the question of whether the "global existence" of a smooth stable solution of problem (1) can be verified by means of a numerical experiment. There is a known argument for bounding a solution's Dirichlet norm (and thus obtaining its full regularity) "locally" via a numerical experiment combined with an a posteriori error estimate. It goes roughly as follows. Suppose the Dirichlet norm of the discrete solution, for a given mesh size h , is found to remain less than some number N_h . Choosing a second number $M > N_h$, the Dirichlet norm of the smooth solution certainly remains less than M on some unknown interval $[0,t_h]$. Using the local error estimate (Proposition 2) which holds on the basis of the assumed bound M , one then obtains an explicit estimate (exponential in time) for the solution's Dirichlet norm on $[0,t_h]$. Equating the right side of this estimate with M one may solve for t_h , or more precisely, a lower bound for t_h , i.e., an interval of time during which M does indeed bound the Dirichlet norm. At best, if the computed numbers N_h remain bounded as $h \to 0$, one finds that $t_h \sim -\log h$, because of the exponential growth of the local error estimate. In other words, to verify existence this way on an interval [0,T] requires a numerical experiment with mesh size $h \sim \exp(-T)$.

The point of Theorem 6 below is to demonstrate that in verifying existence over time intervals of any length, it suffices to work with a single sufficiently small choice of the mesh size, provided the discrete solution is found to be stable as well as of bounded Dirichlet norm.

This raises the question of whether it is possible to verify numerically the stability of a discrete solution. It certainly is not if one has in mind the usual notions of stability, which set a condition to be satisfied by all perturbations, no matter how small. For this reason we introduce, for use as a hypothesis in Theorem 6, another notion of stability which we call "contractive stability to a tolerance". In Theorem 7 it is shown that the question of whether a discrete solution possesses this type of stability can be answered through a fixed, finite amount of computation per unit of time. The question of whether the discrete approximations of an exponentially stable solution inherit the property of being contractively stable to a tolerance is answered affirmatively in Theorem 8. It is shown, moreover, that the stability parameters of the discrete solution are bounded uniformly in h as $h \to 0$, so that the hypotheses of Theorem 6 are necessarily satisfied for all sufficiently small values of h. Together, Theorems 6, 7 and 8 imply that the existence of a stable smooth solution can be verified (at least in principle) through a fixed, finite amount of computation per unit of time. The proofs are supplied in [1]. Below, for simplicity, we define contractive stability to a tolerance relative to the infinite time interval $0 \le t < \infty$ and state our theorems accordingly, the modification to solutions defined on finite time intervals being obvious.

Let u_h be a solution of the discretized Navier-Stokes equations, defined for $t \ge 0$. In analogy with the continuous case, we call w_h a "perturbation" of u_h if $w_h = v_h - u_h$, where v_h is a second discrete solution, starting at some initial time $t_o \ge 0$, with an initial value $v_h(t_o)$ near $u_h(t_o)$. Whenever we speak below of a perturbation w_h, it is to be understood that the associated initial time, initial value, and perturbed solution are denoted by t_o, $w_h(t_o)$ and v_h, respectively.

Definition 4. A solution u_h *of the discretized Navier-Stokes equations (defined for* $t \ge 0$*) is said to be "contractively stable to a tolerance" if there exist positive numbers* δ, ρ, A *and* T, *with* $\rho < \delta$, *such that for any time* $t_o \ge 0$ *and any perturbation* w_h *of* u_h *satisfying* $\|\nabla_h w_h(t_o)\| < \delta$, *there holds*

$$\|\nabla_h w_h(t_o+T)\| < \rho\ , \quad \sup_{[t_o,t_o+T]} \|\nabla_h w_h\| \le A\ .$$

We call ρ the "tolerance", δ the "stability radius" and T the "decay time" of u_h, and A a "Dirichlet bound" for its perturbations.

Theorem 6. Suppose that, for some h, *there is a discrete solution* u_h *which is contractively stable to a tolerance. Then, if* h *is sufficiently small in relation to* $\sup_{t\ge 0}\|\nabla_h u_h\|$, $\|\tilde{\Delta}_h u_h(o)\|$, *and the stability parameters* ρ,δ,T *of Definition 4, there exists a continuous solution* u *of the Navier-Stokes equations satisfying*

$$\sup_{t\ge 0}\|\nabla_h(u-u_h)\| \le 2\rho\ .$$

Further, if $3\rho < \delta$, *and if* h *is small enough, the continuous solution* u *will also be contractively stable to a tolerance.*

The proof that a discrete solution's contractive stability to a tolerance is amenable to numerical verification depends upon discrete solutions enjoying continuous dependence properties analogous to those stated for continuous solutions in Lemmas 1 and 2. As we are dealing abstractly with the discretization of the Navier-Stokes equations, we must assume such properties of continuous dependence. The following then holds.

Theorem 7. Suppose u_h *is a discrete solution satisfying* $\sup_{t\ge 0}\|\nabla_h u_h\| < \infty$. *Then, if* u_h *is contractively stable to a tolerance, this can be verified by checking the decay (to a tolerance in fixed time) of a fixed finite number of test perturbations per unit of time.*

The assurance that discrete solutions approximating an exponentially stable solution will, for all sufficiently small values of h, satisfy the hypotheses of Theorem 6 is provided in our final result. Contractive stability to a tolerance is defined for continuous solutions analogously to Definition 4.

Theorem 8. Let u *and* u_h *be continuous and discrete solutions of problem (1). Suppose that* $\sup_{t\geq 0}\|\nabla u\| < \infty$ *and that* u *is contractively stable to a tolerance, with parameters* ρ, δ, T *and* A. *Then there exist constants* K *and* h_o, *such that*

$$\sup_{t\geq 0}\|\nabla_h(u-u_h)\| \leq \rho + Kh, \tag{26}$$

for all $h < h_o$. *Further, if* $3\rho < \delta$, *then the discrete solution* u_h *is contractively stable to a tolerance for certain fixed values of the stability parameters, for all sufficiently small values of* h.

We have stated this last result for continuous solutions assumed merely to be contractively stable to a tolerance, rather than to possess the stronger property of exponential stability, as we think there is a naturally occuring and important class of flows which possess this weaker stability property without being, in fact, exponentially stable. For example, imagine that Ω is a section of pipe or tubing and let smooth boundary values be prescribed for a flow entering across an upstream section and exiting across a downstream section. Adjusting the rate of flow and the length of the pipe, one may expect to observe incipient turbulence in a flow which is yet, in some sense, stable to larger disturbances. Small perturbations in the nearly uniform upstream flow begin to grow. However, before they grow very large they pass out of Ω across the downstream boundary. Yet, their effect may not decay to zero. Even as they pass downstream they influence the upstream flow; the flow is analytic after all. Their effect might be likened to the introduction of new perturbations upstream, which in their turn will grow, pass downstream, and again create new perturbations upstream. If a larger distrubance is introduced, its effect will decay to the same ambient level of minor disturbances. Another type of example probably occurs in von-Kármán vortex shedding, if there are slight instabilities in the vortices. Still another in the Taylor experiment, when wavy cells appear with slightly turbulent cores. If these flows really are contractively stable to a tolerance, the error estimate to a tolerance (26) applies to their discrete approximations, at least after taking account of time shifts and spatial rotations as was done for exponential stability in section 4.

References

[1] Heywood, J.G. and Rannacher R., Finite Element Approximation of the Nonstationary Navier-Stokes Problem, Part II: Stability of Solutions and Error Estimates Uniform in Time, preprint, Univ. of British Columbia (October 1982).

[2] Heywood, J.G. and Rannacher R., Finite Element Approximation of the Nonstationary Navier-Stokes Problem, Part I: Regularity of Solutions and Second-order Error Estimates for Spatial Discretization, SIAM J. Numer. Anal. 19 (1982) 275-311.

Lecture Notes in Num. Appl. Anal., **5**, 399–407 (1982)
Nonlinear PDE in Applied Science. U.S.-Japan Seminar, Tokyo, 1982

THE EXISTENCE AND THE FINITE ELEMENT APPROXIMATION FOR THE SYSTEM $\Delta u = \Sigma u_j \frac{\partial u}{\partial x_j} + f$

Lin Qun and Jiang Lishang

Institute of Systems Science
Academia Sinica
Beijing, China

Department of Mathematics
Beijing University
Beijing, China

There is no global existence theorem about the nonlinear Navier-Stokes equation until now. However, there is an existence theorem for the linear Stokes equation, and there also exists an existence theorem, as we will prove in this paper, for the nonlinear system

$$(1)\quad \begin{aligned} \Delta u_i &= \sum_1^N u_j \frac{\partial u_i}{\partial x_j} + f_i \quad \text{in} \quad \Omega \subset R^N, \\ u_i &= g_i \quad \text{on} \quad \partial\Omega, \ 1 \le i \le N , \end{aligned}$$

which has the same nonlinearity as the Navier-Stokes equation but without the condition div u = 0. These results mean, as pointed out by R. B. Kellogg, that the difficulty about the Navier-Stokes equation stems from both the condition div u = 0 and the nonlinearity rather than from only one of them.

We will also prove in this paper the convergence theorem for the finite element approximation of the system (1) and some acceleration results.

1. Existence Theorem

The system (1) was discussed by Kiselev and Ladyzenskaya in 1957. It has been pointed by Nirenberg that the proof of existence is incorrect (see MR 20 #6881, by Finn). We first prove that the system (1) always has a classical solution.

<u>Theorem 1</u>. Suppose $\Omega \in A_{2+\alpha}$. Then for any data $f_i \in C_\alpha(\bar{\Omega})$ and $g_i \in C_{2+\alpha}(\partial\Omega)$, the system (1) has a solution $u_i \in C_{2+\alpha}(\bar{\Omega})$, $1 \le i \le N$.

The proof is based on

<u>Lemma 1</u>.

$$\max_{\bar{\Omega}} \Sigma |u_i(x)| \le \max_{\bar{\Omega}} \Sigma |f_i(x)| + \max_{\partial\Omega} \Sigma |g_i(x)| + 2 \max_{\bar{\Omega}} \Sigma |x_i| .$$

Proof. We write, for fixed i,

$$u_i(x) = v(x) + x_i .$$

Hence the ith equation of (1) becomes

$$\Delta v - \sum_1^N u_j \frac{\partial v}{\partial x_j} - v = f_i(x) + x_i \quad \text{in } \Omega ,$$

$$v = g_i(x) - x_i \quad \text{on } \partial\Omega .$$

Now assume for v a positive maximum at $P_o = (x_1^o, \ldots, x_N^o) \in \Omega$ that

$$\frac{\partial v}{\partial x_j} = 0 \quad (1 \le j \le N), \quad \Delta v \le 0 \quad \text{at } P_o .$$

Hence

$$- v(P_o) \ge f_i(P_o) + x_i^o ,$$

$$v(P_o) \le \max_{\bar{\Omega}} |f_i(x)| + \max_{\bar{\Omega}} |x_i| .$$

Similarly, assume for v a negative minimum at $P_o \in \Omega$ so that

$$v(P_o) \ge -\max_{\bar{\Omega}} |f_i(x)| - \max_{\bar{\Omega}} |x_i| .$$

Finally, assume for v an extreme value at $P_o \in \partial\Omega$

$$|v(P_o)| \le \max_{\partial\Omega} |g_i(x)| + \max_{\partial\Omega} |x_i| .$$

Thus in any case we have

$$\max_{\bar\Omega} |u_i(x)| \le \max_{\bar\Omega} |v(x)| + \max_{\bar\Omega} |x_i|$$

$$\le \max_{\bar\Omega} |f_i(x)| + \max_{\partial\Omega} |g_i(x)| + 2 \max_{\bar\Omega} |x_i|$$

and Lemma 1 is proved.

Applying Lemma 1 and the Learay-Schauder theorem we obtain Theorem 1.

The existence theorem for the nonstationary problem corresponding to (1) can be treated in a similar fashion.

2. Finite Element Approximation and Its Acceleration

Consider the simple solution u of (1) defined by

$$(w,\varphi)_1 + \Sigma \left(u_j \frac{\partial w}{\partial x_j} + \frac{\partial u}{\partial x_j} w_j, \varphi\right) = 0 \quad \forall \ \varphi \in \overset{o}{H}_1 \Rightarrow w = 0 ,$$

with scalars

$$(\psi,\varphi) = \Sigma \int \psi_i \varphi_i dx , \quad (\psi,\varphi)_1 = \Sigma \left(\frac{\partial\psi}{\partial x_j}, \frac{\partial\varphi}{\partial x_j}\right) .$$

We have the following approximation theorem.

<u>Theorem 2</u>. Suppose that $N = 3$, $g = 0$, Ω is a convex polygon and $u \in H_2 \cap \overset{o}{H}_1$ is a simple solution of (1). Then for h small enough, the finite element equation

$$(2) \qquad - (u_h,\varphi_h)_1 = \Sigma \left(u_{hj} \frac{\partial u_h}{\partial x_j}, \varphi_h\right) + (f,\varphi_h) \quad \forall \ \varphi_h \in S_h$$

in the piecewise linear trial space S_h has a unique solution $u_h \in S_h$ in a neighborhood of u, which satisfies

$$(3) \qquad \|u_h - u\|_1 \le ch \|u\|_2 , \quad \|u_h - u\|_0 \le ch^2 \|u\|_2 .$$

Furthermore, if $u \in H_3$, the accuracy of energy norm can be raised by computing the following quadratic finite element solutions

$\bar{u}_k \in S_k$,

(4) $$-(\bar{u}_k, \varphi_k)_1 = \Sigma\, (u_{hj} \frac{\partial u_h}{\partial x_j}, \varphi_k) + (f, \varphi_k) \quad \forall\, \varphi_k \in S_k$$

in the piecewise quadratic trial space S_k with a coarser mesh size k (see the figure). We have

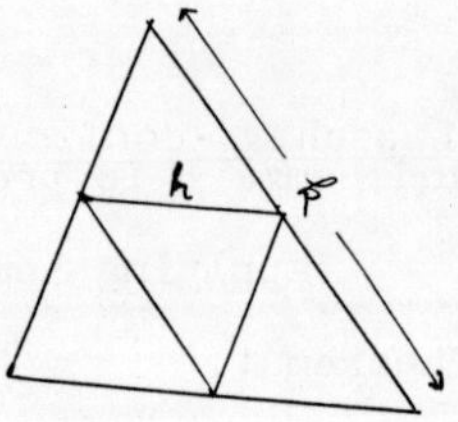

(5) $$\|\bar{u}_k - u\|_1 \le ch^2 .$$

Note that the two finite element equations for u_h and $\bar{u}_k$ have the same degrees of freedom.

Theorem 2 will be proved by the following abstract operator framework.

Consider the abstract operator equation

(6) $$u = Ku ,$$

where K is Fréchet-differentiable in a Banach space E. The following lemma can be found in, for example, G. Alefeld, Beitr Numer. Anal. 6(1977).

Lemma 2. If $(I - K'(u_o))^{-1}$ exists for some $u_o \in E$, K' is Lipschitz-continuous with constant c_1 in the ball $\|u - u_o\| \le r_1$, and

$$\|(I - K'(u_o))^{-1}\| \le c_2, \quad \|(I - K'(u_o))^{-1}(u_o - Ku_o)\| \le c_3 ,$$

$$c_4 = c_1 c_2 c_3 < \frac{1}{2} , \quad r_1 \ge (1 + \sqrt{1 - 2c_4})\, \frac{c_3}{c_4} = r_2 ,$$

then (6) has a unique solution u in the region

$$r_3 = (-1 + \sqrt{1 + 2c_4})\, \frac{c_3}{c_4} \le \|u - u_o\| \le (1 - \sqrt{1 - 2c_4})\, \frac{c_3}{c_4} = r_4$$

and has no solution in the region

$$\|u - u_o\| \le r_3 \quad \text{and} \quad r_4 < \|u - u_o\| < r_2 .$$

Moreover, u is a simple solution: $\exists\, (I - K'(u))^{-1}$.

We now consider the projection equation

(7) $$u_h = P_h K u_h$$

where P_h is a projection with $P_h E = S_h \subset E$.

Lemma 3. Suppose that u is a simple solution of (6), $P_h K$ is Lipschitz-continuous in a neighborhood of u, and

$$\|u - P_h u\| \to 0 \ , \quad \|(I - P_h)K'(u)\| \to 0 \quad (\text{as } h \to 0).$$

Then for h small enough, equation (7) has a unique solution u_h in a neighborhood of u, which satisfies

$$\|u_h - u\| \le c\|u - P_h u\| \ .$$

The Newton iterates for (7) exist and converge quadratically. Moreover

(8) $$\|Ku_h - u\| \le c(\|u_h - u\| + \|K'(u)(I - P_h)\|)\|u_h - u\| \ .$$

Proof. Replacing K with $P_h K$ and chooocing $u_o = u$ in Lemma 2, one obtains the first part of Lemma 3. The estimate (8) can be derived by using the following identities

$$(I - K'(u))(Ku_h - u) = Ku_h - Ku - K'(u)(u_h - u) + K'(u)(I - P_h)(u_h - Ku_h) \ ,$$

$$Ku_h - u_h = Ku_h - Ku + u - u_h \ .$$

We now apply Lemma 3 to the system (1). Let $E = \overset{o}{H}_1$ and K be an operator defined by

$$Ku \in \overset{o}{H}_1$$

$$-(Ku,\varphi)_1 = \Sigma\, (u_j \frac{\partial u}{\partial x_j}, \varphi) + (f,\varphi) \quad \forall\ \varphi \in \overset{o}{H}_1 \ .$$

Let S_h be the piecewise linear trial space with the mesh size h and P_h the standard Ritz-projection defined by

$$P_h u \in S_h \quad \text{and} \quad (P_h u, \varphi_h)_1 = (u, \varphi_h)_1 \quad \forall\ \varphi_h \in S_h \ .$$

Then (1) and (2) can be written as (6) and (7) respectively. And the Fréchet derivative K'(u) will be determined by

$$v = K'(u)w \in \overset{o}{H}_1 ,$$
$$-(v,\varphi)_1 = \Sigma\, (u_j \frac{\partial w}{\partial x_j} + \frac{\partial u}{\partial x_j} w_j, \varphi) \quad \forall\, \varphi \in \overset{o}{H}_1 .$$

We now come to prove

(9) $$\|K'(u)(I - P_h)\|_{1\to 1} \le ch .$$

For this we note

$$\|K'(u)(I - P_h)\|_{1\to 1} = \|(I - P_h)K'(u)^*\|_{1\to 1}$$
$$\|(I - P_h)K'(u)^*\varphi\|_1 \le ch\|K'(u)^*\varphi\|_2$$

where

$$(K'(u)^*\varphi,\psi)_1 = (\varphi, K'(u^*)\psi)_1$$
$$= -\Sigma\, (\frac{\partial u}{\partial x_j}\varphi, \psi_j) + \Sigma\, (\frac{\partial u_j}{\partial x_j}\varphi + u_j \frac{\partial \varphi}{\partial x_j}, \psi) \quad \forall\, \psi \in \overset{o}{H}_1 .$$

Since

$$\|K'(u)^*\varphi\|_2 \le c\|u\|_2\|\varphi\|_1 ,$$

so (9) holds.

By means of Lemma 3 and Nitsche's trick we obtain the estimate (3) and the following estimate

$$\|\bar{u} - u\|_1 \le ch^2\|u\|_2$$

where $\bar{u}$ is the solutions of the uncoupled Poisson equation

(10) $$\Delta\bar{u} = \Sigma\, u_{hj} \frac{\partial u_h}{\partial x_j} + f \quad \text{in } \Omega$$
$$\bar{u} = 0 \quad \text{on } \partial\Omega .$$

The estimate (5) can be derived by the following Brezzi's trick.

First, split the Poisson equation (10) into two parts:

$$- (\bar{u}_1,\varphi)_1 = \Sigma\ (u_j \frac{\partial u}{\partial x_j},\varphi) + (f,\varphi) \quad \forall\ \varphi \in \overset{o}{H}_1 ,$$

$$- (\bar{u}_2,\varphi)_1 = \Sigma\ (u_{hj} \frac{\partial u_h}{\partial x_j} - u_j \frac{\partial u}{\partial x_j},\varphi) \quad \forall\ \varphi \in \overset{o}{H}_1$$

and

$$\bar{u} = \bar{u}_1 + \bar{u}_2 .$$

Correspondingly the quadratic finite element equation (4) can also be splitted into two parts:

$$- (\bar{u}_{1k},\varphi_k)_1 = \Sigma\ (u_j \frac{\partial u}{\partial x_j},\varphi_k) + (f,\varphi_k) \quad \forall\ \varphi_k \in S_k ,$$

$$- (\bar{u}_{2k},\varphi_k)_1 = \Sigma\ (u_{hj} \frac{\partial u_h}{\partial x_j} - u_j \frac{\partial u}{\partial x_j},\varphi_k) \quad \forall\ \varphi_k \in S_k$$

and

$$\bar{u}_k = \bar{u}_{1k} + \bar{u}_{2k} .$$

Then we have

$$\|\bar{u}_k - \bar{u}\|_1 \le \|\bar{u}_{1k} - \bar{u}_1\|_1 + \|\bar{u}_{2k} - \bar{u}_2\|_1$$

$$\le ch^2\|\bar{u}_1\|_3 + ch\|\bar{u}_2\|_2 \le ch^2$$

and (5) follows.

We would like to make the following remark.

Recently Lin Qun and Lu Tao proposed a "splitting extrapolation difference method" for solving the linear elliptic problem

$$\Delta u = \sum_1^N a_j \frac{\partial u}{\partial x_j} + f \quad \text{in } \Omega ,$$

$$u = g \quad \text{on } \partial\Omega$$

in a cubic $\Omega \subset R^N$. The well known difference method consists in replacing the differential equation with the central difference quotient equation, the continuous domain Ω with the discrete mesh $V(h_1, h_2, \ldots, h_N)$ with size $h_1, h_2, \ldots, h_N$ along variables $x_1, x_2, \ldots, x_N$

respectively and replacing the continuous exact solution u with the discrete approximate solution $U(h_1, h_2, \ldots, h_N)$. We have, if u is smooth enough,

$$u - U(h_1, h_2, \ldots, h_N) = \sum_{j=1}^{M} \sum_{i=1}^{N} c_i^{(2j)} h_i^{2j} + O(\sum_{i=1}^{N} h_i^{2M+2}) .$$

The usual global extrapolation method involves the following: Make the homogeneous refinement meshes $V(\frac{h_1}{2}, \frac{h_2}{2}, \ldots, \frac{h_N}{2})$, $V(\frac{h_1}{4}, \frac{h_2}{4}, \ldots, \frac{h_N}{4})$, ... with corresponding difference solutions $U(\frac{h_1}{2}, \frac{h_2}{2}, \ldots, \frac{h_N}{2})$, $U(\frac{h_1}{4}, \frac{h_2}{4}, \ldots, \frac{h_N}{4})$, ... and compute the homogeneous extrapolation solutions

$$HE_1 = \frac{1}{3}(4U(\frac{h_1}{2}, \frac{h_2}{2}, \ldots, \frac{h_N}{2}) - U(h_1, h_2, \ldots, h_N))$$

$$HE_2 = \frac{1}{45}(64U(\frac{h_1}{4}, \frac{h_2}{4}, \ldots, \frac{h_N}{4}) - 20U(\frac{h_1}{2}, \frac{h_2}{2}, \ldots, \frac{h_N}{2}) + U(h_1, h_2, \ldots, h_N))$$

....

Then one has

$$u - HE_1 = O(\Sigma h_i^4) \ ; \ u - HE_2 = O(\Sigma h_i^6) \ ; \ \ldots .$$

Our splitting extrapolation method involves the following: Make the one-variable refinement meshes $V(\frac{h_1}{2}, h_2, \ldots, h_N)$, $V(h_1, \frac{h_2}{2}, \ldots, h_N)$, ..., $V(h_1, h_2, \ldots, \frac{h_N}{2})$; $V(\frac{h_1}{4}, h_2, \ldots, h_N)$, $V(h_1, \frac{h_2}{4}, \ldots, h_N)$, ..., $V(h_1, h_2, \ldots, \frac{h_N}{4})$; ... with the corresponding difference solutions $U(\frac{h_1}{2}, h_2, \ldots, h_N)$, $U(h_1, \frac{h_2}{2}, \ldots, h_N)$, ..., $U(h_1, h_2, \ldots, \frac{h_N}{2})$; $U(\frac{h_1}{4}, h_2, \ldots, h_N)$, $U(h_1, \frac{h_2}{4}, \ldots, h_N)$, ..., $U(h_1, h_2, \ldots, \frac{h_N}{4})$; ... and compute the splitting extrapolation solutions

$$SE_1 = \frac{1}{3}\sum_{i=1}^{N} (4\, U(h_1, \ldots, \frac{h_i}{2}, \ldots, h_N) - U(h_1, \ldots, h_N)) - (N - 1)U(h_1, \ldots, h_N)$$

$$SE_2 = \frac{1}{45}\sum_{i=1}^{N}(64\,U(h_1,\ldots,\frac{h_i}{4},\ldots,h_N) - 20\,U(h_1,\ldots,\frac{h_i}{2},\ldots,h_N)$$

$$+ U(h_1,\ldots,h_N)) - (N-1)U(h_1,\ldots,h_N)$$

....

Then one has

$$u - SE_1 = O(\Sigma\, h_i^4) \;;\; u - SE_2 = O(\Sigma\, h_i^6) \;;\; \ldots .$$

Actually,

$$SE_1 = HE_1 \;,\quad SE_2 = HE_2 \;,\; \ldots$$

in the asymptotic sense. It is easy to see that the splitting extrapolation method will save computational effort in comparison with the homogeneous extrapolation method. We hope that the splitting method will be effective also for the system (1).

Acknowledgments. The authors are greatly indebted to Professors F. Brezzi, J. Frehse, B. Kellogg and G. Strang for their helpful comments.

Lecture Notes in Num. Appl. Anal., **5**, 409–434 (1982)
Nonlinear PDE in Applied Science. U.S.-Japan Seminar, Tokyo, 1982

A hyperbolic model of combustion

Ying Lung-an and Teng Zhen-huan

Department of Mathematics
Peking University
Beijing
CHINA

If fluid flow is accompanied by chemical reaction, then very complicated wave motion phenomena occur. Chapman and Jouguet used a simple and typical model which showed various waves of combustion: strong detonation wave, weak detonation wave, strong deflagration wave, weak deflagration wave, and their critical states, the so-called Chapman-Jouguet detonation wave and deflagration wave [1,2]. Afterwards, many authors have done various works about the structure of these waves and their formative conditions using different kinds of models. More research works have been done in the laboratories and by numerical experiments.

It is an interesting problem how a mathematical model can be applied to these phenomena and one may investigate them by the theory of partial differential equations. Some authors have investigated the travelling wave solutions and some Riemann initial value problems for these problems, but up to now these investigations are not so deep as that for the shock waves.

In this paper we consider a model system of combustion as

$$\left.\begin{aligned} \frac{\partial}{\partial t}(u + qz) + \frac{\partial}{\partial x}f(u) &= \nu\frac{\partial^2 u}{\partial x^2}, \\ \frac{\partial z}{\partial t} &= -K\phi(u)z, \end{aligned}\right\} \tag{1}$$

where constants $q > 0$, $\nu \geq 0$, $K > 0$ represent the binding energy, viscosity and the rate of chemical reaction respectively, u is a lumped variable representing density, velocity and temperature, z is the fraction of unburht gas. Majda [3] has investigated the travelling wave solutions of (1) and explained some interesting phenomena from it, such as strong and weak detonation waves.

The properties of (1) when $\nu = +0$ and $K = +\infty$ are of most interest because the mathematical shock waves and mathematical detonation waves are involved in the solutions at this case. We will prove the global existence of the weak solutions for the initial value problems under some hypotheses. The relationship between system (1) and the reacting fluid dynamic system is just the same as that between Burgers' equation and the fluid dynamic system. But system (1) is much more complicated than Burgers' equation, because first of all it is a system, not a single equation, secondly, because many properties of Burgers' equation, for example, the order principle, are violated here, another example is that there is no "overshot" of shock waves in the solutions of Burgers' equation, while it is just normal with discontinuous solutions of (1). Many difficulties in analysis arise from this.

We will give some hypotheses and two definitions of weak solutions: Problem P and Problem Q, discuss the strong discontinuous curve and the Riemann Problem in the first section, the formulation of Problem Q is stronger than that of Problem P since it determines the state at critical point $u = 0$. We will prove the global existence of Problem P at the second section if, roughly speaking, the initial values are functions with bounded variation. Under an additional hypothesis on the points where the initial value $u_0(x)$ assumes the value 0 (Hypothesis A), we will prove the global existence of Problem Q at the third section.

§1. The definitions of solutions.

We always assume that the function $f(u)$ is sufficiently smooth and $f' > 0$, $f'' > 0$. Function ϕ is defined as

$$\phi(u) = \begin{cases} 0, & u < 0, \\ 1, & u > 0, \end{cases}$$

where $u = 0$ is the "ignition temperature", which is a critical point, we will assume that $\phi(0) = 1$ at the following Problem Q. Clearly $0 \leq z \leq 1$, according to its physical background.

Let $\nu \to +0$, $K \to +\infty$ in system (1), we obtain a formally classical formulation as

$$\frac{\partial}{\partial t}(u + qz) + \frac{\partial}{\partial x} f(u) = 0, \tag{2}$$

$$\left.\begin{aligned} z &= 0, && \text{as} \quad u > 0, \\ \frac{\partial z}{\partial t} &= 0, && \text{as} \quad u < 0. \end{aligned}\right\} \tag{3}$$

The Rankine-Hugoniot condition is also obtained as

$$[u + qz]\sigma = [f]. \tag{4}$$

where [ ] denotes the jump of function, σ is the slope of the discontinuity curve.

If the limit of u, z from the left and right sides of the discontinuous curve are denoted by u^-, z^- and u^+, z^+ respectively, then it is easy to classify the discontinuous curves into five classes:

a) shock waves (abbr. S), either $u^-, u^+ < 0$ or $u^-, u^+ > 0$, and $z^- = z^+$,

b) strong detonation waves (abbr. SD), $u^- > 0$, $u^+ < 0$, and $f'(u^-) > \sigma$,

c) weak detonation waves (abbr. WD), $u^- > 0$, $u^+ < 0$, and $f'(u^-) < \sigma$,

d) Chapman-Jouguet detonation waves (abbr. CJ), $u^- > 0$, $u^+ < 0$, and $f'(u^-) = \sigma$,

e) contact discontinuities (abbr. C), $u^+ = u^-$, $z^+ \neq z^-$, where $\sigma = 0$.

Some other cases are possible, for instance the case when $u^- < 0$, $u^+ > 0$, but we assume that the Lax condition of stability

$$\lambda_i^- \geq \sigma \geq \lambda_i^+, \qquad \text{for } i = 1 \text{ or } 2,$$

is satisfied, where $\lambda_1(u) \equiv 0$, $\lambda_2(u) = f'(u)$, $\lambda_i^- = \lambda_i(u(x - 0, t))$, $\lambda_i^+ = \lambda_i^+(u(x + 0, t))$, then neither the case $u^- < 0$, $u^+ > 0$, nor weak detonation wave are admissible. We will assume that only cases a) b) d) e) are admissible in the following.

There are some other critical cases, for instance $u^+ = 0$ or $u^- = 0$. We will assume in the following that $z = 0$ when $u = 0$, hence we may change $u^- > 0$ or $u^+ > 0$ to $u^- \geq 0$ or $u^+ \geq 0$ at the above inequalities.

For the convenience of following discussion, two auxiliary functions are defined.

Function $u^* = g(u,z)$ is defined by

$$f'(u^*) = \frac{f(u^*) - f(u)}{u^* - u - qz}, \quad u^* \geq u. \tag{5}$$

<u>Lemma</u> 1. u^* exists uniquely and $\frac{\partial g}{\partial u} > 0$, $\frac{\partial g}{\partial z} > 0$.

<u>Proof</u>. Set

$$\varphi(u^*) = \int_{u+qz}^{u^*} \{f'(u^*) - f'(t)\}dt + \int_{u+qz}^{u} f'(t)dt,$$

then (5) is equivalent to $\varphi(u^*) = 0$, $u^* \geq u$. It is easy to verify $\varphi'(u^*) > 0$ and

$$\varphi(u + qz) = \int_{u+qz}^{u} f'(t)dt \leq 0.$$

But $\varphi(+\infty) = +\infty$, therefore there exists a unique $u^* \geq u + qz$ such that $\varphi(u^*) = 0$. But $z \geq 0$, hence $u^* \geq u$.

$\frac{\partial g}{\partial u} > 0$ and $\frac{\partial g}{\partial z} > 0$ can be verified from (5) directly. □

By (5) it is easy to see that SD corresponds to $u^- > g(u^+, z^+)$ and CJ corresponds to $u^- = g(u^+, z^+)$.

The second auxiliary function is $w = \psi(u)$, satisfying

$$w = \begin{cases} u - \dfrac{f(u) - f(0)}{f'(u)}, & u > 0, \\ u, & u \leq 0. \end{cases} \tag{6}$$

It is easy to see that ψ is continuous, monotonous and one-to-one, $\psi(0) = 0$, $\psi'(u) \geq 0$.

<u>Lemma</u> 2. If $u_0 < 0$, $z_1 > z_2 > 0$, $g(u_0, z_2) \geq 0$, then

$$\psi(g(u_0, z_1)) - \psi(g(u_0, z_2)) < q(z_1 - z_2). \tag{7}$$

If $u_1 < u_2 < 0$, $z_0 = 0$, $g(u_1, z_0) \geq 0$, then

$$\psi(g(u_2, z_0)) - \psi(g(u_1, z_0)) < u_2 - u_1. \tag{8}$$

Proof. On the (u, f) plane, the straight lines

$$f - f(u_0) = f'(g(u_0, z_1))(u - u_0 - qz_1),$$
$$f - f(u_0) = f'(g(u_0, z_2))(u - u_0 - qz_2),$$

are the tangent lines of curve $f = f(u)$ by (5). The intersection points of these two lines with horizontal line $f = f(0)$ are

$$w_1 = u_0 + qz_1 + \frac{f(0) - f(u_0)}{f'(g(u_0,z_1))},$$

$$w_2 = u_0 + qz_2 + \frac{f(0) - f(u_0)}{f'(g(u_0,z_2))},$$

respectively, hence

$$w_1 - w_2 = q(z_1 - z_2) + (f(0) - f(u_0))\left\{ \frac{1}{f'(g(u_0,z_1))} - \frac{1}{f'(g(u_0,z_2))} \right\},$$

but

$$g(u_0, z_1) > g(u_0, z_2),$$

by $\frac{\partial g}{\partial z} > 0$, we get

$$w_1 - w_2 < q(z_1 - z_2)$$

by $f'' > 0$. From (5) (6) we know $w_1 = \psi(g(u_0, z_1))$, $w_2 = \psi(g(u_0, z_2))$, therefore (7) is proved. The proof of inequality (8) is similar. □

We consider the initial value problem of (2) (3) with initial values

$$u(x, 0) = u_0(x), \quad z(x, 0) = z_0(x), \tag{9}$$

where z_0 satisfies $0 \leq z_0 \leq 1$ and $z_0(x) = 0$ when $u_0(x) \geq 0$. First of all, we consider the Riemann problem, that is the case of

$$u_0(x) = \begin{cases} u_\ell, & x \le 0, \\ u_r, & x > 0, \end{cases} \qquad z_0(x) = \begin{cases} z_\ell, & x \le 0, \\ z_r, & x > 0, \end{cases}$$

where u_ℓ, u_r, z_ℓ, z_r are constants. There are four cases:

a) $u_\ell \le u_r$. We construct $u(x, t)$ as the solution of equation

$$\frac{\partial u}{\partial t} + \frac{\partial f(u)}{\partial x} = 0, \tag{10}$$

and the initial value. Set $z(x, t) \equiv z_0(x)$, then the solution of this case is obtained. There is a C in the solution.

b) $u_\ell > u_r$, but $u_\ell < 0$. We can construct the solution as case a). There are a C and a S in the solution.

c) $u_\ell > u_r$, $u_\ell \ge 0$, and $u_\ell > g(u_r, z_r)$. Let

$$(u, z) = \begin{cases} (u_\ell, 0), & x \le \dfrac{f(u_\ell) - f(u_r)}{u_\ell - u_r - qz_r} t, \\ (u_r, z_r), & x > \dfrac{f(u_\ell) - f(u_r)}{u_\ell - u_r - qz_r} t. \end{cases}$$

There is a SD, it degenerates to a S when $z_r = 0$.

d) $u_\ell > u_r$, $u_\ell \ge 0$, but $u_\ell \le g(u_r, z_r)$. Let

$$(u, z) = \begin{cases} (u_\ell, 0), & x \le f'(u_\ell)t, \\ ((f')^{-1}(\frac{x}{t}), 0), & f'(u_\ell)t < x < f'(g(u_r, z_r))t, \\ (u_r, z_r), & x > f'(g(u_r, z_r))t. \end{cases}$$

There is a CJ.

Therefore, the Riemann problem is always solvable. But it should be noticed that even the condition of stability is satisfied, the solutions are still not unique. For example, when $u_0(x) \equiv u_0 < 0$, $z_0(x) \equiv z_0 > 0$, $g(u_0, z_0) \ge 0$, besides the trivial solution $u \equiv u_0$ and $z \equiv z_0$, we may also construct a solution as:

$$(u, z) = \begin{cases} (u_0, z_0), & x \le 0, \\ (u_0, 0), & 0 < x \le f'(u_0)t, \\ ((f')^{-1}(\frac{x}{t}), 0), & f'(u_0)t < x \le f'(g(u_0, z_0))t, \\ (u_0, z_0), & x > f'(g(u_0, z_0))t, \end{cases}$$

this solution corresponds to the case when one fires a match in a space filled with combustible gas and oxygen. We conjecture that the solutions obtained in the following are not those solutions of "catastrophe".

For the general initial value problem, $u_0(x)$, $z_0(x)$ are assumed to be bounded measurable functions. Two formulations of weak solutions are given.

Problem P. To find bounded measurable functions $u(x, t)$, $z(x, t)$ defined in $t > 0$ such that for all t and x,

$$\lim_{h \to +0} \frac{1}{h} \int_{x-h}^{x} u(\xi, t)d\xi \equiv u^{-}(x, t) \tag{11}$$

exists, and for any smooth function $\varphi(x, t)$ with compact support on $t \ge 0$,

$$\iint_{t \ge 0} \left\{ \frac{\partial \varphi}{\partial t} (u+qz) + \frac{\partial \varphi}{\partial x} f(u) \right\} dxdt + \int_{-\infty}^{+\infty} \{u_0(x)+qz_0(x)\} \varphi(x, 0)dx = 0 \tag{12}$$

holds, moreover, for any non-negative smooth function $\varphi(x, t)$ with compact support on $t \ge 0$,

$$\iint_{t \ge 0} \frac{\partial \varphi}{\partial t} zdxdt + \int_{-\infty}^{+\infty} z_0(x) \varphi(x, 0)dx \ge 0 \tag{13}$$

holds, and finally such that with

$$v(x, t) = \sup_{0 \le \tau \le t} u^{-}(x, \tau), \tag{14}$$

we have

$$z(x, t) = \begin{cases} 0, & \text{if } v(x, t) > 0, \\ z_0(x), & \text{if } v(x, t) < 0. \end{cases} \tag{15}$$

Problem Q. To find bounded measurable functions $u(x, t)$, $z(x, t)$ defined in

$t > 0$ satisfying (11) (12) and

$$z(x, t) = \begin{cases} 0, & \text{if } v(x, t) \geq 0, \\ z_0(x), & \text{if } v(x, t) < 0, \end{cases} \tag{16}$$

where $v(x, t)$ is defined by (14).

The formulation of Problem Q is stronger than that of Problem P, because (16) implies (13) and determines the state as $u = 0$.

§2. The existence of the solutions of Problem P.

First of all, let us consider a class of special initial values and discuss the properties of the solutions for these special initial value problems.

Lemma 3. If $(-\infty, +\infty)$ consists of a finite number of intervals, and $z_0(x) = \text{constant}$, $u_0(x)$ does not decrease on each of them, then the solution of Problem Q exists.

Proof. Suppose there are N intervals.

When $N = 1$, a solution $u(x, t)$ is constructed as the solution of equation (10) with initial value $u_0(x)$, $u(x, t)$ does not decrease as a function of x for each t. It is sufficient to set $z(x, t) \equiv z_0(x)$.

When $N = 2$, we may suppose the two intervals are $x \leq 0$ and $x > 0$ without losing generality. Let $u_r = u_0(+0)$ and $u_\ell = u_0(-0)$,

$$z_0(x) = \begin{cases} z_r, & x > 0, \\ z_\ell, & x \leq 0. \end{cases}$$

There are four cases (consult with the Riemann problem):

a) $u_\ell \leq u_r$, construct u as the case $N = 1$ and set $z(x, t) \equiv z_0(x)$.

b) $u_\ell > u_r$, but $u_\ell < 0$. Construct $u(x, t)$ with the initial value on $x > 0$ and $x < 0$ separately just like the case $N = 1$, then construct a discontinuity through the origin defined by

$$\frac{dx}{dt} = \frac{f(u(x - 0, t)) - f(u(x + 0, t))}{u(x - 0, t) - u(x + 0, t)}.$$

$u(x + 0, t)$ increases and $u(x - 0, t)$ decreases as t increases, but they are always unequal. We have $z(x, t) \equiv z_0(x)$ in this case.

c) $u_\ell > u_r$, $u_\ell \geq 0$, and $u_\ell > g(u_r, z_r)$. Construct $u(x, t)$ separately like b), then construct a discontinuity defined by

$$\frac{dx}{dt} = \frac{f(u(x - 0, t)) - f(u(x + 0, t))}{u(x - 0, t) - u(x + 0, t) - qz_r}.$$

As t increases, $u(x + 0, t)$, $u(x - 0, t)$ vary as the previous, if $u(x - 0, t) = g(u(x + 0, t_0), z_r)$ at some t_0, then it becomes the case d), i.e. the discontinuous varies from SD to CJ.

d) $u_\ell > u_r$, $u_\ell \geq 0$, but $u_\ell \leq g(u_r, z_r)$.

Step 1. Using the solution of $N = 1$ we obtain the solution on $x \geq f'(u_r)t$ and $x \leq f'(u_\ell)t$, denote them by $u_r(x, t)$ and $u_\ell(x, t)$ respectively.

Step 2. Solve the initial value problem of ordinary differential equation:

$$\frac{dx}{dt} = f'(g(u_r(x, t), z_r)),$$
$$x|_{t=0} = 0.$$

Since $u_r(x, t)$ is continuous on $t > 0$, we obtain a smooth solution $x = x(t)$. Because $g(u_r, z_r) \geq u_\ell$ and the slope of $x(t)$ increases, the curve $x = x(t)$ always lies in region $x \geq f'(u_\ell)t$.

Step 3. Construct a solution

$$u(x, t) = (f')^{-1}(\frac{x}{t}), \quad z = 0$$

on the sector $f'(u_\ell)t < x \leq f'(g(u_r, z_r))t$.

Step 4. Construct characteristics

$$\ell(\tau): \quad x = x(\tau) + f'(g(u_r(x(\tau), \tau), z_r))(t - \tau), \quad t \geq \tau,$$

in region $f'(g(u_r, z_r))t < x \leq x(t)$, then define

$$u = g(u_r(x(\tau), \tau), z_r), \quad z = 0,$$

on $\ell(\tau)$, it is easy to prove $\ell(\tau)$ cover the whole region and (u, z) is a solution, $x = x(t)$ is a CJ.

For the general case when $N \geq 2$, it is easy to prove by induction. □

Remark. If $(u(x, t), z(x, t))$ is the solution obtained by Lemma 3, then through any point (x_0, t_0) there is a characteristic

$$\frac{dx}{dt} = f'(u(x, t)), \quad t < t_0.$$

Identity $u(x, t) \equiv u(x_0, t_0)$ holds on this characteristic. If $u(x_0, t_0) \leq 0$, this line must intersect the x-axis, if $u(x, t) > 0$, it intersects either the x-axis, or a CJ.

Lemma 4. If the solution by Lemma 3 satisfies $u(x, t) \geq -M_0$, and $u(x, t) \leq 0$ at a point (x, t), then for any $\xi < x$, we have

$$\frac{u(x, t) - u(\xi, t)}{x - \xi} \leq \frac{C}{t},$$

where C is a constant depending on function f and constant M_0 only.

Proof. Take any $\xi < x$. If $u(\xi, t) \leq 0$, then by the Remark above, through (ξ, t) there is a downward characteristic which intersects the x-axis at point ξ_1, if x_1 is the intersection point of the characteristic through (x, t) and the x-axis, then

$$x - x_1 = f'(u(x, t))t,$$
$$\xi - \xi_1 = f'(u(\xi, t))t.$$

But these two characteristic do not intersect by the Remark, hence $x_1 \geq \xi_1$. We get

$$\begin{aligned} x - \xi &= x_1 + f'(u(x, t))t - \xi_1 - f'(u(\xi, t))t \\ &\geq f'(u(x, t))t - f'(u(\xi, t))t \\ &= f''(\tilde{u})(u(x, t) - u(\xi, t))t, \end{aligned}$$

where $\tilde{u}$ is a certain mean value, $\tilde{u} \geq -M_0$, so

$$\frac{u(x, t) - u(\xi, t)}{x - \xi} \leq \frac{1}{f''(\tilde{u})t} .$$

If $u(\xi, t) > 0$, then the above inequality is obvious because $u(x, t) \leq 0$. □

Lemma 5. The solution by Lemma 3 satisfies

$$\inf_x u_0(x) \leq u(x, t) \leq \max\{\sup_x u_0(x), g(0, 1)\}.$$

Lemma 6. The solution by Lemma 3 satisfies

$$0 \leq z(x, t) \leq 1.$$

The above two lemmas are obvious because of the structure of the solution.

Lemma 7. $\text{var}\ \psi(u(\cdot, t)) \leq C\{\text{var}\ \psi \circ u_0 + \text{var}\ z_0\}$,

where C is a constant depending on q, function f and $\sup|u(x, t)|$ only, $\psi \circ u_0$ is the composition of ψ and u_0.

Proof. For a given t, $u(x, t)$ is piecewise monotonous, $z(x, t)$ is piecewise constant, therefore the left and right limit $u^-(x, t)$, $z^-(x, t)$ and $u^+(x, t)$, $z^+(x, t)$, exists. Let x_i be the discontinuous points such that $z^-(x_i, t) < z^+(x_i, t)$, $u^-(x_i, t) < 0$ $(i = 1,2,\cdots)$, y_i be the discontinuous points such that $0 = z^-(y_i, t) < z^+(y_i, t)$, $u^-(y_i, t) \geq 0$ $(i = 1,2,\cdots)$, set

$$F(t) = \text{var}\ \psi(u(\cdot, t)) + 2q\sum_i\{z^+(x_i, t) - z^-(x_i, t)\}$$
$$+ 2\sum_i \max\{\psi(g(u^+(y_i, t), z^+(y_i, t))) - \psi(u^-(y_i, t)), 0\}.$$

Let us prove that $F(t) \leq F(0)$.

Firstly we compare $F(0)$ and $F(+0)$. Each term in F does not change locally at the interior points of every intervals and the discontinuous points for cases a) b) c). For case d), when $t = +0$, the jump of $\psi(u(x, t))$ at the discontinuous points are

$$2\psi(g(u^+(y_i, 0), z^+(y_i, 0))) - \psi(u^-(y_i, 0)) - \psi(u^+(y_i, 0))$$
$$= 2\{\psi(g(u^+(y_i, 0), z^+(y_i,0))) - \psi(u^-(y_i, 0))\} + \{\psi(u^-(y_i, 0)) - \psi(u^+(y_i,0))\},$$

hence

$$F(+0) = \text{var}\ \psi(u(\cdot,+0)) + 2q\sum_i\{z^+(x_i, +0) - z^-(x_i, +0)\} = F(0).$$

There are only a finite number of discontinuous curves and the numbers of intersection points are finite, we may assume that the lowest intersection point is at $t = t_0$. Next, we consider $F(t)$ as $t < t_0$. The third term in $F(t)$ disappears and the second term keeps invariable in this case, therefore we will consider the variation of $\psi \circ u$ only. For the continuous points of u, because u is constant along characteristic, the local variation keeps invariable along the characteristic. We will consider the influence of S, SD, CJ to the local variation only.

The local variation decreases for S and SD, hence they does not cause an increase of $F(t)$.

For a CJ curve $x(t)$, set $b = x(t)$, construct a characteristic through (b, t) which intersects the x-axis at $x = c$, construct a tangent line of $x(t)$ at $x = x(0)$, which is denoted by $x_1(\tau)$, let $a = x_1(t)$ as $\tau = t$, then

$$u(a, t) = u^-(x(0), +0).$$

By Lemma 2

$$\begin{aligned}\psi(u^-(b, t)) &= \psi(g(u^+(b, t), z^+(b, t)))\\ &= \psi(g(u(c, 0), z^+(x(0), 0)))\\ &\le \psi(g(u^+(x(0), 0), z^+(x(0), 0))) + \{u(c, 0) - u^+(x(0), 0)\}\\ &= \psi(u(a, t)) + \{u(c, 0) - u^+(x(0), 0)\}.\end{aligned}$$

Hence

$$\begin{aligned}&\{\psi(u^-(b, t)) - \psi(u(a, t))\} + \{\psi(u^-(b, t)) - \psi(u^+(b, t))\}\\ &\le \psi(u^-(b, t)) - u^+(x(0), 0))\end{aligned}$$

$$\leq \{\psi(u^-(x(0), +0)) - \psi(u^+(x(0), 0))\} + \{\psi(u(c, 0)) - \psi(u^+(x(0), 0))\},$$

therefore

$$F(t) \leq F(+0).$$

The discontinuous curves intersect each other at $t = t_0$, we denote by SD $\to$ C that a SD catchs up with a C. The variance from $F(t_0 - 0)$ to $F(t_0)$ is considered in the following cases:

SD $\to$ C or CJ $\to$ C: we denote by z^+ and z^- the values of z from the right and left sides of C, by u^+ and u^- the values of u from the right and left sides of SD or CJ, then $u^- \geq g(u^+, z^-)$.

If $z^+ < z^-$, then there is a new SD at $t = t_0$. The last two terms of $F(t)$ have no contribution, the local variation does not change.

If $z^+ > z^-$, then it corresponds to a point x_i, if $u^- > g(u^+, z^+)$, then there is also a SD, the local variation does not change, and $F(t)$ decreases.

If $z^+ > z^-$ and $u^- \leq g(u^+, z^+)$, then one term $2q\{z^+ - z^-\}$ in $F(t)$ will be eliminated and one term $2\{\psi(g(u^+, z^+)) - \psi(u^-)\}$ will be added from $t = t_0 - 0$ to $t = t_0$, but by Lemma 2

$$\psi(g(u^+, z^+)) - \psi(g(u^+, z^-)) \leq q(z^+ - z^-)$$

hence

$$\psi(g(u^+, z^+)) - \psi(u^-) \leq q(z^+ - z^-).$$

$F(t)$ does not increase either.

There are other cases such as S $\to$ C, S $\to$ SD, S $\to$ CJ, SD $\to$ S, CJ $\to$ S, it is easy to check $F(t)$ is invariable under these cases.

There is no difference between the analysis starting from $F(t_0)$ and from $F(0)$. By

$$F(0) \leq C\{\text{var } \psi \circ u_0 + \text{var } z_0\},$$

we obtain

$$F(t) \leq C\{\text{var } \psi \circ u_0 + \text{var } z_0\},$$

therefore

$$\text{var } \psi(u(\cdot, t)) \leq C\{\text{var } \psi \circ u_0 + \text{var } z_0\}. \qquad \square$$

Lemma 8. $\text{var } z(\cdot, t) \leq \text{var } z_0$.

Proof. Each interval where $z = 0$ extends as t increases, while z does not vary at the rest part, thus this conclusion follows. $\square$

Lemma 9. $\int_{-\infty}^{+\infty}|\psi(u(x, t)) - \psi(u(x, \tau))|dx \leq C|t - \tau|\{\text{var } \psi \circ u_0 + \text{var } z_0\}$,

where C depends on q, function f and $u_m = \sup|u(x, t)|$ only.

Proof. We may assume that $t > \tau$ without losing generality, construct a downward characteristic through point (x, t), it intersects either the x-axis or a CJ by the Remark of Lemma 3. If it is a CJ, then let the point move down along this CJ, if this CJ stops at a point (x_1, t_1), then continue this procedure by constructing a characteristic through (x_1, t_1), after a finite number of steps, this moving point will arrive at a point (x', τ) which lies at the same horizontal line as (x, τ).

Firstly we estimate $|\psi(u(x, t)) - \psi(u(x', \tau))|$, it is sufficient to consider the variance of u along CJ only because u does not change its value along characteristics. If $\widehat{P_1P_2}$ is a CJ arc mentioned before, we construct downward characteristics through P_1, P_2, they intersect the τ-horizontal line at points Q_1, Q_2 because $u^+(P_1)$ and $u^+(P_2)$ are negative, we construct perpendicular lines through P_1 and P_2 which intersect the τ-horizonted line at R_1 and R_2, then

$$\begin{aligned}&|\psi(u^-(P_2)) - \psi(u^-(P_1))| = |\psi(g(u^+(P_2), z^+(P_2))) - \psi(g(u^+(P_1), z^+(P_1)))|\\ &= |\psi(g(u(Q_2), z(R_2))) - \psi(g(u(Q_1), z(R_1)))|.\end{aligned}$$

By Lemma 2

$$|\psi(u^-(P_2)) - \psi(u^-(P_1))| \le |u(Q_2) - u(Q_1)| + q|z(R_2) - z(R_1)|.$$

There is no overlap among the $\overline{Q_1Q_2}$'s and among the $\overline{R_1R_2}$'s formed by the $\widehat{P_1P_2}$'s, hence

$$|\psi(u(x, t)) - \psi(u(x', \tau))| \le \mathrm{var}\{\psi(u(\cdot, \tau))|\ [x', x]\} + q\mathrm{var}\{z(\cdot, \tau)|\ [x', x]\}.$$

Secondly by applying the triangular inequality, we get

$$|\psi(u(x, t)) - \psi(u(x, \tau))| \le 2\ \mathrm{var}\{\psi(u(\cdot, \tau))|[x',x]\} + q\ \mathrm{var}\{z(\cdot, \tau)|\ [x', x]\},$$

taking $K = f'(u_m)$, we obtain

$$\begin{aligned}&|\psi(u(x, t)) - \psi(u(x, \tau))|\\ &\le 2\ \mathrm{var}\{\psi(u(\cdot, \tau))|\ [x - K|t - \tau|, x]\} + q\ \mathrm{var}\{z(\cdot, \tau)|\ [x - K|t - \tau|, x]\},\end{aligned}$$

if measure μ_u and μ_z are the partial derivatives of $\psi(u(x, \tau))$ and $z(x, \tau)$ with respect to x for fixed τ, then

$$|\psi(u(x, t)) - \psi(u(x, \tau))| \le \max(2, q)\int_{x-K|t-\tau|}^{x} d(|\mu_u| + |\mu_z|),$$

$$\begin{aligned}&\int_{-\infty}^{\infty} |\psi(u(x, t)) - \psi(u(x, \tau))|dx\\ &\le \max(2, q)\int_{-\infty}^{+\infty} dx\int_{x-K|t-\tau|}^{x} d(|\mu_u| + |\mu_z|)\\ &= \max(2, q)\int_{-\infty}^{+\infty} d(|\mu_u| + |\mu_z|)\int_{x-K|t-\tau|}^{x} dx\\ &= K\max(2, q)|t - \tau|\{\mathrm{var}\ \psi(u(\cdot, \tau)) + \mathrm{var}\ z(\cdot, \tau)\}.\end{aligned}$$

The conclusion follows from Lemma 7 and Lemma 8. □

Lemma 10. $\int_{-\infty}^{+\infty} |z(x, t) - z(x, \tau)|dx \le f'(u_m)|t - \tau|\ \mathrm{var}\ z_0$.

Proof. $z(x, t)$ varies on SD and CJ only, the slope of which is not greater than $f'(u_m)$, hence in the sense of distributions,

$$\int_{-\infty}^{+\infty} \left|\frac{\partial z}{\partial t}\right|dx \le f'(u_m)\int_{-\infty}^{+\infty} \left|\frac{\partial z}{\partial x}\right|dx = f'(u_m)\ \mathrm{var}\ z(\cdot, t).$$

By Lemma 8 the conclusion holds. □

Finally, we prove

Theorem 1. If $\psi(u_0(x))$ and $z_0(x)$ are functions with bounded variation, then the solution of Problem P exists, and $\psi \circ u$, $z \in BV$ ([4]).

Proof. We may assume that $u_0(x)$ is left continuous, and we may assume that $z_0(x)$ is left continuous on these points where $u_0(x) < 0$. For any integer n, we can define a function $v_0^{(n)}(x)$, such that $(-\infty, +\infty)$ consists of a finite number of intervals, $v_0^{(n)}(x)$ is constant on each interval, $|v_0^{(n)}(x) - \psi(u_0(x))| < \frac{1}{n}$. If $v_0^{(n)}(x) \geq 0$ on some interval, then set $z_0^{(n)}(x) \equiv 0$ on it, if $v_0^{(n)}(x) < 0$ on some interval, then we define $z_0^{(n)}(x)$ on this interval such that $z_0^{(n)}(x)$ are piecewise constants on a finite number of intervals, and $|z_0^{(n)}(x) - z_0(x)| < \frac{1}{n}$. Let $u_0^{(n)}(x) = \psi^{-1}(v_0^{(n)}(x))$. It is easy to see that as $n \to \infty$, $u_0^{(n)}(x)$ converges to $u_0(x)$ uniformly and $z_0^{(n)}(x)$ converges to $z_0(x)$ pointwise, and it is easy to make the variation of $\psi \circ u_0^{(n)}$, $z_0^{(n)}$ bounded uniformly. By definition, $z_0^{(n)}(x) = 0$ if $u_0^{(n)}(x) \geq 0$.

Problem Q with the initial value $u_0^{(n)}(x)$ and $z_0^{(n)}(x)$ has a solution $u_n(x, t)$ and $z_n(x, t)$ by Lemma 3, and the estimation in Lemmas 5-10 holds. We take a subsequence of $\{n\}$ such that for all $M > 0$ and $T > 0$, the corresponding subsequences of $\{\psi(u_n(x, t))\}$ and $\{z_n(x, t)\}$ converge in space $C([0, T]; L_1(-M, M))$, we still denote these subsequences by $\{\psi \circ u_n\}$ and $\{z_n\}$ for convenience. Denote by $w(x, t)$ and $z(x, t)$ the limit functions. They belong to space BV. Let $u(x, t) = \psi^{-1}(w(x, t))$. We change the value of $w(x, t)$ and $z(x, t)$ on a null measure set such that for every t, $w(x, t)$ and $z(x, t)$ are left continuous and the variation of them is bounded, then $u(x, t)$ is left continuous too. $u_n(x, t)$ and $z_n(x, t)$ are also left continuous by the Proof of Lemma 3.

Now we prove that $(u(x, t), z(x, t))$ is the solution of Problem P. Clearly it satisfies (12) and (13) because $(u_n(x, t), z_n(x, t))$ are solutions. We have to verify (15). Let

$$v_n(x, t) = \sup_{0 \leq \tau \leq t} u_n(x, \tau).$$

For any $t > 0$, we take a subsequence of $\{(u_n, z_n)\}$ again such that $\psi(u_n(x,t))$ and $z_n(x, t)$ converge to w and z almost everywhere as the functions of independent variable x, the subsequence is still denoted by $\{(u_n, z_n)\}$. u_n also converges almost everywhere by the continuity of ψ^{-1}. The set of points where $\{(u_n, z_n)\}$ does not converge to (u, z) is denoted by N_1.

Suppose that $x \bar{\in} N_1$. If $v(x, t) > 0$, there exists $t_1 \leq t$ such that $u(x, t_1) > 0$ by (14). By the left continuity, there are $\varepsilon > 0$ and $h > 0$, such that $u(\xi, t_1) \geq \varepsilon$ for $\xi \in [x - h, x]$. But $\psi(u_n(x, t_1))$ converges to $w(x, t_1)$ in L_1 norm, we take a subsequence such that it converges almost everywhere, if ξ is a point where it converges, then for sufficiently large n, $u_n(\xi, t_1) > 0$, hence $v_n(\xi, t) > 0$, $z_n(\xi, t) = 0$ by (15). Let $n \to \infty$, we get $z(\xi, t) = 0$ only if $\xi \bar{\in} N_1$. Therefore $z(\xi, t) = 0$ holds almost everywhere on $[x - h, x]$. Thus $z(x, t) = 0$ by the left continuity of z.

If $v(x, t) < 0$, we prove that there is a subsequence such that $v_n(x, t) < 0$. If not, then $v_n(x, t) \geq 0$ for all sufficiently large n. That is, for $\varepsilon > 0$ and every n, there exists a $\tau_n < t$, such that $u_n(x, \tau_n) \geq -\varepsilon$. If τ is a accumulation point of $\{\tau_n\}$. We take a subsequence, still denoted by $\{\tau_n\}$, such that $\tau_n \to \tau$ as $n \to \infty$. If $\tau > 0$, then for sufficiently large n, $\tau_n \geq \frac{\tau}{2}$. We have

$$u_n(\xi, \tau_n) \geq - \varepsilon - \frac{2C}{\tau} (x - \xi),$$

for all $\xi \leq x$ by Lemma 4. We take $|\xi - x| < \tau\varepsilon/2C$, then we get $u_n(\xi, \tau_n) \geq -2\varepsilon$ uniformly with respect to n. $\{\psi \circ u_n\}$ converges to w in L_1 norm uniformly with respect to t, hence

$$u(\xi, \tau) \geq -2\varepsilon, \quad 0 \leq x - \xi \leq \tau\varepsilon/2C,$$

almost every where. By the left continuity

$$u(x, \tau) \geq -2\varepsilon.$$

Hence $v(x, t) \geq -2\varepsilon$, but ε is arbitrary, hence $v(x, t) \geq 0$, which contradicts

$v(x, t) < 0$. If $\tau = 0$, then we may construct a characteristic or a curve consisting of piecewise characteristics and CJ, which intersects the x-axis at ξ_n, $\xi_n \in [x - f'(u_m)\tau_n, x]$, and $u_0^{(n)}(\xi_n) \geq -\varepsilon$. $\xi_n \to x$ as $n \to \infty$, we get $u_0(x) \geq -\varepsilon$ by the left continuity of $u_0(x)$ and the uniform convergence of $u_0^{(n)}$, but ε is arbitrary, hence $u_0(x) \geq 0$, so $v(x, t) \geq 0$, it is also a contradiction. Therefore, there is a subsequence such that $v_n(x, t) < 0$, hence $z_n(x, t) = z_0^{(n)}(x)$. Because $x \overline{\in} N_1$,

$$z(x, t) = \lim_{n\to\infty} z_n(x, t) = z_0(x).$$

Therefore (15) holds for almost every x. But t is arbitrary, thus (15) holds almost everywhere. We can change the value of $z(x, t)$ on a null set such that (15) holds everywhere. □

§3. The existence of the solutions of Problem Q.

First of all, let us introduce a definition and a hypothesis.

Definition. It is said that $u_0(x)$ assumes the value a at point x, if one of the following holds:

$$u_0(x - 0) = a;$$

$$u_0(x + 0) = a;$$

$$u_0(x + 0) > u_0(x - 0),\ a \in (u_0(x - 0), u_0(x + 0)).$$

Lemma 11. Suppose $(u(x, t), z(x, t))$ is the solution obtained by Theorem 1, if $u(x_0, t_0) \leq 0$ at a point (x_0, t_0), then the straight line, which is called the characteristic,

$$x - x_0 = f'(u(x_0, t_0))(t - t_0), \quad t \leq t_0,$$

has the following two properties:

a) $u_0(x)$ assumes the value $u(x_0, t_0)$ at point $x = x_0 - f'(u(x_0, t_0))t_0$ which is the intersection of this characteristic and x-axis;

b) if $u(x_0, t_0) \neq u(x_1, t_1)$ and $u(x_1, t_1) \leq 0$, then the downward characteristics through points (x_0, t_0) and (x_1, t_1) do not intersect on $t > 0$.

Proof. For any $h > 0$, $\varepsilon > 0$, there is an integer n and $\xi_n \in [x_0 - h, x_0]$, such that

$$|u_n(\xi_n, t_0) - u(x_0, t_0)| < \varepsilon,$$

because if not, then there were $h > 0$, $\varepsilon > 0$, such that

$$|u_n(\xi, t_0) - u(x_0, t_0)| \geq \varepsilon, \quad \xi \in [x_0 - h, x_0],$$

for sufficiently large n, which contradicts the L_1 convergence of $\psi \circ u_n$ and the left continuity of u.

According to the above property, there exists a subsequence of $\{u_n\}$, still denoted by $\{u_n\}$, and a series of points $\xi_n \to x$ as $n \to \infty$, such that

$$u_n(\xi_n, t_0) \longrightarrow u(x_0, t_0), \quad \text{as} \quad n \longrightarrow \infty.$$

We construct a downward characteristic of u_n through (ξ_n, t), if $u_n(\xi_n, t_0) \leq 0$, it intersects the x-axis, the intersection point is $x_n = \xi_n - f'(u_n(\xi_n, t_0))t_0$ and $u_0^{(n)}(x)$ assumes the value $u_n(\xi_n, t_0)$ at x_n, if $u_n(\xi_n, t_0) > 0$, we can construct a piecewise characteristic and CJ curve as in Lemma 9 which intersects the x-axis, $u_0^{(n)}(x)$ assumes a value at the intersection point x_n which is nonnegative and not greater than $u_n(\xi_n, t_0)$, hence $x \in [\xi_n - f'(u_n(\xi_n, t_0))t_0, \xi_n - f'(0)t_0]$. Let $n \to \infty$, $x_n \to x_0 - f'(u(x_0, t_0))t_0$. $u_0(x)$ assumes the value $u(x_0, t_0)$ at this point because $\{u_0^{(n)}\}$ converges uniformly and the left and right limit of $u_0(x)$ exists at each point, thus a) is proved.

As for b), we can take a subsequence $\{u_n\}$ and two series of points $\{\xi_n\}$ and $\{\xi_n'\}$, such that $\xi_n \to x_0$, $\xi_n' \to x_1$, $u_n(\xi_n, t_0) \to u(x_0, t_0)$, $u_n(\xi_n', t_1) \to u(x_1, t_1)$ as $n \to \infty$. $u_n(\xi_n, t) \neq u_n(\xi_n', t)$ for sufficiently large n, at least one of them is negative, therefore the characteristics or piecewise characteristic and CJ curves through points (ξ_n, t_0) and (ξ'_n, t_1) do not intersect on $t > 0$. These two families of curves converge to their limit positions as $n \to \infty$, which do not intersect on $t > 0$ either. □

We make the following hypothesis on the initial values:

Hypothesis A. If $u_0(x)$ assumes the value 0 at a point x, then there is an interval $[b, c] \ni x$ such that $u_0(x) \geq 0$ on (b, c).

Lemma 12. For the solution obtained by Theorem 1, if there are $t > t_0 > 0$ and $x \in (-\infty, +\infty)$, such that $v(x, t) = 0$, $z(x, t) > 0$, and t_0 is the supremum of those τ satisfying

$$v(x, \tau) < 0, \quad \tau \in [0, t_0),$$

then $u(x, t_0) = 0$.

Proof. Because t_0 is a supremum, there are only two possibilities:

a) there is a series $\tau_n \to t_0$, such that $u(x, \tau_n) \to 0$;

b) $u(x, t_0) = 0$.

If possibility a) holds, and if τ_n decreases montonously, $u(x, \tau_n) \leq 0$ because $v(x, t) = 0$. We construct a downward characteristic of u through the point (x, τ_n), which intersects the x-axis at $x_n = x - f'(u(x, \tau_n))\tau_n$. $x_n \to x_0 = x - f'(0)t$ as $n \to \infty$. If $u(\xi, t_0) \leq 0$ for some $\xi < x$, then a characteristic can be constructed from point (ξ, t_0) to the x-axis, let the intersection point be ξ'. (ξ, t_0) is on the left side of characteristic through (x, τ_n) for sufficiently large n, by Lemma 11, $\xi' \leq x_n$. Let $n \to \infty$, we get $\xi' \leq x_0$, hence the slope

$$f'(u(\xi, t_0)) = \frac{\xi - \xi'}{t_0} \geq \frac{\xi - x_0}{t_0},$$

that is

$$u(\xi, t_0) \geq (f')^{-1}\left(\frac{\xi - x_0}{t_0}\right).$$

The above inequality still holds if $u(\xi, t_0) \geq 0$. Let $\xi \to x$, u is left continuous, hence

$$u(x, t_0) \geq (f')^{-1}\left(\frac{x - x_0}{t_0}\right)$$

i.e. $u(x, t_0) \geq 0$. But $u(x, t_0) \leq 0$, hence $u(x, t_0) = 0$.

If τ_n increases monotonously, because $u(x, t_0) \leq 0$, we can construct a characteristic through point (x, t_0) and intersect the x-axis at point x', $x' \geq x_0 = x - f'(0)t_0$. But we have $x' \leq x_n$ by Lemma 11 for any n. $x_n \to x_0$ as $n \to \infty$, therefore $x' = x_0$, we also get $u(x, t_0) = 0$.

Therefore we get $u(x, t_0) = 0$ at any case. □

We know by Lemmas 11 and 12 that $u_0(x)$ assumes the value 0 at point x_0. Moreover, we can prove

<u>Lemma</u> 13. Under the conditions of Lemma 12, if Hypothesis A holds and $u_0^{(n)}(x) \geq u_0(x)$, $\lim_{n\to\infty} z_n(x, t) = z(x, t)$, $x_0 = x - f'(0)t_0$, then there is a constant $h > 0$, such that $u_0(\xi) \geq 0$ for every $\xi \in (x_0, x_0 + h)$, and there is always a point ξ in $(x_0 - \delta, x_0)$, such that $u_0(\xi) < 0$ for any $\delta > 0$.

<u>Proof</u>. We prove it by contradiction. If the conclusion were false, then there would be a $\delta > 0$, such that $u_0(\xi) \geq 0$ for $\xi \in (x_0 - \delta, x_0)$. We take δ sufficiently small, such that $\delta \leq f'(0)(t - t_0)$. Take an arbitrary $\tau \in [t_0, t_0 + \delta/f'(0)]$, then construct a characteristic of u through (x, τ). Since $u(x, \tau) \leq 0$, this characteristic intersects the x-axis at a point ξ. By Lemma 11, $\xi \leq x_0$. If $\xi \in [x_0 - \delta, x_0]$, then we get $u(x, \tau) \geq 0$ from $u_0(\xi) \geq 0$. If $\xi < x_0 - \delta$, then from the slope

$$\frac{x - \xi}{\tau} \geq \frac{x - x_0 + \delta}{t_0 + \delta/f'(0)} = \frac{f'(0)t_0 + \delta}{t_0 + \delta/f'(0)} = f'(0),$$

we also get $u(x, \tau) \geq 0$. But $u(x, \tau) \leq 0$, hence $u(x, \tau) = 0$. Because $z(x, t) > 0$ and $\lim_{n\to\infty} z_n(x, t) = z(x, t)$, $z_n(x, t) > 0$ for sufficiently large n. But (u_n, z_n) is the solution of Problem Q, so $v_n(x, t) < 0$ by (16), that is,

$u_n(x, \tau) < 0$ for all $\tau \le t$. As $\tau \in [t_0, t_0 + \delta/f'(0)]$, we construct a characteristic of u_n through (x, τ) intersecting the x-axis at point ξ. Because $u_0^{(n)}(x) \ge 0$ on $(x_0 - \delta, x_0)$, $\xi \ge x_0$. We fix one point $t_1 \in (t_0, t_0 + \delta/f'(0))$, then the slope of characteristic satisfies

$$\frac{x - \xi}{\tau} \le \frac{x - x_0}{t_1} < \frac{x - x_0}{t_0} = f'(0)$$

for $\tau \ge t_1$. Hence

$$u_n(x, \tau) \le -\eta < 0, \quad \tau \in [t_1, t_0 + \delta/f'(0)],$$

for all n. We take a $\beta \in (t_1, t_0 + \delta/f'(0))$, then $u(x, \beta) = 0$. In the same way as in the proof of Lemma 11, we may take a subsequence $\{u_n\}$ and a sequence of points $\xi_n \to x$ as $n \to \infty$, such that

$$u_n(\xi_n, \beta) \longrightarrow 0, \quad \text{as } n \longrightarrow \infty.$$

We construct the downward characteristic of u_n through (ξ_n, β), then the slope tends to $f'(0)$ as $n \to \infty$. We construct a downward straight line through the point $(x, t_0 + \delta/f'(0))$ with slope $f'(-\eta)$, a straight line through the point (x, β) with slope $f'(0)$, then the two lines intersect, (see Figure 1). Put

$$b = \frac{f'(-\eta)(t_0 + \delta/f'(0) - \beta)}{f'(0) - f'(-\eta)}.$$

Take β close enough to point $t_0 + \delta/f'(0)$, such that $b < \beta$, then the downward characteristics of u_n through $(x, t_0 + \delta/f'(0))$ and (ξ_n, β) would intersect on $t > 0$ for sufficiently large n. This is a contradiction.

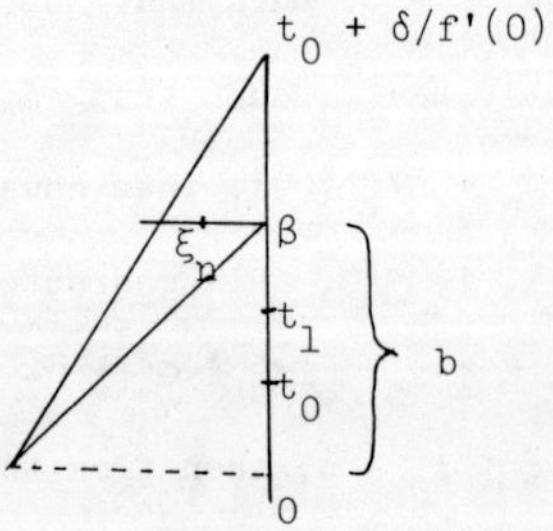

FIGURE 1

We have proved that there is always a point ξ in $(x_0 - \delta, x_0)$, such that $u_0(\xi) < 0$ for any $\delta > 0$. By Hypothesis A and the fact that $u_0(x)$ assumes the value 0 at point x_0, the only possibility is that there is a constant $h > 0$, such that $u_0(\xi) \ge 0$ for every $\xi \in (x_0, x_0 + h)$. □

Remark. Clearly the condition $\lim_{n\to\infty} z_n(x, t) = z(x, t)$ at the above lemma can be relaxed to the effect that this limit holds only for a subsequence.

By the convergence of sequence $\{z_n(x, t)\}$ proved in Theorem 1, we can take a subsequence from it again, still denoted by $\{z_n(x, t)\}$, such that it converges to $z(x, t)$ almost everywhere on $t \geq 0$. We define N_1 as a set of (x, t) such that any subsequence of $\{z_n(x, t)\}$ does not converge to $z(x, t)$. It is obvious that N_1 is a null measure set.

Lemma 14. If Hypothesis A holds and $u_0^{(n)}(x) \geq u_0(x)$, there are $t > t_0 > 0$, $x \in (-\infty, +\infty)$ and $s > s_0 > 0$, $y \in (-\infty, +\infty)$ such that t, t_0, x and s, s_0, y satisfy the conditions of Lemma 12 respectively, and $(x, t) \bar{\in} N_1$, $(y, s) \bar{\in} N_1$. Set

$$x_0 = x - f'(0)t_0, \quad y_0 = y - f'(0)s_0$$

then $x_0 \neq y_0$, if $x \neq y$.

Proof. We may assume that $x < y$. Take a subsequence of $\{(u_n, z_n)\}$ such that $\{z_n\}$ converges to z at point (x, t). We obtain $z_0(x) > 0$ from $z(x, t) > 0$. Thus $u_0(x) < 0$.

If $u_0(x + 0) > 0$, then there is a constant $\delta > 0$, such that $u_0(\xi) \geq 0$ for $\xi \in (x, x + \delta)$, hence $u_0^{(n)}(\xi) \geq 0$.

If $u_0(x + 0) < 0$, by the same reason, there is a constant $\delta > 0$, such that $u_0(\xi) < 0$ for $\xi \in (x, x + \delta)$, hence $u_0^{(n)}(\xi) < 0$.

If $u_0(x + 0) = 0$, then u_0 assumes the value 0 at point x, but $u_0(x) < 0$, by Hypothesis A, there is also a $\delta > 0$, such that $u_0^{(n)}(\xi) \geq 0$ for $\xi \in (x, x + \delta)$.

It is known that $u_n(x, \tau) < 0$ for $\tau \in [0, t]$ and sufficiently large n, and that $u_0^{(n)}$ does not change its sign on interval $(x,x+\delta)$. Therefore $u_n(\xi,\tau)$ is a generalized solution of equation (10) on domain $\Omega = \{(\xi, \tau);\ x < \xi < x + \delta,\ 0 < \tau < t\}$ with the value on $\tau = 0$ and $\xi = x$ as its initial value and boundary value, $\frac{\partial z_n}{\partial t} = 0$, so is the limit function $u(\xi, \tau)$. Therefore through any point (ξ, τ) in Ω we can always construct a downward characteristic of u. If it intersects line $\xi = x$, then $u \leq 0$. We can continue this characteristic to the x-axis. Let the intersection point be ξ'. By Lemma 13, $\xi' \bar{\in} (x_0, x_0 + h)$. Because if $\xi' \in (x_0, x_0 + h)$, then $u(\xi, \tau) \geq 0$, hence $u(\xi, \tau) = 0$. If (x, t_0) is on this characteristic, then $\xi' = x_0$. If (x, t_0) is at the left side of it, then, by $u(x, \tau_1) < 0$ $(\forall \tau_1 < t_0)$, $u(\xi, \tau) \geq 0$ is impossible. If (x, t_0) is at the right side of it, then by Lemma 11, $\xi' \leq x_0$. Therefore, all contradict $\xi' \in (x_0, x_0 + h)$.

Let ξ' be the intersection point of characteristic through point (ξ, τ) and the x-axis. If $\xi' \geq x_0 + h$, then from the slope of it we get

$$u(\xi, \tau) \leq (f')^{-1}\left(\frac{\xi - x_0 - h}{\tau}\right) .$$

If $\xi' \leq x_0$, from the slope we get

$$u(\xi, \tau) \geq (f')^{-1}\left(\frac{\xi - x_0}{\tau}\right) .$$

Let $(\xi, \tau) \to (x, t_0)$, then the right hand sides have the limit $(f')^{-1}\left(\frac{x-x_0-h}{t_0}\right)$ and $(f')^{-1}\left(\frac{x-x_0}{t_0}\right) = 0$ respectively. Therefore there is a neighborhood of (x, t_0) such that

$$(f')^{-1}\left(\frac{\xi - x_0 - h}{\tau}\right) \leq -2\theta < 0,$$

$$(f')^{-1}\left(\frac{\xi - x_0}{\tau}\right) \geq -\theta,$$

then a discontinuity curve $\xi = x(\tau)$ through point (x, t_0) is generated [5]. By the Rankine-Hugoniot condition we know the slope of this discontinuity is

$$x'(\tau) = \frac{f(u(x(\tau) - 0, \tau)) - f(u(x(\tau) + 0, \tau))}{u(x(\tau) - 0, \tau) - u(x(\tau) + 0, \tau)}$$

$$\leq \frac{f(0) - f(-2\theta)}{2\theta} < f'(0).$$

The points which lie at the neighborhood of the line

$$\xi - x = f'(0)(\tau - t_0) \tag{17}$$

are at the right side of $x(\tau)$, hence the values of u on these points satisfy $u(\xi, \tau) \leq -2\theta$.

Now we prove $x_0 \neq y_0$. If not, then line (17) would coincide with line

$$\xi - y = f'(0)(\tau - s_0), \tag{18}$$

because they intersect the x-axis at the same point and their slopes are equal. But when $\xi \in (x, x + \delta)$, at the neighborhood of line (17) we always have $u(\xi, \tau) \leq -2\theta$, but $u \equiv 0$ on line (18), this is a contradiction. □

Finally, we prove the existence theorem.

Theorem 2. If $\psi(u_0(x))$ and $z_0(x)$ are functions with bounded variation and Hypothesis A holds, then the solutions of Problem Q exist, and $\psi \circ u, z \in BV$.

Proof. There are at most countably many points x_0 which satisfy the conclusion of Lemma 13. By Lemma 14, there is at most one x which corresponds to an x_0 such that the condition of Lemma 12 holds. Hence there are at most countably many such points x. Denote the set of them by N_x. Set $N_2 = N_x \times [0, +\infty)$, then N_2 is a null measure set on the (x, t) plane.

We take $u_0^{(n)}(x) \geq u_0(x)$ and obtain the solution $u(x, t), z(x, t)$ of Problem P by Theorem 1. We prove that $u(x, t)$ and $z(x, t)$ also satisfy equation (16) in the case of $v(x, t) = 0$. Taking an arbitrary $x \in (-\infty, +\infty)$, we prove there is at most one t satisfying

$$(x, t) \overline{\in} N_1 \cup N_2, \ z(x, t) > 0, \ v(x, t) = 0. \tag{19}$$

If not, there would be t and t_1 satisfying the above condition. We may assume that $t > t_1$. If t_0 is the supremum of those τ satisfying $v(x, \tau) < 0$, $\tau \in [0, t_0)$, then $t_0 \leq t_1$. By Lemma 13, x corresponds to an x_0 which satisfies the conclusion of Lemma 13, hence $(x, t) \in N_2$. This is a contradiction.

Therefore the set of points which satisfy (19) is of measure zero, which is denoted by N_3.

$N_1 \cup N_2 \cup N_3$ is a null measure set, we define the value of $z(x, t)$

according to (16) on this set, then (16) is satisfied everywhere. The obtained $u(x, t)$, $z(x, t)$ is the solution of Problem Q. □

References

[1] Williams, F.A., Combustion Theory, Addison-Wesley, Reading, Mass., 1965.

[2] Courant, R. and Friedrichs, K.O., Supersonic Flow and Shock Waves (Interscience Publishers, Inc., New York, 1948).

[3] Majda, A., A qualitative model for dynamic combustion, SIAM J. Appl. Math. 41, 1 (1981) 70-93.

[4] Volpert, A.I., The space BV and quasilinear equations, Math. USSR Sb., 2 (1967) 257-267.

[5] Dafermos, C.M., Characteristics in hyperbolic conservation laws, a study of the structure and the asymptotic behavior of solutions, in Nonlinear Analysis and Mechanics, Vol. 1, Pitman, London, 1977.

Lecture Notes in Num. Appl. Anal., **5**, 435–457 (1982)
Nonlinear PDE in Applied Science. U.S.-Japan Seminar, Tokyo, 1982

Boundary Value Problems for Some Nonlinear Evolutional Systems of Partial Differential Equations

Zhou Yu-lin

Department of Mathematics
Peking University
Beijing

CHINA

In the paper, the boundary value problems for the nonlinear systems of the Schrödinger type, the pseudo-parabolic type and the pseudo-hyperbolic type of partial differential equations are considered. The generalized global solutions and the classical global solutions for the boundary value problems of these nonlinear systems are obtained.

§1. Systems of Schrödinger Type.

The nonlinear Schrödinger equations

$$u_t - iu_{xx} + \beta|u|^p u = 0 \tag{1.1}$$

and the nonlinear Schrödinger systems

$$u_t - iu_{xx} + u(\alpha|u|^2 + \beta|v|^2) = 0,$$

$$v_t - iv_{xx} + v(\alpha|u|^2 + \beta|v|^2) = 0 \tag{1.2}$$

of complex valued functions [1-4], regarded as the systems of real value functions (of real parts and imaginary parts) are contained in the general system

$$u_t - A(t)u_{xx} = f(u) \tag{1.3}$$

as simple special cases, where u and $f(u)$ are N-dimensional vector valued functions, $A(t)$ is a nonsingular and nonnegatively definite matrix. In the problems of the theoretical physics, chemical reactions etc., it is very often that there appear the equations and systems of such kind. For the systems of form (1.3) of higher order, the periodic boundary problems and the initial value problems have been studied in [5-7] and the generalized global solutions and the classical global solutions are obtained.

Now in the present section, we are going to consider the first boundary value problems

$$u(0, t) = u(\ell, t) = 0,$$
$$u(x, 0) = u_0(x) \tag{1.4}$$

in the rectangular domain $Q_T = \{0 \leqslant x \leqslant \ell, 0 \leqslant t \leqslant T\}$ for the systems (1.3) of the Schrödinger type of second order, where $u_0(x)$ is an initial vector valued function.

Let us take the approximate semilinear parabolic system

$$u_t - A(t)u_{xx} - \varepsilon u_{xx} = f(u) \tag{1.5}$$

where $\varepsilon > 0$. Firstly we establish the solutions for the problem (1.5), (1.4). And next we get the solution for the degenerate problem (1.3), (1.4) by passing to limit as $\varepsilon \to 0$.

As a consequence of the result in [8], we have the following lemma for the case of the linear parabolic systems.

Lemma 1.1. Suppose that for the linear parabolic systems

$$u_t - A(x, t)u_{xx} + B(x, t)u_x + C(x, t)u = f(x, t) \tag{1.6}$$

and the boundary condition (1.4), hold the following assumptions.

(1) The $N \times N$ coefficient matrices $A(x, t)$, $B(x, t)$ and $C(x, t)$ are measurable and bounded and $A(x, t)$ is positively definite.

(2) The free term vector valued function $f(x, t)$ is quadratic integrable in Q_T.

(3) The initial vector valued function $u_0(x)$ belongs to $W_2^{(1)}(0, \ell)$.

Then the boundary value problem (1.6), (1.4) has a unique solution $u(x, t) \in L_\infty((0, T); W_2^{(1)}(0, \ell)) \cap W_2^{(2,1)}(Q_T)$, satisfying the estimating relation

$$\sup_{0\leq t\leq T} \| u(\cdot,t)\|_{W_2^{(1)}(0,\ell)} + \| u_t\|_{L_2(Q_T)} + \| u_{xx}\|_{L_2(Q_T)}$$

$$\leq K_1\{\| u_0\|_{W_2^{(1)}(0,\ell)} + \| f\|_{L_2(Q_T)}\} , \tag{1.7}$$

where K_1 is a constant.

Theorem 1.1. Suppose that the coefficient matrix $A(t)$ is bounded, the Jacobi derivative matrix $\frac{\partial f(u)}{\partial u}$ of the vector valued function $f(u)$ is semibounded, i.e., there exists a constant b such that for any N-dimensional vectors $\xi \in \mathbb{R}^N$, $(\xi, \frac{\partial f(u)}{\partial u}\xi) \leq b(\xi, \xi)$ holds, for all $|u| < \infty$ and the initial vector valued function $u_0(x) \in W_2^{(1)}(0, \ell)$. Then the boundary problem (1.5), (1.4) has a unique global solution $u(x, t) \in L_\infty((0, T); W_2^{(1)}(0, \ell)) \cap W_2^{(2,1)}(Q_T)$, where $\varepsilon > 0$.

The proof of the existence of the approximate solutions $u_\varepsilon(x, t)$ is based on the fixed point theorem treatment as similarly used in [9, 10]. The method of integral estimations for the proof of this theorem is very similar to the way of the estimations needed in the limiting process of the approximate solutions to the solution of degenerate problem.

Thus we have a set of approximate solutions $\{u_\varepsilon(x, t)\}$ for the nondegenerate problem (1.5), (1.4), where $\varepsilon > 0$.

Taking the scalar product of the vector u_ε and the system (1.5) and integrating the resulting relation for x in the interval $[0, \ell]$, we get

$$\frac{1}{2}\frac{d}{dt}\|u(\cdot,t)\|^2_{L_2(0,\ell)} - \int_0^\ell(u(x,t), A(t)u_{xx}(x,t))dx - \varepsilon\int_0^\ell(u(x,t),u_{xx}(x,t))dx$$

$$= \int_0^\ell(u(x,t), f(u(x,t)))dx.$$

By making use of the boundary condition (1.4), the second and the third terms of the left hand side of the above equality take the forms

$$-\int_0^\ell (u(x,t),\ A(t)u_{xx}(x,t))dx = \int_0^\ell (u_x(x,t),\ A(t)u_x(x,t))dx \geqslant 0,$$

$$-\int_0^\ell (u(x,t),\ u_{xx}(x,t))dx = \|u_x(\cdot,t)\|^2_{L_2(0,\ell)}$$

respectively. By virtue of the semiboundedness of the Jacobi derivative matrix $\frac{\partial f(u)}{\partial u}$, the last term of the above equality can be written as

$$\int_0^\ell (u(x,t),\ f(u(x,t)))dx \leqslant (b+\tfrac{1}{2})\|u(\cdot,t)\|^2_{L_2(0,\ell)} + \tfrac{1}{2}\ell|f(0)|^2.$$

Then the above mentioned equality becomes

$$\frac{d}{dt}\|u(\cdot,t)\|^2_{L_2(0,\ell)} + 2\varepsilon\|u_x(\cdot,t)\|^2_{L_2(0,\ell)} \leqslant (2b+1)\|u(\cdot,t)\|^2_{L_2(0,\ell)} + \ell|f(0)|^2.$$

By means of Gronwall's lemma, the following lemma holds.

Lemma 1.2. Under the conditions of Theorem 1.1, the approximate solutions $\{u_\varepsilon(x,\ t)\}$ of the problem (1.5), (1.4) have the estimation

$$\sup_{0\leqslant t\leqslant T} \|u_\varepsilon(\cdot,t)\|_{L_2(0,\ell)} + \sqrt{\varepsilon}\ \|u_{\varepsilon x}(\cdot,t)\|_{L_2(0,\ell)} \leqslant K_2, \tag{1.8}$$

where K_2 is independent of $\varepsilon > 0$ and directly dependent on $\ell|f(0)|^2$. When $f(0) = 0$ is a zero vector or the system (1.5) is homogeneous, K_2 is also independent of $\ell > 0$.

In order to estimate the derivative $u_{\varepsilon x}(x,\ t)$, we make the scalar product of u_{xx} with the system (1.5) and integrate the resulting relation for x in interval $[0,\ \ell]$ by parts. Then we have

$$\frac{d}{dt}\|u_x(\cdot,t)\|^2_{L_2(0,\ell)} + 2\varepsilon\|u_{xx}(\cdot,t)\|^2_{L_2(0,\ell)} < 2b\|u_x(\cdot,t)\|^2_{L_2(0,\ell)} \tag{1.9}$$

where the system (1.5) is assumed to be homogeneous, i.e., $f(0) = 0$. In fact, by virtue of the boundary condition (1.4),

$$u_t(0,\ t) = u_t(\ell,\ t) = 0,$$

then we have

$$\int_0^\ell (u_{xx}, u_t)dx = -\frac{1}{2}\frac{d}{dt}\|u_x(\cdot, t)\|^2_{L_2(0,\ell)}.$$

On the other hand,

$$\int_0^\ell (u_{xx}, f(u))dx = (u_x(x,t), f(u(x,t)))\Big|_{x=0}^{x=\ell} - \int_0^\ell (u_x, \frac{\partial f(u)}{\partial u}u_x)dx.$$

Under the assumption $f(0) = 0$, this becomes

$$-\int_0^\ell (u_{xx}, f(u))dx = \int_0^\ell (u_x, \frac{\partial f(u)}{\partial u} u_x)dx \leqslant b\|u_x(\cdot, t)\|^2_{L_2(0,\ell)}.$$

From the inequality (1.9), we have the following lemma.

Lemma 1.3. For the homogeneous system (1.5), i.e., $f(0) = 0$, under the assumptions of Theorem 1.1, the approximate solutions $\{u_\varepsilon(x, t)\}$ have the estimation

$$\sup_{0\leqslant t\leqslant T} \|u_{\varepsilon x}(\cdot, t)\|_{L_2(0,\ell)} + \sqrt{\varepsilon}\,\|u_{\varepsilon xx}\|_{L_2(Q_T)} \leqslant K_3, \tag{1.10}$$

where K_3 is independent of $\varepsilon > 0$ and $\ell > 0$.

Differentiating the system (1.5) with respect to x, making the scalar product of the resulting relation and u_{xxx}, then integrating for x in interval $[0, \ell]$, we have

$$\int_0^\ell (u_{xxx}, u_{xt})dx - \int_0^\ell (u_{xxx}, A(t)u_{xxx})dx - \varepsilon\int_0^\ell (u_{xxx}, u_{xxx})dx$$
$$= \int_0^\ell (u_{xxx}, f(u)_x)dx. \tag{1.11}$$

On the lateral boundaries $x = 0, \ell$ of the rectangular domain Q_T, the relations

$$u_{xx}(0, t) = u_{xx}(\ell, t) = 0 \tag{1.12}$$

follow immediately from the system (1.5). In fact, on account of the nonsingularity of the matrix $A(t)$, the inverse matrix $A_\varepsilon^{-1}(t)$ of the matrix $A_\varepsilon(t) = A(t) + \varepsilon E$ is bounded for $\varepsilon > 0$ and $0 \leqslant t \leqslant T$, then the system (1.5) can be expressed as

$$A_\varepsilon^{-1}(t)u_t - u_{xx} = A_\varepsilon^{-1}(t)f(u).$$

Thus the conditions (1.12) are obviously available. Also

$$\int_0^\ell (u_{xxx}, f(u)_x)dx \leqslant C_1 \| u_{xxx}(\cdot, t)\|^2_{L_2(0,\ell)} + C_2 ,$$

where C_1 and C_2 are constants dependent on K_1, K_2 and are independent of $\varepsilon > 0$ and $\ell > 0$.

<u>Lemma</u> 1.4. Besides the conditions of Lemma 1.3, assume that $f(u)$ is twice continuously differentiable and $u_0(x) \in W_2^{(2)}(0, \ell)$. The approximate solutions $\{u_\varepsilon(x, t)\}$ satisfy the inequality

$$\sup_{0\leqslant t\leqslant T} \| u_{xx}(\cdot, t)\|_{L_2(0,\ell)} + \| u_t(\cdot, t)\|_{L_2(0,\ell)} \leqslant K_4, \tag{1.13}$$

where K_4 is independent of $\varepsilon > 0$ and $\ell > 0$.

By means of the above estimations we can construct the global solution of problem (1.3), (1.4) from the set of approximate solutions $\{u_\varepsilon(x, t)\}$. Under the assumptions of Lemma 1.4, $\{u_\varepsilon(x, t)\}$ is uniformly bounded in the functional space $L_\infty((0, T); W_2^{(2)}(0, \ell)) \cap W_2^{(1)}((0, T); L_2(0, \ell))$ for $\varepsilon > 0$. Then $\{u_\varepsilon(x, t)\}$ and $\{u_{\varepsilon x}(x, t)\}$ are uniformly bounded in the space of Hölder continuous functions for $\varepsilon > 0$. It can be selected from $\{u_\varepsilon(x, t)\}$, a sequence $\{u_{\varepsilon_i}(x, t)\}$, that there exists a vector valued function $u(x, t)$, such that when $i \to \infty$, $\varepsilon_i \to 0$, the sequences $\{u_{\varepsilon_i}(x, t)\}$ and $\{u_{\varepsilon_i x}(x, t)\}$ are uniformly convergent to $u(x, t)$ and $u_x(x, t)$ respectively in Q_T. Hence it is clear that $\{f(u_{\varepsilon_i}(x, t))\}$ uniformly converges to $f(u(x, t))$ and also $\{u_{\varepsilon_i xx}(x, t)\}$ and $\{u_{\varepsilon_i t}(x, t)\}$ converge weakly to $u_{xx}(x, t)$ and $u_t(x, t)$ respectively. For any test function $\psi(x, t)$, there is the integral relation

$$\iint_{Q_T} \psi[u_{\varepsilon t} - A(t)u_{\varepsilon xx} - \varepsilon u_{\varepsilon xx} - f(u_\varepsilon)]dxdt = 0. \tag{1.14}$$

From the estimation formular (1.10), we know that

$$|\iint_{Q_T} \psi\varepsilon u_{\varepsilon xx} dxdt| \leqslant \sqrt{\varepsilon}\, \|\psi\|_{L_2(Q_T)} \|\sqrt{\varepsilon}\, u_{\varepsilon xx}\|_{L_2(Q_T)}$$

tends to zero as $\varepsilon_i \to 0$. Therefore passing to the limit for the (1.4), we get

$$\iint_{Q_T} \psi[u_t - A(t)u_{xx} - f(u)]dxdt = 0. \tag{1.15}$$

This means that $u(x, t)$ satisfies (1.3) almost everywhere, i.e., $u(x, t)$ is a generalized global solution of the boundary value problem (1.4) for the degenerate system (1.3) of the Schrödinger type.

Theorem 1.2. Under the conditions of Lemma 1.4, the boundary value problem (1.4) for the system (1.3) of Schrödinger type has a unique global solution $u(x, t) \in L_\infty((0, T); W_2^{(2)}(0, \ell)) \cap W_\infty^{(1)}((0, T); L_2(0, \ell))$.

Since the estimations given in the last three lemmas are all independent of the width $\ell > 0$ of the rectangular domain Q_T, by taking the limiting process for $\ell \to \infty$, we can obtain the solution of the boundary value problem

$$\begin{aligned} &u(0, t) = 0, && 0 \leqslant t \leqslant T; \\ &u(x, 0) = u_0(x), && 0 \leqslant x < \infty \end{aligned} \tag{1.16}$$

in the infinite domain $Q_T^* = \{0 \leqslant x < \infty, \ 0 \leqslant t \leqslant T\}$ for the system (1.3) of Schrödinger type.

Theorem 1.3. Suppose that all conditions of Lemma 1.4 hold with the replacement of $\ell > 0$ by ∞. Then the boundary value problem (1.16) in Q_T^* for the system (1.3) of Schrödinger type has a unique global solution $u(x, t) \in L_\infty((0, T); W_2^{(2)}(0, \infty)) \cap W_\infty^{(1)}((0, T); L_2(0, \infty))$.

By the similar way it is not difficult to obtain the calssical global solutions and the smooth global solutions of the boundary value problem (1.4) in Q_T and (1.16) in Q_T^* for the nonlinear systems of the generalized Schrödinger type.

§2. Systems of Pseudo-parabolic Type.

Recently many authors have paid great attention to the study of the linear and nonlinear pseudo-parabolic equations. The nonlinear pseudo-parabolic equations often occur in practical research, such as the so-called BBM equations

$$u_t + f(u)_x = u_{xxt} \tag{2.1}$$

for long waves in nonlinear dispersion systems[11-13], the equations in the cooling process according to two-temperature of heat conduction, the equations for filtration of fluids in the broken rocks, the equations of Sobolev-Galpern type and so forth[14-20]. These equations contain the differential operator $u_t - u_{xxt}$ as their mean part. Some fairly general family of nonlinear pseudo-parabolic systems[21, 22], which contain all above mentioned equations as simple special cases are considered, such as the systems

$$u_t + (-1)^M A u_{x^{2M}t} = f(u, u_x, \dots, u_{x^{2M}}) \tag{2.2}$$

with the special form of the right hand part

$$f_j = \sum_{m=1}^{M} (-1)^m D_x^{m+1} \frac{\partial F}{\partial p_{j,m-1}} + \sum_{m=1}^{M} (-1)^m D_x^m \frac{\partial G}{\partial p_{j,m-1}} + h_j(u), \quad j = 1, 2, \dots, N, \tag{2.3}$$

where u and $h(u)$ are N-dimensional vector valued functions, $F(u, u_x, \dots, u_{x^{M-1}})$ and $G(u, u_x, \dots, u_{x^{M-1}})$ are smooth functions, $p_{j,m} = u_{j x^m}$ ($j = 1, \dots, N$; $m = 0, 1, \dots, M - 1$), and A is a $N \times N$ positively definite constant matrix. The generalized global solutions and the classical global solutions of the periodic boundary problems and Cauchy problems for the systems (2.2), (2.3) are obtained in [21].

In this section, firstly we are going to consider the questions of a priori estimation for the linear pseudo-parabolic systems, then we turn to study the problems for nonlinear pseudo-parabolic systems.

For the linear pseudo-parabolic systems

$$(-1)^M u_t + A(x, t) u_{x^{2M}t} + \sum_{k=0}^{2M} B_k(x, t) u_{x^{2M-k}} = f(x, t), \tag{2.4}$$

let us consider the boundary value problem

$$u_{x^k}(0, t) = u_{x^k}(\ell, t) = 0 \quad (k = 0, 1, \dots, M - 1),$$

$$u(x, 0) = \varphi(x), \tag{2.5}$$

where u is a N-dimensional unknown vector valued function, $f(x, t)$ is a N-dimensional quadratic integrable in Q_T vector valued function, $\varphi(x)$ is a N-dimensional initial vector valued function, satisfying the homogeneous boundary conditions, $A(x, t)$ is a $N \times N$ symmetric positively definite matrices, $B_k(x, t)$ $(k = 0, 1,\dots, 2M)$, $A(x, t)$ and $A_t(x, t)$ are measurable and bounded matrices. .

Taking the scalar product of $u_{x^{2M}}$ with the linear system (2.4) and integrating the resulting relation with respect to x in interval $[0, \ell]$, we get

$$(-1)^M\int_0^\ell(u_{x^{2M}}, u_t)dx + \int_0^\ell(u_{x^{2M}}, A(x,t)u_{x^{2M}t})dx$$

$$+ \int_0^\ell \sum_{k=0}^{2M}(u_{x^{2M}}, B(x,t)u_{x^{2M-k}})dx = \int_0^\ell(u_{x^{2M}}, f(x,t))dx. \tag{2.6}$$

Since $A(x, t)$ is a symmetric matrix,

$$(u_{x^{2M}}, A(x,t)u_{x^{2M}t}) = \frac{1}{2}(u_{x^{2M}}, A(x,t)u_{x^{2M}})_t - \frac{1}{2}(u_{x^{2M}}, A_t(x,t)u_{x^{2M}}).$$

From the interpolation formulae

$$\| u_{x^{2M-k}}(\cdot,t)\|^2_{L_2(0,\ell)} \leq C_1\|u_{x^{2M}}(\cdot,t)\|^2_{L_2(0,\ell)} + C_2\|u(\cdot,t)\|^2_{L_2(0,\ell)}$$

$$(k = 0, 1,\dots, 2M)$$

and the relation

$$\| u(\cdot,t)\|_{L_2(0,\ell)} \leq C_3\|u_{x^M}(\cdot,t)\|_{L_2(0,\ell)}$$

obtained directly from the homogeneous boundary condition (2.5), the above equality becomes

$$\frac{d}{dt}\|u_{x^M}(\cdot,t)\|^2_{L_2(0,\ell)} + \frac{d}{dt}\int_0^\ell(u_{x^{2M}}, Au_{x^{2M}})dx$$

$$\leq \| f(\cdot,t)\|^2_{L_2(0,\ell)} + C_4\|u_{x^M}(\cdot,t)\|^2_{L_2(0,\ell)} + C_5\|u_{x^{2M}}(\cdot,t)\|^2_{L_2(0,\ell)}. \tag{2.7}$$

On account of the positive definiteness of the matrix $A(x, t)$, we get the estimation relation

$$\sup_{0\leq t\leq T} \|u(\cdot,t)\|_{W_2^{(2M)}(0,\ell)} \leq K_1\{\|f\|_{L_2(Q_T)} + \|\varphi\|_{W_2^{(2M)}(0,\ell)}\}. \tag{2.8}$$

Again making the scalar product of vector $u_{x^{2M}t}$ with the linear system (2.4), we get the relation

$$(-1)^M(u_{x^{2M}t}, u_t) + (u_{x^{2M}t}, Au_{x^{2M}t}) + \sum_{k=0}^{2M} (u_{x^{2M}t}, B_k u_{x^{2M-k}}) = (u_{x^{2M}t}, f).$$

Then integrating this equality in the rectangular domain Q_T, it follows that

$$\|u_{x^M t}\|^2_{L_2(Q_T)} + a\|u_{x^{2M}t}\|^2_{L_2(Q_T)}$$

$$\leq C_6\{\|f\|_{L_2(Q_T)} + \|u\|_{L_2((0,T);\, W_2^{(2M)}(0,\ell))}\}\|u_{x^{2M}t}\|_{L_2(Q_T)},$$

where in the derivation, the boundary relations

$$u_{x^k t}(0, t) = u_{x^k t}(\ell, t) = 0 \qquad (k = 0, 1, \ldots, M - 1) \tag{2.9}$$

have been used and a is the least eigenvalue of the positively definite matrix $A(x, t)$. Hence there is the estimation relation

$$\|u_t\|_{L_2((0,T);\, W_2^{(2M)}(0,\ell))} < K_2\{\|f\|_{L_2(Q_T)} + \|\varphi\|_{W_2^{(2M)}(0,\ell)}\}. \tag{2.10}$$

By means of the method of continuation of parameter, it follows from the obtained a priori estimations (2.8) and (2.10), the boundary value problem (2.5) of the linear pseudo-parabolic system (2.4) has the solution in the functional space $W_2^{(1)}((0, T);\, W_2^{(2M)}(0, \ell))$. Since the problem (2.4), (2.5) is linear, the uniqueness of the solution is an immediate consequence of the estimation relations.

Theorem 2.1. Suppose that the linear pseudo-parabolic system (2.4) and the boundary value condition (2.5) satisfy the following conditions.

(1) $A(x, t)$ is a $N \times N$ symmetric positively definite matrix and is differentiable with respect to t.

(2) $B_k(x, t)$ $(k = 0, 1, \ldots, 2M)$, $A(x, t)$ and $A_t(x, t)$ are $N \times N$ measurable and bounded matrices.

(3) The N-dimensional vector valued function $f(x, t)$ is quadratic integrable in Q_T.

(4) The N-dimensional initial vector valued function $\varphi(x) \in W_2^{(2M)}(0, \ell)$ equals to zero together with the derivatives of order up to $M - 1$ at the ends of segment $[0, \ell]$.

Then the boundary value problem (2.4), (2.5) has a unique solution (x, t) $\in Z(Q_T) = W_2^{(1)}((0, T); W_2^{(2M)}(0, \ell))$ and the estimation relation

$$\|u\|_{Z(Q_T)} < K_3 \{\|f\|_{L_2(Q_T)} + \|\varphi\|_{W_2^{(2M)}(0,\ell)}\} \tag{2.11}$$

holds.

Let us turn to condiser in the rectangular domain Q_T, the nonlinear pseudo-parabolic system of partial differential equations

$$(-1)^M u_t + A(x,t)u_{x^{2M}t} = B(x, t, u, u, \ldots, u_{x^{2M-1}})u_{x^{2M}}$$

$$+ g(x, t, u, u_x, \ldots, u_{x^{2M-1}}), \tag{2.12}$$

where u and g are two N-dimensional vector valued function, $A(x, t)$ is a $N \times N$ sysmmetric positively definite matrix with bounded derivative $A_t(x, t)$ with respect to t. Matrix $B(x, t, u, u_x, \ldots, u_{x^{2M-1}})$ is semibounded, i.e., for $(x, t) \in Q_T$ and for any $u, u_x, \ldots, u_{x^{2M-1}} \in \mathbb{R}^N$, there is a constant b, such that for any N-dimensional vector $\xi \in \mathbb{R}^N$

$$(\xi, B(x, t, u, u_x, \ldots, u_{x^{2M-1}})\xi) \leqslant b(\xi, \xi). \tag{2.13}$$

Assume that g is a term of lower degree, which means that

$$|g(x, t, u, u_x, \ldots, u_{x^{2M-1}})| \leq K_4\{\sum_{k=0}^{2M-1} |u_{x^k}| + f_0(x, t)\} \tag{2.14}$$

where $f_0(x, t) \in L_2(Q_T)$ and K_4 is a constant.

In order to prove the existence of the global solution of the homogeneous boundary value problem (2.5) for the system (2.12), we take the functional space $G = L_\infty((0, T); W_\infty^{(2M-1)}(0, \ell))$ as the base space for the fixed point theorem treatment.

For every $v \in G$, we construct a N-dimensional vector valued function u defined as the unique solution of the boundary value problem (2.5) for the linear pseudo-parabolic system

$$(-1)^M u_t + A(x, t)u_{x^{2M}t} = \lambda B(x, t, v, v_x, \ldots, v_{x^{2M-1}})u_{x^{2M}} + \lambda g(x, t, v, v_x, \ldots, v_{x^{2M-1}}) \tag{2.15}$$

with a parameter $0 \leq \lambda \leq 1$. It can be easily seen that all conditions of Theorem 2.1 are available, so $u(x, t)$ is uniquely defined and $u(x, t) \in W_2^{(1)}((0, T); W_2^{(2M)}(0, \ell))$.

The correspondence of v to u defines a functional mapping $T_\lambda : G \to G$ of the base space G into itself, where $0 \leq \lambda \leq 1$ is a parameter. For every $v \in G$, the image $T_\lambda v = u$ belongs to $Z \subset G$. Since the imbedding mapping $Z \hookrightarrow G$ is compact, for every $0 \leq \lambda \leq 1$, the mapping $T_\lambda : G \to Z \hookrightarrow G$ is completely continuous.

Let M be a bounded set of G. For any $v \in M \subset G$ and any $0 \leq \lambda, \overline{\lambda} \leq 1$, there are $T_\lambda v = u_\lambda$ and $T_{\overline{\lambda}} v = u_{\overline{\lambda}}$. The difference vector $w = u_\lambda - u_{\overline{\lambda}}$ satisfies the linear pseudo-parabolic systems

$$(-1)^M w_t + Aw_{x^{2M}t} = \lambda B(x,t,v,\ldots,v_{x^{2M-1}})w_{x^{2M}} + (\lambda-\overline{\lambda})B(x,t,v,\ldots,v_{x^{2M-1}})u_{\overline{\lambda}x^{2M}} + (\lambda-\overline{\lambda})g(x,t,v,\ldots,v_{x^{2M-1}}),$$

$$w_{x^k}(0, t) = w_{x^k}(\ell, t) = 0 \qquad (k = 0, 1, \ldots, M - 1),$$

$$w(x, 0) = 0.$$

It follows immediately that the estimation

$$\| u_\lambda - u_{\bar{\lambda}} \|_G \leqslant C_t \, | \lambda - \bar{\lambda} | \tag{2.16}$$

holds, which means that for any bounded subset M of G, the mapping $T_\lambda : M \to G$ is uniformly continuous for $0 \leqslant \lambda \leqslant 1$.

When $\lambda = 0$, for any $v \in G$, $T_0 v = u_0$ is a fixed vector.

Now we turn to consider the a priori estimations of the solution of the boundary value problem (2.5) for the nonlinear pseudo-parabolic system

$$(-1)^M u_t + A(x,t) u_{x^{2M} t} = \lambda B(x,t,u,u_x,\ldots,u_{x^{2M-1}}) u_{x^{2M}}$$
$$+ \lambda g(x,t,u,u_x,\ldots,u_{x^{2M-1}}) \tag{2.17}$$

with parameter $0 \leqslant \lambda \leqslant 1$.

Taking the scalar product of the vector $u_{x^{2M}}$ with the above system (2.17) and integrating the resulting relation in the rectangular domain $Q_t (0 \leqslant t \leqslant T)$, we have

$$(-1)^M \iint_{Q_t} (u_{x^{2M}}, u_t) dxdt + \iint_{Q_t} (u_{x^{2M}}, A u_{x^{2M} t}) dxdt$$

$$= \lambda \iint_{Q_t} (u_{x^{2M}}, B u_{x^{2M}}) dxdt + \lambda \iint_{Q_t} (u_{x^{2M}}, g) dxdt.$$

Since B is semibounded, there is

$$\iint_{Q_t} (u_{x^{2M}}, B u_{x^{2M}}) dxdt \leqslant b \, \| u_{x^{2M}} \|^2_{L_2(Q_t)}.$$

For the last term of the above equality, we have

$$\iint_{Q_t}(u_{x^{2M}}, g)dxdt \leqslant \frac{1}{2}\|u_{x^{2M}}\|^2_{L_2(Q_t)} + \frac{1}{2}\iint_{Q_t}(g, g)dxdt$$

$$\leqslant C_8\{\|u\|^2_{L_2(Q_t)} + \|u_{x^{2M}}\|^2_{L_2(Q_t)} + \|f_0\|^2_{L_2(Q_t)}\}.$$

By the procedure similar to that used in previous section, we have the estimation relation

$$\sup_{0\leqslant t\leqslant T}\|u(\cdot, t)\|_{W_2^{(2M)}(0,\ell)} \leqslant K_5\{\|f_0\|_{L_2(Q_T)} + \|\varphi\|_{W_2^{(2M)}(0,\ell)}\}, \tag{2.18}$$

where K_5 is a constant independent of the parameter $0 \leqslant \lambda \leqslant 1$. It follows that all possible solutions of the nonliear problem (2.17), (2.5) are uniformly bounded for $0 < \lambda < 1$ in the base space $G = L_\infty((0, T); W_\infty^{(2M-1)}(0, \ell))$.

Therefore the boundary value problem (2.5) for the nonlinear pseudo-parabolic system (2.12) has at least one global solution $u(x, t) \in W_2^{(1)}((0, T), W_2^{(2M)}(0,\ell))$.

Theorem 2.2. Suppose that the nonlinear pseudo-parabolic system (2.12) and the boundary conditions (2.5) satisfy the following assumptions.

(1) $A(x, t)$ is a $N \times N$ symmetric positively definite matrix and has bounded derivative $A_t(x, t)$ with respect to t.

(2) $B(x, t, u, u_x, \ldots, u_{x^{2M-1}})$ is a semibounded matrix valued continuous function of variables $(x, t) \in Q_T$ and $u, u_x, \ldots, u_{x^{2M-1}} \in \mathbb{R}^N$.

(3) $g(x, t, u, u_x, \ldots, u_{x^{2M-1}})$ is a N-dimensional vector valued continuous function satisfying the relation

$$|g(x, t, u, u_x, \ldots, u_{x^{2M-1}})| \leqslant K_4\{\sum_{k=0}^{2M-1}|u_{x^k}| + f_0(x, t)\}, \tag{2.14}$$

where $f_0(x, t) \in L_2(Q_T)$ and K_4 is a constant.

(4) $u_0(x) \in W_2^{(2M)}(0, \ell)$ satisfies the homogeneous boundary conditions (2.5).

Then the problem (2.12), (2.5) has a unique global vector valued solution $u(x, t) \in W_2^{(1)}((0, T); W_2^{(2M)}(0, \ell))$.

The uniqueness of the solution can be obtained by usual estimation of the difference vector valued function of two given generalized global solutions.

The results of the classical global solutions and the smooth global solutions of the boundary value problems (2.5) for the nonlinear pseudo-parabolic system (2.12) can be obtained by the similar way.

In the case of the system

$$u_t + (-1)^M A(x, t) u_{x^{2M}t} = f(u, u_x, \ldots, u_{x^{2M}}) \tag{2.19}$$

with special right hand side (2.3), the boundary value problem (2.5) can be discussed by the method used above. Similarly we have the following result.

<u>Theorem</u> 2.3. Suppose that the system (2.19), (2.3) and the boundary conditions (2.5) satisfy the following conditions.

(1) $A(x, t)$ is a $N \times N$ symmetric positively definite matrix and has bounded derivative $A_t(x, t)$ with respect to t.

(2) $F(p_0, p_1, \ldots, p_{M-1})$ is $M + 1$ times continuously differentiable with respect to all its variables $p_0, p_1, \ldots, p_{M-1} \in \mathbb{R}^N$. $G(p_0, p_1, \ldots, p_{M-1})$ is M times continuously differentiable. The Hessian matrix H of the function F is semibounded, i.e., for any NM-dimensional vector $\xi \in \mathbb{R}^{MN}$ $\xi = (\xi_{m\ell})$ $(m = 1, \ldots, M;\ \ell = 1, \ldots, N)$ such that

$$\sum_{j,\ell=1}^{N} \sum_{m,s=1}^{M} \frac{\partial^2 F}{\partial p_{jm-1}\, \partial p_{\ell s-1}} \xi_{jm} \xi_{\ell s} \leqslant b \sum_{\ell=1}^{N} \sum_{m=1}^{M} \xi_{\ell m}^2 ,$$

where b is a constant for $p_0, p_1, \ldots, p_{M-1} \in \mathbb{R}^N$.

(3) $h(u)$ is a N-dimensional continuous vector valued function satisfying the relation

$$(u, h(u)) \leqslant C(u, u) + d,$$

where C and d are constants.

(4) $\varphi(x) \in W_2^{(2M)}(0, \ell)$ satisfies the homogeneous boundary conditions (2.5).

Then the problem (2.19), (2.3), (2.5) has a unique global vector valued solution $u(x, t) \in W_2^{(1)}((0, T); W_2^{(2M)}(0, \ell))$.

§3. Systems of Pseudo-hyperbolic Type.

In the study of the practical problems in physics, mechanics, biology etc., such as the forced vibration of plane boundary layer, the transfer of the bio-electric signal in animal nervous systems, the linear and nonlinear equations with the principal part $u_{tt} - u_{xxt}$ of pseudo-hyperbolic type often appear. A lot of authors have paid much attention to consider various problems for the linear and nonlinear pseudo-hyperbolic equations[23-28]. In [7,29], the quasi-linear systems of pseudo-hyperbolic type of higher order

$$u_{tt} + (-1)^M A u_{x^{2M}t} = f(x, t, u, u_x, \dots, u_{x^M}, u_t) \tag{3.1}$$

with the special right hand side

$$f_j = \sum_{m=0}^{[\frac{M}{2}]} (-1)^{m+1} D_x^m \left(\frac{\partial F}{\partial p_{mj}} \right) + g_j, \quad j = 1, 2, \dots, N \tag{3.2}$$

are considered, where u and $g(x, t, u, \dots, u_{x^M}, u_t)$ are the N-dimensional vector valued functions, A is a $N \times N$ symmetric positively definete constant matrix, $F(u, u_x, \dots, u_{x^{[\frac{M}{2}]}})$ is a smooth non-negative function and g is the term of lower degree. The generalized global solutions and the classical global solutions of the periodic boundary problems and the initial value problems for the systems (3.1) with special right hand side (3.2) are obtained.

In this section we are going to consider the boundary value problems for the nonlinear systems of pseudo-hyperbolic type of higher order. First of all we will talk about the linear case for the use of further investigation.

Suppose that in the rectangular domain Q_T, the general linear pseudo-hyperbolic systems

$$(-1)^M u_{tt} + \sum_{k=0}^{2M} A_k(s,t) u_{x^{2M-k}t} + \sum_{k=0}^{2M} B_k(x,t) u_{x^{2M-k}} = f(x,t) \tag{3.3}$$

and the boundary value conditions

$$u_{x^k}(0, t) = u_{x^k}(\ell, t) = 0 \quad (k = 0, 1, \dots, M - 1),$$

$$u(x, 0) = \varphi(x),$$

$$u_t(x, 0) = \psi(x). \tag{3.4}$$

Here we assume that the coefficient matrices $A_k(x, t)$ and $B_k(x, t)$ $(k = 0, 1, \dots, 2M)$ are measurable and bounded in Q_T. $A_0(x, t)$ is a $N \times N$ positively definite matrix and $B_0(x, t)$ is a symmetric positively definite matrix having bounded derivative $B_{0t}(x, t)$ with respect to t. The N-dimensional free term vector valued function $f(x, t)$ is quadratic integrable in Q_T. The initial vector valued functions $\varphi(x) \in W_2^{(2M)}(0, \ell)$ and $\psi(x) \in W_2^{(M)}(0, \ell)$ and they vanish together with their derivatives of order up to $M - 1$ at the ends of the interval $[0, \ell]$.

Taking the scalar product of the vector $u_{x^{2M}t}$ and the linear system (3.3) and integrating in $Q_t (0 \leqslant t \leqslant T)$, we get

$$(-1)^M \iint_{Q_t} (u_{x^{2M}t}, u_{tt}) dxdt + \sum_{k=0}^{2M} \iint_{Q_t} (u_{x^{2M}t}, A_k u_{x^{2M-k}t}) dxdt$$

$$+ \sum_{k=0}^{2M} \iint_{Q_t} (u_{x^{2M}t}, B_k u_{x^{2M-k}}) dxdt = \iint_{Q_t} (u_{x^{2M}t}, f) dxdt. \tag{3.5}$$

Using the boundary relations

$$u_{x^k t^2}(0, t) = u_{x^k t^2}(\ell, t) = 0 \quad (k = 0, 1, \dots, M - 1) \tag{3.6}$$

obtained directly from the homogeneous boundary conditions (3.4), we have

$$(-1)^M \iint_{Q_t} (u_{x^{2M}t}, u_{tt}) dxdt = \frac{1}{2} \|u_{x^M t}(\cdot, t)\|^2_{L_2(0,\ell)} - \frac{1}{2} \|\psi^{(M)}\|^2_{L_2(0,\ell)} \cdot$$

Since $A_0(x, t)$ is positively definite, there is $a_0 > 0$, such that

$$\iint_{Q_t} (u_{x^{2M}t}, A_0(x, t)u_{x^{2M}t})dxdt \geqslant a_0\|u_{x^{2M}t}\|^2_{L_2(Q_t)}.$$

Because $B_0(x, t)$ is symmetric positively definite and $B_{0t}(x, t)$ is bounded, there is

$$\iint_{Q_t} (u_{x^{2M}t}, B_0 u_{x^{2M}})dxdt$$

$$= \frac{1}{2}\iint_{Q_t} (u_{x^{2M}}, B_0 u_{x^{2M}})_t \, dxdt - \frac{1}{2}\iint_{Q_t} (u_{x^{2M}}, B_{0t} u_{x^{2M}})dxdt$$

$$\geqslant \frac{1}{2} b_0\|u_{x^{2M}}(\cdot,t)\|^2_{L_2(0,\ell)} - C_1\|\varphi^{(2M)}\|^2_{L_2(0,\ell)} - C_2\|u_{x^{2M}}\|^2_{L_2(Q_t)}.$$

If $B_{0t}(x, t)$ is a non-positively definite matrix, $B_0(x, t)$ can be non-negatively definite. Hence it can be proved that the equality (3.5) may be replaced by the inequality

$$\|u_{x^M t}(\cdot,t)\|^2_{L_2(0,\ell)} + a_0\|u_{x^{2M}t}\|^2_{L_2(Q_t)} + b_0\|u_{x^{2M}}(\cdot,t)\|^2_{L_2(0,\ell)}$$

$$\leqslant C_3\{\|u_{x^M t}\|^2_{L_2(Q_t)} + \|u_{x^{2M}}\|^2_{L_2(Q_t)} + \|\varphi\|^2_{W_2^{(2M)}(0,\ell)} + \|\psi\|^2_{W_2^{(M)}(0,\ell)} + \|f\|^2_{L_2(Q_t)}\}.$$

By the similar way, we can obtain the following theorem.

Theorem 3.1. Suppose that the linear pseudo-hyperbolic system (3.3) and the boundary value problem (3.4) satisfy the following assumptions.

(1) $A_0(x, t)$ is a $N \times N$ positively definite matrix.

(2) $B_0(x, t)$ is a $N \times N$ symmetric positively definite matrix. When $B_{0_t}(x, t)$ is non-positively definite, $B_0(x, t)$ is symmetric non-negatively definite.

(3) Matrices $A_k(x, t), B_x(x, t)$ $(k = 0, 1, \ldots, 2M)$ and $B_{0_t}(x, t)$ are all measurable and bounded in Q_T.

(4) $f(x, t)$ is quadratic integrable in Q_T.

(5) $\varphi(x) \in W_2^{(2M)}(0, \ell)$, $\psi(x) \in W_2^{(M)}(0, \ell)$ and they vanish together with all

of their derivatives of order up to $M - 1$ at the ends of the interval $[0, \ell]$. Then the boundary value problem (3.4) for the linear pseudo-hyperbolic system (3.3) has a unique generalized global vector valued solution $u(x, t) \in Z = W_\infty^{(1)}((0, T); W_2^{(M)}(0, \ell)) \cap W_2^{(1)}((0, T); W_2^{(2M)}(0, \ell)) \cap W_2^{(2)}((0, T); L_2(0, \ell))$. There is estimation

$$\sup_{0 \leq t \leq T} \|u(\cdot, t)\|_{W_2^{(2M)}(0,\ell)} + \sup_{0 \leq t \leq T} \|u_t(\cdot, t)\|_{W_2^{(M)}(0,\ell)}$$

$$+ \|u\|_{W_2^{(1)}((0,T);\, W_2^{(2M)}(0,\ell))} + \|u_{tt}\|_{L_2(Q_T)}$$

$$\leq K_1 \{ \|\varphi\|_{W_2^{(2M)}(0,\ell)} + \|\psi\|_{W_2^{(M)}(0,\ell)} + \|f\|_{L_2(Q_T)} \}. \tag{3.7}$$

Now we turn to consider a nonlinear pseudo-hyperbolic system

$$(-1)^M u_{tt} + A(x,t,u,u_x,\ldots,u_{x^{2M-1}},u_t,u_{xt},\ldots,u_{x^{M-1}t})u_{x^{2M}t}$$

$$= f(x,t,u,u,\ldots,u_{x^{2M-1}},u_t,u_{xt},\ldots,u_{x^{M-1}t}) \tag{3.8}$$

Suppose that $A \equiv A(x, t, p, q) \equiv A(x, t, p_0, p_1, \ldots, p_{2M-1}, q_0, q_1, \ldots, q_{M-1})$ is a positively definite matrix valued function of variables $(x, t) \in Q_T$ and $3M$ vector variables p_k $(k = 0, 1, \ldots, 2M - 1)$ and q_h $(h = 0, 1, \ldots, M - 1)$ of dimension N. For the sake of brevity we assume that N-dimensional vector valued function $f(x, t, p, q)$ is the term of lower degree of the system (3.8), it means that for $(x, t) \in Q_T$ and $p \in \mathbb{R}^{2MN}$, $q \in \mathbb{R}^{MN}$,

$$|f| \leq K_2 \{ \sum_{k=0}^{2M-1} |p_k| + \sum_{h=0}^{M-1} |q_h| + f_0(x, t) \} \tag{3.9}$$

where $f_0(x, t) \in L_2(Q_T)$.

We construct the solution of the boundary value problem (3.4) for the non-linear pseudo-hyperbolic system (3.8) also by the method of the fixed point technique. Let us take $G = L_\infty((0, T); W_\infty^{(2M-1)}(0, \ell)) \cap W_\infty^{(1)}((0,T); W_\infty^{(M-1)}(0,\ell))$

as the base space. Now we define the functional mapping $T_\lambda : G \to G$ of the base space into itself with parameter $0 \leqslant \lambda \leqslant 1$ as follows. For every $v \in G$, we take $u = T_\lambda v$ satisfying the following linear pseudo-hyperbolic system

$$(-1)^M u_{tt} + A(x, t, v, v_x, \dots, v_{x^{2M-1}}, v_t, v_{xt}, \dots, v_{x^{M-1}t})u_{x^{2M}t}$$

$$= \lambda f(x, t, v, v_x, \dots, v_{x^{2M-1}}, v_t, v_{xt}, \dots, v_{x^{M-1}t}) \tag{3.10}$$

and the boundary conditions (3.4).

By the compactness of imbedding $Z \hookrightarrow G$ and the estimation relation (3.7), it can easily seen that for every $0 \leqslant \lambda \leqslant 1$, the mapping T_λ is completely continuous and for any bounded subset M of G, the mapping $T_\lambda : M \to G$ is uniformly continuous with respect to $0 \leqslant \lambda \leqslant 1$.

In order to prove the existence of the solution for the problem (3.8), (3.4), it remains to get the uniform boundedness for λ in G of all possible solutions of the boundary value problems for the nonlinear pseudo-hyperbolic system with parameter $0 \leqslant \lambda \leqslant 1$

$$(-1)^M u_{tt} + A(x, t, p, q)u_{x^{2M}t} = \lambda f(x, t, p, q). \tag{3.11}$$

Making the scalar product of vector $u_{x^{2M}t}$ and the system (3.11) and then integrating the resulting reaction in $Q_t (0 \leqslant t \leqslant T)$, we can obtain that all possible solutions of problem (3.11), (3.4) are uniformly bounded in G for $0 \leqslant \lambda \leqslant 1$.

<u>Theorem</u> 3.2. Suppose that the nonlinear pseudo-hyperbolic system (3.8) and the boundary value problem (3.4) satisfy the following assumptions.

(1) $A(x, t, p_0, p_1, \dots, p_{2M-1}, q_0, q_1, \dots, q_{M-1})$ is a $N \times N$ positively definite matrix valued continuous function of variables $(x, t) \in Q_T$ and $3M$ vector variables $p_k (k = 0, 1, \dots, 2M - 1)$ and $q_h (h = 0, 1, \dots, M - 1)$ of dimension N.

(2) $f(x, t, p_0, p_1, \dots, p_{2M-1}, q_0, q_1, \dots, q_{M-1})$ is a N-dimensional vector

valued continuous function, having the properties (3.9).

(3) $\varphi(x) \in W_2^{(2M)}(0, \ell)$ and $\psi(x) \in W_2^{(M)}(0, \ell)$ are two N-dimensional initial vector valued functions, vanishing together with their derivatives of order up to $M - 1$ at the ends of the interval $[0, \ell]$.

Then the problem (3.8), (3.4) has a unique generalized global solution $u(x, t) \in Z$.

Similarly we can obtain the results for the boundary value problems of the pseudo-hyperbolic systems (3.1) with special right term (3.2). Also by the similar methods, we may obtain the results for the classical and smooth global solutions for the above mentioned problems.

References

[1] Scott, A.C., Chu, F.Y.F. and Mclaughlin, D.W., The soliton, a new concept in applied science, Proc. IEEE, 61 (1973) 1443-1483.

[2] Lions, J., Quelques methods de resolution des problems aux limites nonlinearies (Paris, 1969).

[3] Ablowitz, M.J., Lectures on the inverse scattering transform, Studies in Appl. Math., 58 (1978) 17-94.

[4] Guo Bo-ling, The global solution for some systems of nonlinear Schrödinger equations, Proceedings 1980 Beijing DD-Symposium, vol. 3, 1227-1246.

[5] Zhou Yu-lin and Fu Hong-yuan, The Periodic boundary problems for the nonlinear systems of generalized Schrödinger type of higher order, Acta Mathematica Scientia, 1 (1981) 156-164 (in Chinese).

[6] Zhou Yu-lin and Fu Hong-yuan, Initial value problems for the semilinear systems of generalized Schrödinger type of higher order, Proceedings 1980 Beijing DD-Symposium, vol. 3, 1713-1729.

[7] Zhou Yu-lin and Fu Hong-Yuan, Global solutions for semilinear hyperbolic and pseudo-hyperbolic systems of higher order, (preprint presented to DD3-Sympo-

sium, Changchu , China, 1982).

[8] Кружков, С.Н., Квазилинейные параьолические уравнения и системы с двумя независимыми переменными, Труды семинара нм. И. Г. Петррвского, Вып. 5 (1979), 217-272.

[9] Zhou Yu-lin and Guo Bo-ling, The periodic boundary problems and Cauchy problems for the systems of generalized Korteweg-de Vries type of higher order, (to appear), (preprint presented to DD3-Symposium, Changchun, China, 1982).

[10] Zhou Yu-lin and Guo Bo-ling, The Solvability of Cauchy problem for the quasilinear degenerate parabolic systems $z_t = z \times z_{xx} + f(x, t, z)$, (preprint present to DD3-Symposium, Changchun, China, 1980).

[11] Benjamin, T.B., Bona, J.L. and Mahong, J.J., Model equations for long waves in nonlinear dispersion system, Philos. Trans. Roy. Soc. London, ser.A, 272 (1972) 47-78.

[12] Bona, J.L. and Bryant, P.J., A mathematical model for long waves generated by wavesmakers in nonlinear dispersion systems, Proc. Cambridge Philos. Soc. 73 (1973) 391-405.

[13] Medeiros, L.A. and Miranda, M.M., Weak solutions for a nonlinear dispersive equation, J. Math. Anal. Appl., 59 (1977) 432-441.

[14] Colton, D., Pseudo-parabolic equations in one space variable, J. Diff. Equations, 12 (1972) 559-565.

[15] Twing, T.W., A cooling process according to two-temperature theory of heat conduction, J. Math. Anal. Appl., 45 (1974) 23-31.

[16] Rundel, W., The construction of solutions to pseudo-parabolic equations in noncylinderical domain, J. Diff. Equations, 27 (1978) 394-404.

[17] Ton, B.A., Nonlinear evolution equations of Sobolev-Galpern type, Math. Z., 151 (1976) 219-233.

[18] Davis, P.L., A quasilinear parabolic and a related third order problem, J. Math. Anal. Appl., 40 (1972) 327-335.

[19] Showalter, R.E. and Ting, T.W., Pseudo-parabolic partial differential equations, SIAM J. Math. Anal., 1 (1970) 1-26.

[20] Showalter, R.E., Weak solutions of nonlinear evolution equations of Sobolev-Galpern type, J. Diff. Equations, 11 (1972) 252-265.

[21] Zhou Yu-lin and Fu Hong-yuan, Periodic boundary problems and initial value problems for nonlinear pseudo-parabolic systems of higher order, Scientia Sinica, (1982) 116-124 (in Chinese).

[22] Zhou Yu-lin and Fu Hong-yuan, Cauchy problems of a class of semilinear degenerate evolution systems, Kexue Tongbao, 27 (1982) 590-593.

[23] Pao, C.V., A mixed initial boundary value problem arising in neurophysiology, J. Math. Anal. Appl., 52 (1975) 105-119.

[24] Гоявдберг, З.А., О параметрическом усилении стояцих волн в жидкости, ДАН СССР, 201 (1971) 304-306.

[25] Davis, P.L., A quasilinear hyperbolic and related third order equations, J. Math. Anal. Appl., 51 (1975) 596-606.

[26] Caughey T.K. and Ellison, J., Existence, uniqueness and stability of solutions of a class of nonlinear partial differential equations, J. Math. Anal. Appl., 51 (1975) 1-32.

[27] Andrews, G., On the existence of the equation $u_{tt} = u_{xxt} + (\sigma(u_x))_x$, J. Diff. Equations, 35 (1980) 200-231.

[28] Ebihara, Y., On some nonlinear evolution equations with the strong dissipation, J. Diff. Equations, 30 (1979) 149-164.

[29] Zhou Yu-lin and Fu Hong-yuan, The periodic boundary problems and the Cauchy problems for the quasilinear systems of pseudo-hyperbolic type of higher order, (to appear, in Chinese).

DIRECTORY OF PARTICIPANTS
OF
US-JAPAN SEMINAR '82 IN APPLIED ANALYSIS

GUESTS OF SEMINAR

Professor Tetsuichi Asaka
Science University of Tokyo
(Emeritus Professor of University of Tokyo)

Dr. Tatuo Simizu
Laboratory of Shimizu Construction Co. Ltd.

Professor Shoji Tanaka
University of Tokyo

Dr. Hajimu Yoneguchi
Nippon UNIVAC Sōgō Kenkyusho, Inc.

FOREIGN PARTICIPANTS

(US Delegates)

Ronald J. DiPerna	Duke University
Tosio Kato	University of California, Berkeley
Robert V. Kohn	Courant Institute of Mathematical Sciences, New York University
Alan C. Newell	The University of Arizona
George Papanicolaou	Courant Institute of Mathematical Sciences, New York University
Gilbert Strang	Massachusetts Institute of Technology
Hans F. Weinberger	University of Minnesota

(Special Participants from Third Countries)

John G. Heywood	The University of British Columbia
Lin Qun	Institute of Systems Science, Academia Sinica
Ying Lung-an	Peking University
Zhou Yu-lin	Peking University

JAPANESE PARTICIPANTS

Rentaro Agemi	Hokkaido University
Kiyoshi Asano	Kyoto University
Hiroshi Fujii	Kyoto Sangyo University
Hiroshi Fujita	University of Tokyo
Daisuke Fujiwara	Tokyo Institute of Technology
Isamu Fukuda	Kokushikan University
Yoshikazu Giga	Nagoya University
Ei-Ichi Hanzawa	Hokkaido University
Masayoshi Hata	Kyoto University
Imsik Hong	Nihon University
Yuzo Hosono	Kyoto Sangyo University
Atsushi Inoue	Tokyo Institute of Technology
Hitoshi Ishii	Chuo University
Nobutoshi Itaya	Kobe University of Commerce
Masayuki Ito	Hiroshima University
Seizô Itô	University of Tokyo
Tatsuo Itoh	University of Tokyo
Takao Kakita	Waseda University
Hideo Kawarada	University of Tokyo
Shuichi Kawashima	Nara Women's University
Fumio Kikuchi	University of Tokyo
Hikosaburo Komatsu	University of Tokyo
Yukio Kōmura	Ochanomizu University
Yoshio Konishi	University of Tokyo
Takeshi Kotake	Tohoku University
ShigeToshi Kuroda	University of Tokyo
Kyûya Masuda	Tohoku University
Hiroshi Matano	Hiroshima University
Akitaka Matsumura	Kyoto University
Akihiko Miyachi	University of Tokyo
Isao Miyadera	Waseda University
Sadao Miyatake	Kyoto University
Tetsuhiko Miyoshi	Kumamoto University
Sigeru Mizohata	Kyoto University
Ryuichi Mizumachi	Tohoku University
Hiroko Morimoto	Meiji University

Katsuya Nakashima	Waseda University
Yoshimoto Nakata	Science University of Tokyo
Takaaki Nishida	Kyoto University
Yasumasa Nishiura	Kyoto Sangyo University
Hisashi Okamoto	University of Tokyo
Shin Ozawa	University of Tokyo
Mikio Sato	Kyoto University
Yasuko Sato	Ryukyu University
Norio Shimakura	Kyoto University
Taira Shirota	Hokkaido University
Takashi Suzuki	University of Tokyo
Masahisa Tabata	The University of Electro-Communications
Izumi Takagi	Tokyo Metropolitan College of Aeronautical Engineering
Hiroki Tanabe	Osaka University
Seiji Ukai	Osaka City University
Teruo Ushijima	The University of Electro-Communications
Shigehiro Ushiki	Kyoto University
Masaya Yamaguti	Kyoto University
Kiyoshi Yoshida	Kumamoto University
Kôsaku Yosida	University of Tokyo

INSTITUTIONAL PARTICIPANTS

AISIN SEIKI Co., Ltd.
AISIN-WARNER Ltd.
BUN-EIDO
Energy Research Laboratry, Hitachi, Ltd.
FACOM HITAC
FUJIFACOM CO.
Fujitsu Ltd.
FUKUMOTO-SHOIN, Ltd.
Hitachi Ltd., Central Research Laboratory
Hitachi Ltd., Software Works
Hitachi Ltd., System Development Laboratory
Hitachi Software Engineering Co., Ltd.
HOKUSHIN ELECTRIC WORKS Ltd.
IBM Japan, Ltd.
Institute of Japanese Union of Scientists and Engineers
Ishikawajima-Harima Heavy Industries Co., Ltd.
Japan Advanced Numerical Analysis. Inc.
Japan Process Development Co., Ltd.
Japanese Standards Association
Kajima Co.
Kawasaki Steel Co., Chiba Works
KENBUN SHOIN Co., Ltd.
Kyoei Information Processing Service Center Ltd.
Maeda Construction Co., Ltd.
Mitsubishi Central Research Laboratory
Mitsubishi Heavy Industries Co., Ltd.
Mitsubishi Research Institute Inc.
N.C.R. Japan Ltd.
NIPPON BUSINESS CONSULTANT Co., Ltd.
Nippon Electric Company (C & C Systems Research Laboratories)
Nippon Sheet Glass Co.
Nippon UNIVAC Kaisha, Ltd.
Oki Electric Industry Company, Ltd.
PANAFACOM Ltd.
Shimizu Construction Co., Ltd.
Surugadai Gakuen

TOKYO SHUPPAN Co., Ltd.
Tokyo Shoseki Co., Ltd.
Toshiba Research and Development Center
YAZAKI Corporation
YOYOGI SEMINAR

COMMITTEES AND STAFFS

1. Coordinators

Japanese Coordinator

Hiroshi Fujita	University of Tokyo

U.S. Coordinators

Peter D. Lax	Courant Institute of Mathematical Scicences, New York University
Gilbert Strang	Massachusetts Institute of Technology

2. Local Organizing Committee

Hiroshi Fujita*	Univ. of Tokyo
SigeToshi Kuroda	Univ. of Tokyo
Shigeru Mizohata	Kyoto Univ.
Masaya Yamaguti	Kyoto Univ.
Kôsaku Yosida	Emeritus Professor of Univ. of Tokyo

3. Scientific Committee

Hiroshi Fujii	Hiroshi Fujita*
Hideo Kawarada	Takaaki Nishida
Teruo Ushijima	Masaya Yamaguti

4. Executive Committee

Hiroshi Fujii	Hiroshi Fujita*
Asako Hatori	Hideo Kawarada
Katsuya Nakashima	Teruo Ushijima

5. Working Committee

Akihiko Miyachi	Hisashi Okamoto
Shin Ozawa	Takashi Suzuki*
Kunihiko Takase	Masahiro Yamamoto

* Chairman or Chief